POWER LINES

POWER LINES

THE HUMAN COSTS OF AMERICAN ENERGY IN TRANSITION

Sanya Carley and David Konisky

THE UNIVERSITY OF CHICAGO PRESS

Chicago and London

The University of Chicago Press, Chicago 60637
The University of Chicago Press, Ltd., London
© 2025 by The University of Chicago
Published 2025
Printed in the United States of America

34 33 32 31 30 29 28 27 26 25   1 2 3 4 5

ISBN-13: 978-0-226-82562-5 (cloth)
ISBN-13: 978-0-226-84366-7 (ebook)
DOI: https://doi.org/10.7208/chicago/9780226843667.001.0001

Library of Congress Cataloging-in-Publication Data

Names: Carley, Sanya, author. | Konisky, David M., author.
Title: Power lines : the human costs of American energy in
transition /
Sanya Carley and David Konisky.
Description: Chicago : The University of Chicago Press, 2025. |
Includes bibliographical references and index.
Identifiers: LCCN 2025004402 | ISBN 9780226825625 (cloth) |
ISBN 9780226843667 (ebook)
Subjects: LCSH: Energy development—Economic aspects—United
States. | Power resources—Economic aspects—United States. |
Clean energy investment—United States.
Classification: LCC HD9502.U52 C3727 2025 | DDC 333.790973—
dc23/eng/20250208
LC record available at https://lccn.loc.gov/2025004402

♾ This paper meets the requirements of ANSI/NISO Z39.48-1992
(Permanence of Paper).

Authorized Representative for EU General Product Safety
Regulation (GPSR) queries: **Easy Access System Europe**—
Mustamäe tee 50, 10621 Tallinn, Estonia, gpsr.requests@
easproject.com
Any other queries: https://press.uchicago.edu/press/contact.html

For our families

CONTENTS

Preface ix

1. An American Injustice 1

2. Sacrifice Zones 23

3. Beaten, Broken, Forgotten 55

4. Life Without Energy 79

5. Where New Technologies Don't Go 105

6. Backyards and Ballots 129

7. The Life Cycle of an Injustice 163

8. The Uneasy, Uneven Future 201

Acknowledgments 227

Notes 229   References 273   Index 311

PREFACE

Scientists at the US National Oceanic and Atmospheric Administration determined that 2023 was the warmest year recorded since global records began in 1850. They further declared that the ten warmest years in the 174-year record have all occurred during the last decade. As the Earth warms, we are experiencing extreme weather everywhere and with more regularity. From persistent drought in the western United States to wildfires in southern Europe to massive flooding in Pakistan, the effects of climate change are here and getting worse. If the world continues on its current emissions trajectory, temperatures will continue to rise, perhaps to 2 or 3 degrees Celsius above pre-industrial levels, or more. The consequences will be dire, making our collective efforts to address climate change urgent and critical.

It is a gross understatement to say that making progress to mitigate climate change is challenging. Beyond the technical and economic obstacles, the political will to bring about the necessary changes to how we use energy is uncertain and tenuous. The result of the 2024 United States presidential election is only the most recent reminder. In the United States, the lack of coherent policy to decarbonize the economy results in, not surprisingly, incoherent energy use. As just one illustration, in the past few years, the United States has generated both record levels of electricity from clean, renewable-energy sources like wind and solar and record levels of production of fossil fuels, such as oil and natural gas. In 2023, the United States produced more oil in a year than any nation, ever.

For the United States to do its part to reduce the greenhouse gas emissions causing climate change—the United States emits more than all other nations except China and is responsible for a quarter of total historical emissions—the country needs to hasten its clean-energy transition. The good news is that progress has already been made. Over the past couple of decades, energy utility companies have retired nearly four hundred coal-fired power plants. Solar energy and wind are increasingly common sources of electricity generation. Ten years ago, it was a relatively rare occurrence to see someone driving down the road in an electric vehicle. Today, electric vehicles are commonplace. Rooftop solar panels are more and more common and increasingly affordable because of decreases in costs, federal tax incentives, and a broadening array of purchasing and lease options. And people are beginning to reduce energy use in their homes by putting in more efficient appliances and smart meters, and by starting a process of electrification through the replacement of natural-gas furnaces and air conditioners with electric heat pumps.

The benefits of the clean-energy transition are well known and extend well beyond efforts to lessen the impacts of climate change. Transformation of the energy system will bring with it new jobs for individuals and new economic opportunities for communities. Innovations in technology will spur macroeconomic growth and create new business opportunities for entrepreneurs and investors. The move away from burning fossil fuels will result in improved environmental quality, which means fewer people will be exposed to harmful air pollution and contaminated water.

These benefits will be large and widespread but not universal. The energy transition is a process, and one that will neither be linear nor smooth. Not everyone will enjoy the same opportunities to access new jobs and technologies, not all communities will experience growth from new investments, and not all people will experience the same improvements to their quality of life.

Ten years ago, we started to think about the equity and justice implications of the clean-energy transition, and specifically we began to identify and better understand the individuals, households, and communities who might be vulnerable to the public policies and the economic and social shocks that will accompany the transformation of the US energy system. Over the subsequent decade, our collective research, in collaboration with amazing colleagues, has taken us in many directions. This book represents an effort to synthesize our findings and thinking.

We wrote this book before the 2024 presidential election and the return of Donald J. Trump to the White House. President Trump made it clear during the campaign and through his early months in office that he views addressing climate change and hastening the clean-energy transition as impediments to economic prosperity and US interests. We disagree. The shift away from a fossil fuel–based economy may temporarily slow, but it will not be reversed, no matter the hyperbolic political rhetoric or shifts in policy. Moreover, the unevenness and unfairness of the clean-energy transition is likely to intensify, which we think only underscores the themes we emphasize here.

This book brings attention to the types of people and places that are likely to be disproportionately affected in a negative way by the US energy transition. Our choice to focus more on the problems and less on the solutions was deliberate and does not mean that we are pessimistic about the future. Rather, it reflects a belief that it is essential to document current and future hardships as a means to draw attention to how we might better develop and target public policies to enhance resilience to the coming changes.

We recognize that some may question our focus on the challenges of the energy transition rather than its positive dimensions, especially given our positionality as academics, who have the luxury of thinking about these issues from our "ivory-tower" offices and who are currently living in neighborhoods and communities that have largely avoided many of the historical injustices associated with our energy system. We understand this criticism, yet also hope that people understand our intent, which is to shine light on the challenges that lie ahead in an effort to create more inclusive and just policies. To this end, we offer some broad ideas about paths forward in the final pages of the book, and we hope that they will spark additional conversations and optimism that pursuing energy justice as an aspirational goal will produce better outcomes for more people, especially those historically disadvantaged by our energy system.

Sanya Carley
Philadelphia, Pennsylvania

David Konisky
Bloomington, Indiana

CHAPTER 1

An American Injustice

In February 2021, a series of severe winter storms propelled by a polar vortex hit the southern United States. Texas was ground zero. In its wake of snow, ice, sleet, and bitterly cold temperatures, the state's electric-power grid experienced a state of emergency. The extreme temperatures froze transmission lines, and the conditions led to a spike in electricity demand, while electricity supply—from natural gas power plants to wind turbines—became crippled. The state's electric-grid transmission operator, the Electric Reliability Council of Texas (ERCOT), initiated a series of rolling blackouts across the state in an effort to keep the electricity system from complete collapse.[1]

Millions of Texans were affected—at one point over five million were without power. Several million faced days of brutally cold temperatures without any electricity and, for many, a lack of running water due to frozen or burst pipes.[2]

In the aftermath of the storm, details about just how grim the circumstances were became clear. Many Texan households were unable to recover clean, running water for days. Food supplies were scarce, as many grocery stores had to close because they could not keep perishable food due to intermittent power.[3] And many Texans would soon learn that they had astronomical electricity bills for the electricity they consumed during the storm.[4] Several utility companies also went bankrupt.[5]

Most devastating of all was the death toll. The Texas Department of State Health Services put the official number at 246 fatalities,[6] with some accounts of over seven hundred deaths.[7] The deaths spanned seventy-seven counties and included people as young as infants and as old as a hundred years or more.[8] Many of these fatalities were caused by hypothermia due to exposure to the extreme cold temperatures. Such was the case for Cristian Pavon Peneda, an eleven-year-old boy who lived in a mobile home in Houston.[9] Carrol Anderson, a seventy-five-year-old Vietnam War veteran, died of hypothermia when he went out to his truck for his extra oxygen tank.[10] Others died trying to cope with the loss of energy. Some died from carbon monoxide poisoning when running the heat in their car while sitting in the garage, or while operating a generator indoors. Others died from house fires caused by using the fireplace as a desperately needed source of heat.[11]

As with past natural disasters, such as Hurricane Katrina, the impacts of the winter storm and its aftermath were not experienced equally. One study of service disruptions using satellite data on nighttime lights found that areas with more people of color were much more likely to experience blackouts.[12] Other studies found disparities across income and race in the orchestrated power disconnections. A study using cell phone network data showed that places where the ERCOT disconnected regions more frequently and for longer durations were more likely to be lower income and inhabited by racial and ethnic minorities.[13]

Beyond blackouts, Texans who lacked the resources to adapt to the cold weather conditions were most vulnerable. People without cars, backup generators, or fireplaces were unable to stay warm and keep the lights on. People without stored food and water suffered, and they were not able to replenish their supplies when local grocery stores closed. Those with weaker pipes experienced more pipe bursts. Those without insurance were less able to recover from the water damage and therefore more prone to long-term problems like mold. And those without savings were unable to absorb the extra cost of electricity—sometimes in the thousands of dollars, due to the storm-induced price spikes in Texas's deregulated market—and avoid the long-term threat of electricity disconnection as a result of delinquent payments.

The 2021 winter storm brought Texas, one of the most energy-rich places in the world, to its knees. However, for many Americans, accounts like the ones from Texas are at once devastating and not altogether surprising. The rise in climate change–induced disasters over the last several decades has produced an expectation—and perhaps an acceptance—that some communities' losses of life and home are an unavoidable aspect of modern life. According to a recent accounting by the US National Centers for Environmental Information, between 1980 and 2023, there were 372 weather and climate disasters. The economic damages from these events added up to $2.63 trillion, and they resulted in more than sixteen thousand deaths.[14] These disasters have accelerated in the last decade. In 2023 alone, there were over thirty extreme weather events that each generated over $1 billion in damages.[15]

But, for many Americans, what transpired in Texas was also unsurprising for a different reason: many US households routinely lack reliable and affordable energy in their homes. Each year, one in four US households suffers from what is referred to as energy insecurity or energy poverty, and millions of households have their service disconnected by their utility when they are unable to come up with the funds to pay their energy bills.[16] We know from evolving research on the topic that there are extreme disparities in who experiences energy insecurity. African American households, for example, are three times more likely to have their power shut off.[17] These disparities relate to how much people spend on energy and the conditions of their homes; it is well documented that people of color, even controlling for income, spend more on energy services as a share of their income and more commonly live in dwellings that are inefficient. And energy insecurity is not just a material hardship, it is also associated with adverse physical and emotional well-being, particularly for children, and can lead to forced displacement through evictions. For many, the type of power outages and their consequences experienced during the storm in Texas are a regular way of life.

A thousand miles northeast of Texas, in Appalachia, the work of supplying energy to the public is producing its own set of social and economic challenges. In recent decades, competition from natural gas due to advances in hydraulic fracturing and the increased deployment of renewable energy has drastically reduced demand for coal as a source of electricity generation. Federal air-pollution control regulations have also raised the costs of operating coal-fired power plants, which has contributed to their economic vulnerability. Although the US federal government has largely failed to impose direct limits on carbon emissions, the threat of carbon restrictions has also driven markets away from coal. The changes within the electricity sector have been swift and deep. From 2001 to 2020, net generation of electricity from coal fell by over 60 percent.[18] Not surprisingly, the diminished demand has had devastating consequences for coal communities in Appalachia.

The past decade has seen numerous mine closures and coal company bankruptcies, and with it losses in vital tax revenues for local communities. Coal miner employment has also hit record lows. A decade ago, there were seventy-five thousand coal mining jobs across the United States, many providing salaries above the national median; in 2024, the number of such jobs had fallen by more than 40 percent.[19] And, because coal country often lacks economic diversification, there are not many alternative jobs for those who have lost their economic livelihoods with the decline of coal mining.

The shift to cleaner sources of energy means that similar transitions are occurring in other communities reliant on fossil fuels to engine their local economies. The disproportionate impact on these communities is itself a type of energy-related disparity. The conditions in West Virginia, Kentucky, and other parts of Appalachia are befalling coal communities elsewhere in the United States, such as Wyoming and Montana, and they also affect communities that have historically relied on coal-fired power plants for local employment and tax revenue.

Far away from coal mines are the factories that process fossil fuels, many of which are located in communities where the residents are low-income people and people of color. Perhaps the best-known example in the United States is the stretch of the Mississippi River between Baton Rouge and New Orleans, Louisiana, unaffectionately known as "Cancer Alley." This dense, industrial expanse is home to more than 150 oil refineries, chemical factories, and fossil-fuel processing and storage facilities, and it is among the most polluted regions of the nation. This same area is now increasingly targeted for carbon capture and storage projects, despite opposition from communities that have long objected to the heavy presence of industrial polluters.

Living alongside—sometimes literally across the street from—the refinery crackers, flare towers, and storage tanks are families, children, retirees, and others. Often unable to afford to live elsewhere, or living in homes that existed before their unwanted industrial neighbors came to town, people are stuck with the noise, nuisance, and pollution of these facilities. And with these things are the adverse health effects that come with living in close proximity to large sources of particulates, toxic chemicals, and other pollutants that these facilities release into the air and water. The modern-day environmental-justice movement in the United States emerged from these very communities, communities sometimes referred as "sacrifice zones,"[20] where residents fight back against the fossil fuel companies that despoil their homes and communities.

A Changing Climate and Energy System

Accelerating climate disruptions and a growing pressure for energy systems to change in response will profoundly impact these frontline communities, as well as all of society, for years to come. According to estimates from the Intergovernmental Panel on Climate Change (IPCC), the leading global scientific authority on climate change, in absence of drastic measures to mitigate greenhouse gas emissions, the Earth will warm at least 3 degrees Celsius (5.4 degrees Fahrenheit) above pre-industrial levels by 2100, with an increased incidence of extreme weather events, including heat waves, wildfires, tropical storms and hurricanes, and flooding.[21] The United States has already warmed by 1.4 degrees Celsius (1.2 degrees Fahrenheit), exceeding global rates of temperature rise, and the National Climate Assessment, a report produced by leading US experts on climate, predicts that the average US temperature will increase by at least 2.2 degrees Celsius (4 degrees Fahrenheit) by 2050.[22]

This overall rise in temperature will yield more days that are above comfortable temperature thresholds. One assessment estimates that with 2 degrees Celsius of warming, much of the contiguous United States will experience twenty or more days above 35 degrees Celsius (95 degrees Fahrenheit), with closer to forty or fifty days in southeastern states.[23] These conditions will also increase the number of heat waves and decrease the number of cold spells. These temperature changes will be especially prevalent in Alaska, home to massive glaciers that are sustaining the Earth's natural cycles. These glaciers shed about seventy-five billion tons of ice into the ocean as water each year. While these conditions are particularly dangerous to the ecosystems and species within Alaska, threatening the survival of many Native Alaskan communities, they also expose land that has been frozen for thousands of years, which may release more carbon dioxide into the atmosphere and potentially release bacteria that have been frozen and may threaten human health.[24]

Heat waves are more prevalent than they were previously, especially in large cities across the world. These conditions are exacerbated by the urban heat island effect, which is when highly concentrated amounts of buildings, pavement, and hard surfaces retain heat in a localized area or region.[25] Recent examples of urban heat waves include the city of Phoenix, Arizona, which experienced its hottest summer on record in 2023, with an average temperature of 97 degrees Fahrenheit. Among the records set in Phoenix that summer include thirty-one consecutive days of at least 100 degrees Fahrenheit from June 30 to July 30, including seventeen straight days with temperatures at or above 115 degrees Fahrenheit.[26]

To combat climate change, households, communities, and countries alike need to invest in both mitigation and adaptation strategies. Mitigation strategies are those that seek decarbonization—for example, by replacing fossil fuel infrastructure with renewable-energy or energy-efficient infrastructure. Adaptation strategies are those that build our capacity to adjust to new climate conditions and also those that build our resilience—so that a changing climate is less able to cause mass destruction. Both types of strategies will require massive investments in new infrastructure, programming, and social support; they will also require changes in collective behavior around how we, as a society, use energy. The costs of inaction would be greater.

Decarbonization has begun in the United States and elsewhere across the world. Many energy producers and consumers are starting to transition away from heavy reliance on fossil fuel–based energy sources and toward an increased use of lower-carbon, more efficient energy technologies, including renewable resources. Unlike their fossil-fuel counterparts, renewable sources have low or zero "fuel" costs. However, while costs have come down a lot in recent years,[27] substantial associated capital costs are still required to build renewable-energy systems. Renewable energy also often requires new infrastructure that connects it to the electric grid, both on the macro scale (think of new transmission and distribution lines to connect wind facilities) and on the micro scale (think of the equipment needed to connect a residential solar panel). Greater use of these technologies is necessary to achieve decarbonization goals, but they require expenses that are typically passed along to consumers.

All these requirements are on top of the other investments that are needed to upgrade electric grids. Some studies put these costs at around $4.5 trillion for the United States and $14 trillion globally.[28] These are just one type of cost, however. Other costs, some much harder to quantify, will come as part of the large-scale transformation of the energy system.

Those Left Behind

Confronting the climate crisis requires massive energy-system changes, and at all scales—from the electric grid to the appliances we use in our homes and everything in between. Any such transformation comes with disruptions, and there is every reason to believe that the disruptions associated with the current energy transition will be severe, especially for the people and communities on the front lines, such as those who already face energy insecurity and those whose economic livelihoods are dependent on legacy fossil fuels.

Energy systems have always produced winners and losers, and the current shift away from fossil fuels toward a mix of low-carbon, more efficient, and advanced technologies will be no different. For those with high income, geographic mobility, power of influence, and advanced skills, the clean-energy transition will provide opportunities to modernize their use of energy and to more easily adapt to economic and social changes that come with the transition. For many others, the clean-energy transition will create harms and losses. Among the numerous examples include a loss of jobs for fossil fuel workers; loss of revenue for communities that have historically relied on revenue from fossil fuel extraction and use; residential-energy insecurity for a growing number of Americans who cannot afford to pay for critical household-energy services; lack of access to technologies such as solar power, electric vehicles, and smart meters; and lack of inclusion in energy-related decision-making processes, among other challenges.

These energy-related problems spill over to economic and social well-being, and they cannot be easily separated from broader issues of deepening income inequality, food and housing insecurity, community deindustrialization, the opioid epidemic, political extremism, and broadening feelings of economic and social isolation and lack of trust and confidence in institutions. In this way, the emerging challenges of the energy transition are part of an interconnected web of challenges and vulnerabilities that characterize an unequal and unjust American society.

The core argument of this book is that the clean-energy transition will leave many individuals and communities behind. Many people, including those historically marginalized, will be made worse off. Although policymakers, energy companies, advocates, and others rarely talk about the energy transition as a process that explicitly creates harms or losses, such harms and losses are an inevitable outcome of making large energy transformations. To make matters worse, the patterns of harms and losses are likely to calcify existing inequalities and intensify other cleavages that already plague American society, and other societies elsewhere as well.

The energy transition stands to squeeze traditionally underserved communities through unaffordable energy and continued health and nuisance burdens from living in close proximity to energy production facilities. The energy transition is also poised to sacrifice the livelihoods of those employed in the legacy production of energy and in adjacent industries reliant on their use (e.g., makers of internal-combustion-engine cars).

The use of fossil fuels, of course, comes with its own severe problems, which should not be diminished or ignored. The point is that the transition will continue to elevate the "haves" at the sacrifice of the "have-nots."

An irony, of course, is that the clean-energy transition is both vital and urgent if we are to address the climate crisis. Decarbonization is not a choice but a necessity. Replacing fossil fuels with cleaner, renewable alternatives will also generate innumerable public health and economic benefits to society, which raises the question: Is it possible to pursue decarbonization in a way that does not create disproportionate burdens for some people and communities?

Concerns about potential disproportionate burdens related to the clean-energy transition have led some scholars, activists, and policymakers to start talking about "energy justice." Energy justice is the pursuit of a world in which people are centered in energy and climate solutions, and in which the quality and dignity of all lives are valued—especially the lives of those who have been historically oppressed or marginalized.

Notions of energy justice commonly emphasize the distributional, procedural, and historical dimensions of energy systems and energy policymaking. Distributional justice considers how the benefits and burdens of the energy system are spread across different populations; its objective is for no single population to bear a disproportionate share of the burdens and for all populations to have access to the benefits. Procedural justice considers the access and opportunities that individuals and communities have to participate in energy decision-making processes; its objective is for everyone's voice to be heard and respected. Recognition justice addresses the need to account for past harms and to fully consider the many structural and societal dynamics that keep certain populations disadvantaged.

Much of the earliest academic literature and broader conversation about energy justice was theoretical and abstract, often disconnected from the lived experience of low-income households, coal miners, autoworkers, and others who will be first and most directly affected by the energy transition. More generally, the concepts of energy justice and equity are ignored completely across a range of academic work and in much of policymaking. As a matter of practice, for example, most economists judge policies and programs in utilitarian terms— that is, they ask whether more people are better off than harmed by a given course of action. Historically, most economic studies have largely overlooked or omitted a consideration of the distributional or procedural consequences of such actions, let alone made any space to account for historical burdens. Policymakers talk about how achieving net-zero emissions will ultimately result in economic growth and job creation, but they often do not consider who will benefit and who will be left behind. In other words, a focus on efficiency, markets, and economic growth often comes at the neglect of the people who are adversely affected by energy decisions.

Similar blind spots about technological progress afflict much of the academic literature and public-policy practice. Scholars and practitioners alike have historically focused on the difficult technical challenges—of shifting the energy foundation of the economy from fossil fuels to other sources that are less damaging to the environment and to human well-being. Such technological solutions take the form of solar energy, wind energy, battery storage, biomass, electric vehicles, advanced nuclear power, geothermal operations, and—potentially— carbon capture and storage, as well as residential technologies such as efficient appliances, heat pumps, and smart thermostats. This focus on technology tends to dominate broader discussions about how to respond to the climate crisis and stifles further consideration of social justice and human well-being. Energy scholar Jennie Stephens refers to the framing of climate action as isolated to technological solutions as "climate isolationism."[29]

Yet energy technologies and systems are fundamentally rooted and here to serve human activity, which means that the pace and success of the energy transition depends on whether and how society embraces clean-energy technologies, whether everyone can access and use them, and whether they produce disruptions to human and social activities that are manageable or—alternatively—destructive. And for this energy transition to be a "just" one, the United States must pursue decarbonization in a way that, at bare minimum, does not exacerbate the economic, racial, and other inequities that currently scar energy systems nor create new inequities.

In recent years, there has been growing attention to energy justice issues. A burgeoning scholarly literature emphasizes the need to consider the distributional consequences of energy systems, as well as how more people can be included in the energy decisions that affect them. Politicians and activists (even if disingenuous or cynical) often highlight those who have been left behind, or will be left behind, by changes to energy markets and policies. Think of recent Republican politicians appealing to coal miners affected by reduced demand for coal and Democratic and Republican politicians alike appealing to autoworkers who feel threatened by the emergence of electric vehicles.

From a policy standpoint, there has also been some progress toward putting in practice the principles and objectives of energy justice. The Biden administration created the Office of Energy Justice and Equity with the US Department of Energy; directed federal climate, energy, and environmental investments to disadvantaged communities as part of its Justice40 Initiative; and designed new programs under laws such as the Inflation Reduction Act of 2022 to improve access to energy technologies to historically underserved groups, among other items. Several state governments have their own similar initiatives. More broadly, the political discourse around energy, climate, and environmental issues unquestionably has increasingly incorporated ideas about equity and justice. In our view, these efforts reflect sincere and serious efforts to, adapting Martin Luther King Jr.'s famous saying, bend the arc of the clean-energy transition toward justice.

However, like many environmental-justice initiatives before them,[30] many (if not most) of these efforts are administrative and discretionary, and they are subject to the whims of subsequent presidents, governors, and other leaders who may come to office with different ideas and priorities. The switch from President Biden to President Trump is just the latest manifestation of the enormous shifts in federal policy on energy and climate policy. In addition, most of the actions taken to date are incremental and insufficiently scaled to the problem. Moreover, while new technologies are often portrayed in pollyannaish ways as "win-wins," their accessibility is often limited. Take just one example, heat pumps, which are increasingly promoted as an effective way to partially electrify people's homes. The use of a heat pump in one's home can reduce the use of fossil fuels compared to natural-gas furnaces, as well as reduce energy bills because of their relative efficiency. However, heat pumps are expensive purchases out of reach to many low-income households, and they are completely unavailable to renters who do not choose their heating and cooling appliances.

While many scholars, energy thinkers, and politicians are right to aspire to energy justice, the reality is that energy *injustice* is both the status quo and the likely future. None of these observations is to suggest that we should not put in place policies, programs, and reforms to address energy-based inequities and injustices—we absolutely should. But we also should not put our heads in the sand and ignore or underplay the magnitude of the challenge, which is that the clean-energy transition will be messy and uneven across both time and space, and it will create new hardships for many people and communities that we must address.

This book presents stories of individuals and communities who are on the front lines of the energy transition, focusing on the many types of harms and losses that will occur or already are occurring. We conceive of frontline individuals and communities broadly, to include those people and places who will likely experience near-term adverse effects of the energy transition. In so doing, we group coal miners and auto assembly workers who face uncertain economic prospects alongside low-income households who struggle to pay energy bills and gain access to new technologies and agricultural communities where residents are increasingly being asked to take on new renewable-energy infrastructure that disrupts their rural landscapes. We also group them with environmental-justice communities who have been disproportionately affected by the historical and current consumption of fossil fuels and who are now being targeted with new energy infrastructure that may continue or exacerbate these burdens.

In drawing connections across these groups and communities, we are not equating their experiences or suggesting that their future harms and losses will be of the same severity as those who have historically been disproportionately and adversely affected by the current energy system. Their experiences have been and will continue to be unique in both cause and effect, and our grouping them together here does not imply in any way a normative assertion that they are owed the same thing from a justice perspective. Rather, our goal is to find and emphasize commonalities as a way to understand the widespread vulnerabilities to the energy transition.

In telling these stories, we also highlight an underappreciated paradox: Those who have the greatest potential to benefit from decarbonization of the economy (e.g., cleaner air and water; reduced impacts from climate change; access to new technologies; jobs and economic development opportunities) are among the least likely to actually capture them.

Alongside our central argument, we emphasize three additional themes: There is an uneven geography of costs and benefits of this transition; current energy systems often pose impossible trade-offs for frontline communities; and those who are most disadvantaged and poised to benefit from change are largely powerless and often unable to engage in energy decision-making or opportunities.

First, in the domain of energy, the costs and benefits of energy choices are rarely experienced in the same geographic space. Those who benefit are often completely different populations, both geographically and also by sociodemographic characteristics, than those who bear the burden. In some cases, certain populations bear all the localized burdens, while the rest of the country or world enjoys the lion's share of the benefits. These spatial transboundary inequities will be a prominent component of the clean-energy transition, but they will also be difficult to address.[31]

Second, individuals and communities on the front lines of the changing energy systems face difficult and complex trade-offs. The oft-used expression of facing a decision that puts one between a rock and a hard place aptly applies here. Individuals often must choose between providing an economic livelihood for their families and the cost of degrading their environment; relocating their family for new employment after their job disappears or commuting long distances to avoid relocation; profiting off mineral and earth deposits or preserving their ancestral land and environments; staying warm in their home and risking utility disconnection or taking drastic measures to avoid turning up the heat to keep bills low. Yet as a society, we often do not recognize, let alone seek to mitigate, the physical burden and mental toll that these decisions have on frontline individuals and communities; nor do we acknowledge that many of these burdens and trade-offs are concentrated geographically and compound over time.

Energy injustice is a problem reinforced by economic, social, political, and institutional factors. Energy injustices affect people's health and well-being, the dignity of their living conditions, their children's future opportunities, their ability to retire with savings, and the social fabric of their surrounding communities. And energy injustices require that people and families weigh difficult trade-offs and make unimaginable decisions, such as to "heat or eat" or to uproot their families and leave their homes and communities behind in search of new employment.

Third, the communities and populations facing the most burdens from energy systems are literally and often figuratively without power. Millions of Americans and hundreds of millions of people across the world lack access to essential energy services. In the case of most Americans, this type of energy poverty most often occurs because of an inability to afford energy services—that is, people cannot afford to pay their electricity or natural-gas bill and are then disconnected from service by their utility provider. This problem is both compounded by and exacerbates housing insecurity, since the inability to pay one's energy bill contributes to housing precarity and homelessness.

Many who face energy poverty, as well as other frontline populations, however, are also without power in a figurative sense. Across the world, and in the United States as well, people lack the power and agency to rewrite their own relationship—and often their communities' relationship—with the world's energy systems, a relationship marked by historical oppression, social and economic alienation, and abandonment. These individuals lack access to decisions about energy that affect them. They also lack access to clean-energy technologies and services, access to jobs within clean-energy industries, and the opportunity to participate in the decision-making processes that determine the direction and pace of the energy transition.

We argue in this book that inequities and injustices in the energy system are inevitable, which is not to say that they cannot at least be partially mitigated with deliberate action. Inequalities are human-made, and righting them requires deliberate and coordinated human response. The energy transition is *not* a movement toward reliable energy for all or the making right of legacy economic problems tied to energy production, unless society chooses to make it so. Centering people and communities in the transformation of an energy economy presents an opportunity to think about the clean-energy transition in a different way—and to moderate the injustices to come. Among other things, it means talking about and designing energy and climate policy to explicitly promote racial justice and reduce income inequality. It also means thinking about compensating or otherwise providing for those who will lose out from the transition, such as coal miners, oil rig operators, and line workers at automobile assembly plants. And it means assisting low- to middle-income people so they too can access and enjoy the benefits of new technologies.

Our objective in this book is to establish a baseline of understanding about how energy systems of the past, present, and future are not only fraught with inequities but are poised to perpetuate such inequities. We think it is vital that we fully recognize the challenges that lie ahead. This book is about these challenges.

Outline of the Book

Following this introduction, we proceed to chapter 2, where we present a description of the environmental injustices that have been generated by the current fossil-fuel-based energy system. Fossil fuels have powered the US economy for generations, enabling economic growth and prosperity for many, and providing affordable and accessible energy to most Americans. But it has also inflicted harm to millions of fellow citizens. Centering people in discussions of the energy transition first requires acknowledging and reconciling past harms. Chapter 2 provides an overview of the historic burdens that the US energy system has placed on millions of Americans, most often people of color and those with lower incomes.

We present findings from extensive research that has cataloged the disproportionate burdens low-income communities and communities of color have experienced from the fossil-fuel-powered economy. We structure this review to emphasize that burdens exist across the life cycle of the energy system—from extraction, through distribution and processing, to consumption and waste disposal. Examples from energy extraction include coal mining (e.g., black lung disease, contaminated water, permanently altered landscapes), uranium mining on Navajo land, and fracking in rural Pennsylvania and elsewhere. Illustrations from distribution and processing include the siting of pipelines (e.g., Dakota Access) and major infrastructure (e.g., highways) and oil refineries and other petrochemical facilities in Louisiana and Texas. Regarding consumption, the pollution from burning coal and natural gas for electricity and gasoline for cars and trucks has exposed Americans in urban and rural areas alike to harmful air pollution, particularly people who live in close proximity to power plants and major roadways. With respect to waste, communities of color have been targeted for the disposal of the byproducts of energy use, such as coal ash in places like Uniontown, Alabama.

In chapter 3, we turn to the topic of fossil fuel workers and communities and their changing circumstances amid the energy transition. From the coal fields of Wyoming and West Virginia to the oil fields of Alaska and West Texas, communities around the United States, as well as elsewhere around the world, are dependent on the extraction of fossil fuels for their economic livelihoods and, in some cases, their social identity. As the country decarbonizes and shifts away from these sources of energy, some fossil fuel communities are already experiencing economic decline, with spillover effects to their social and cultural fabrics. In some cases, generations of families have worked in these industries, and entire local economies are reliant on the tax revenues from these incumbent energy sources to fund schools, roads, social services, and community events. To date, the largest shocks have come to coal country, but future shocks will hit oil and gas communities as well. Although the fortunes of these communities have long ebbed and flowed through boom-and-bust cycles, absent focused government and other efforts, coal and oil communities are poised to experience a permanent bust.

We describe this important implication of the US energy transition incorporating ideas from the literature on a "just transition." We highlight our own research in Appalachia, which included interviews and focus groups with current and former coal miners and people working to help these communities survive the transition. We present personal narratives from these communities, including stories about how some individuals within coal communities feel used and abandoned by the rest of the country as consumption of coal recedes; how the transition away from coal inflicts both personal and community identity confusion; and the plight of specific individuals in their attempts to find new employment opportunities. We also discuss the implications of shifting energy markets on other fossil fuel communities, including gas and oil communities, in the near future.

In chapter 4, we turn to the topic of energy insecurity. Decarbonization of the US economy will be expensive, requiring trillions of dollars in new public and private investment in infrastructure. Replacing the fossil-fuel-based energy system to one reliant on cleaner, renewable sources of energy will require retiring coal and natural-gas power plants and deploying wind and solar farms with large battery-storage capacity. It will also necessitate upgrading and expanding the electricity grid to move power across the country in new ways. Homes, offices, and commercial buildings will need energy efficiency upgrades, and drivers of electric cars and trucks will require an extensive charging infrastructure to facilitate an electrified transportation sector. A failure to invest in these upgrades will alternatively place costs on grid, home, and road repairs. Either way, the price tag is significant.

We focus our attention on one specific manifestation of higher costs: household energy insecurity. An energy insecure household is one that is unable to pay for electricity and other critical residential-energy services. Households that cannot afford these services face not just inconvenience and discomfort but significant risks to their health and well-being. The costs of the energy transition are likely to exacerbate the problem, as utility companies pass their investment costs to their customers. And climate change will make matters even worse, as people demand more energy to keep their homes comfortable given temperature extremes, especially those who own or rent homes that are old, drafty, and have poor insulation.

The chapter synthesizes our research on household energy insecurity, presenting results from surveys of low-income US households and interviews with people who have experienced hardship from losing access to electricity. We present evidence that among low-income households there are large disparities in energy insecurity; homes of people of color, people with compromised health, and people with children are particularly vulnerable to utility disconnection. Beyond an understanding of the incidence of energy insecurity, we also explore the ways people cope both financially and behaviorally when they struggle to pay their bills. An additional layer to the challenge of energy insecurity is that people are often distrustful of their utility companies based on their past experiences and encounters, and federal programs to help people through these challenges are woefully underfunded—and as a result, many people are left feeling helpless, hopeless, and ultimately without power.

In chapter 5, we address the question of who owns new clean-energy technologies and why that matters. At the foundation of the energy transition is new infrastructure including the innovation and the development, commercialization, and eventual adoption of new technologies. The widespread use of technologies such as solar panels, battery storage systems, electric vehicles, smart meters, and heat pumps is essential if the world is to decarbonize its economies. However, this new infrastructure is often expensive and inaccessible to many Americans. Electric vehicles not only cost more than their gas-powered counterparts, they also require access to charging stations. Solar panels require a roof, typically in a resident-owned home, creating obstacles to those who rent. Both technologies require that specialist retail and service companies (e.g., dealerships and solar installers) be located nearby and willing to serve all customers, yet there is plenty of evidence that, even today, not all dealerships carry electric vehicles and not all regions have solar installers, as is often the case in rural communities. Smart meters require that the owners have the time and inclination to monitor rates and the flexibility to shift the time of their energy use, a luxury that people struggling with multiple jobs and other obligations may not enjoy.

Government policies to incentivize the adoption of new technologies often favor those who are already wealthy. Tax credits for solar and electric vehicles tend to flow to the financially well off and those who pay taxes, as do rebates—since they only go to people who are able to pay the expensive up-front costs of these technologies. Government spending on charging stations tends to follow spending on electric vehicles, again reinforcing the stream of government resources that flow to the wealthy at the exclusion of others.

There are also impacts to those who cannot access the more advanced, efficient, and low-carbon technologies. When more households install solar panels on their roofs and no longer consume as much utility-provided electricity nor pay a sufficient share of the utility's fixed costs, these costs are spread over a smaller number of non-solar customers—so utility-provided rates rise. When enough households switch away from natural gas to electric HVAC and appliances, the gas companies will similarly have to spread the fixed costs of their distribution systems across a smaller number of customers: those who could not afford to switch. In short, these technologies create compounding advantages for the wealthy, at the expense or neglect of those who cannot pay. In this chapter, we explore these market dynamics and the underlying conditions of inequities.

In chapter 6, we turn to the challenge of siting new energy infrastructure. The transition away from fossil fuels to cleaner sources of energy requires shifts in the composition of the economy. To do so in a timeframe to meaningfully reduce the worst impacts of climate change requires massive new investments in energy infrastructure. Despite a multi-decade effort to start on a path toward decarbonization, in 2024 about two-thirds of the US economy remained powered by fossil fuels. Any serious decarbonization attempt will require the deployment of wind, solar, and other zero-emission sources of electricity generation, the buildout of new transmission lines and other infrastructure (e.g., battery storage) to bring this power from sites of generation to where it is used in homes and business, and potentially new facilities and pipelines to capture and store carbon dioxide.

In this chapter we analyze some of the challenges that complicate the siting of new energy infrastructure. Like many other aspects of the energy transition, these challenges are not simply technical. Rather, they are economic, social, and political. We focus on three interrelated challenges. First, the siting of new energy infrastructure will create geographically uneven costs and benefits. In particular, projects often create benefits for some but concentrate the costs on others. Second, people often mistrust the companies and government agencies that propose to bring new technologies to their communities. This mistrust sows doubt about the purported benefits of proposed projects. And third, the deeply polarized state of US politics further complicates the siting of new energy infrastructure. Americans disagree about the urgency and policy path toward decarbonization, and this disagreement shapes the discourse about energy infrastructure broadly and with respect to specific projects. When coupled with a fragmented political system and the decentralized nature of energy decision-making, the obstacles in front of any specific energy project can be formidable.

In chapter 7, we focus on a specific technology, the electric vehicle, and explore sources of social and environmental exploitation and injustice across the entire life cycle of the technology. In this cradle-to-grave narrative, we highlight four case studies. Beginning with the extraction of raw materials and resources, we feature cobalt mining in the Democratic Republic of the Congo and the complicated and in some cases brutal human conditions in these mines. Turning to the production of the electric vehicle, we discuss automobile working communities in the Midwest, and specifically in Detroit, Michigan, and Lordstown, Ohio, where decades of boom-and-bust cycles, feelings of abandonment, and economic scandals have cast a shadow over the evolving transition toward electric vehicles. We share our interviews with autoworkers who express dread about the impending move away from the cars that have provided them, their families, and their communities with an identity and economic security. We then turn to the consumption side of the electric vehicle life cycle with a focus on the American electric vehicle consumer. Following the discussion in chapter 5 about access to clean-energy technologies, the story for electric vehicles is no different. Yet the barriers for access to the electric vehicle are particularly pronounced, especially for traditionally underserved and marginalized communities. Finally, we turn to the disposal of waste with a feature on electronic waste fields surrounding a yam market in Agbogbloshie, Ghana, a region so scarred by toxic and burning debris that it was featured in a national movie production as a dystopian inferno fueled by collective energy greed.

These four case studies highlight how a single technology can be fraught with geographic disparities that appear at all stages across its life cycle. These geographic disparities span oceans, regions, and neighborhoods. Regions that bear the worst of the burdens—in the case of the electric vehicle, such regions include locations of mineral extraction and battery disposal, respectively—are marked by environmental degradation, social and economic stratification, and human exploitation. These regions are energy sacrifice zones due to the heavy concentration of environmental injustices that occur there, and they are present in order for others to benefit from the opportunities provided by the energy commodities produced or destroyed there. At other stages in the electric-vehicle life cycle, workers and consumers both express sentiments of abandonment: by those who hold the power to make energy decisions and by the inaccessibility of this clean-energy technology.

In the final chapter, we connect the main arguments and themes of the book, showing how the basic challenges of the US clean-energy transition extend across all parts of our energy system. We then present a broad blueprint—built on the core principles of energy justice—that can guide policy, community, and industry priorities going forward. Our intent is neither to set out a prescriptive plan full of specific policy proposals nor to make normative claims about what individuals or communities on the front lines of the transition are owed. Rather, our goal is to emphasize a general approach that centers justice considerations in energy decision-making and refocuses attention on the people and communities affected most by these decisions. The path forward is not easy or linear, but with care, kindness, and empathy for those on the front lines of the energy transition, a more just and equitable energy future is possible.

CHAPTER 2

Sacrifice Zones

On a warm November day in 2021, nine months after Winter Storm Uri devastated Texas, US Environmental Protection Agency (EPA) administrator Michael Regan sat on a porch of a small home on Lavender Street in Houston, Texas. Lavender Street is in Houston's Fifth Ward, a neighborhood about two miles northeast of downtown that is composed almost entirely of people of color, and where over 80 percent of residents identify as Hispanic or Black. The median household income of the Fifth Ward in 2022 was $42,660, which is approximately $20,000 less than that for all of Houston.[1] Sandra Edwards and her neighbors told Regan about the high cancer rate on her block, pointing to a nearby, abandoned industrial site where creosote was produced. Creosote is a carcinogenic chemical used to preserve wood. Edwards told Regan, "In the summertime, you can sit in this yard and watch the vapors come up. It's not Lavender Street anymore. We all know, this is Plume Street."[2]

Regan's visit to Houston came on day five of what the EPA called a "Journey to Justice" tour, in which the EPA head participated in listening sessions in communities across three southeastern states—Louisiana, Mississippi, and Texas. According to an agency press release, the tour's purpose was for Regan to spotlight enduring environmental-justice concerns in historically marginalized communities and to hear firsthand from people suffering from the severe impacts of pollution.[3] In his visits to these communities, Regan heard about underinvestment in water infrastructure, inattention to Superfund sites, and concerns about air pollution from oil refineries.[4]

Regan's "Journey to Justice" tour had special significance and some irony. Regan is the first Black man to lead the federal EPA, and only the second person of color—the first was Lisa Jackson during the Barack Obama administration. He is a son of the South, growing up in Goldsboro, North Carolina, which is less than a hundred miles from Warren County, North Carolina, which many consider to be the birthplace of the modern US environmental-justice movement.[5] Regan is the son of two public servants—his mother a nurse and his father an agricultural extension agent—and the first EPA administrator to have graduated from a historically Black college and university, the North Carolina Agricultural and Technical State University. Prior to leading the EPA, Regan served as the secretary of the North Carolina Department of Environmental Quality, where among other accomplishments he created the state's first advisory board on environmental justice and equity.[6] President Joe Biden's appointment of Regan as EPA administrator had symbolic importance, representing part of a broad effort to fill key leadership positions with new leaders, particularly women and people of color, who would bring with them clear commitments to achieve environmental justice.[7] Among the other high-profile appointments were Deb Haaland, the first Native American to lead the US Department of Interior, and Brenda Mallory, the first African American to lead the White House Council on Environmental Quality.

Juxtaposed against this renewed effort to bring environmental justice to the forefront of federal government leadership and decision-making were the communities that EPA Administrator Regan visited during his tour, many of which are part of what has long been known as "Cancer Alley," an eighty-five-mile industrial stretch that runs along the Mississippi River between Baton Rouge and New Orleans, Louisiana. The communities in this area represent a long legacy of pollution and struggle. These are places where historically marginalized Americans experience the burdens of the nation's energy choices. Cancer Alley is home to 150 petrochemical plants and oil refineries, operated by some of the largest companies in the world, including Exxon, Chevron, DuPont, and Shell. This area is a central cog of the US fossil fuel economy, and it is also among the most polluted areas of the country. According to data compiled by the EPA, companies in Louisiana release more total toxic chemicals into the environment per square mile than all other states, except Nevada.[8]

For communities in Cancer Alley, some rural and some urban, very little has changed for generations. They are places where pollution coincides with poverty; they are neighborhoods where people of color have been forced to live due to racist zoning and redlining practices[9] that have restricted housing options for people of color to the least desirable parts of town, often alongside areas zoned for heavy industrial pollution. These communities, and their northern counterparts such as the South Side of Chicago and Detroit's 48217 zip code, dubbed "Michigan's most toxic zip code,"[10] are the same types of communities that Robert Bullard, widely heralded as the father of environmental justice, profiled thirty years ago in his pathbreaking 1990 book on environmental injustice, *Dumping in Dixie*.[11] And they are the same communities that Yale Professor Dorceta Taylor refers to as "toxic communities"[12] and environmental-justice researcher Steve Lerner as "sacrifice zones"[13]—places where society dumps its pollution and waste on those without power, voice, or ability to escape.

The "sacrifice zones" of Cancer Alley provide just one example of places on the front lines of America's fossil-fuel-powered economy. People live, work, learn, and play alongside fossil fuels throughout the energy system—from sites of oil and natural gas extraction; to pipelines that distribute fuels for processing; to the oil refineries and chemical plants where fossil fuels are processed and manufactured; to the major roadways, highways, and ports where cars, trucks, and ships combust fossil fuels; to the places where we dispose of energy waste. Although parts of the fossil fuel economy are everywhere, from the East to West Coast and from northern to southern states, the worst effects are often concentrated on the doorsteps of the most vulnerable Americans.

Before we turn our attention in the coming chapters to those on the front lines of the US clean-energy transition, it is important to recognize the enormous inequities and injustices of the current energy system. The United States has benefited in profound ways from fossil fuels, especially during periods of plentiful and cheap coal, gas, and oil. It is no exaggeration to say that much of the economic prosperity of the post–World War II period that the United States and other high-income countries around the world have enjoyed is due to ample and inexpensive energy that fossil fuels have provided. Americans are periodically reminded of their reliance on a dependable and cheap supply of fossil fuels when there is a disruption. The oil crises of the 1970s are perhaps the most infamous illustration, but examples abound, including the spikes in oil and natural gas prices in 2022 that followed Russia's expanded invasion of Ukraine.

In recent decades, however, there has been a broad realization that reliance on fossil fuels has also come with a steep price. The production and use of fossil fuels has resulted in severe environmental degradation, a changed climate, and immense harm to human health and well-being. Much has been done historically to address these problems. Federal environmental laws such as the Clean Air Act and Clean Water Act have compelled firms in the fossil fuel sector and beyond to reduce their environmental impacts, and undoubtedly air and water quality are better for it. However, the benefits of these and other federal laws have not resulted in the same level of protection for everyone. For example, while air quality has improved throughout the country, there remain important disparities; places that were the most polluted fifty years ago are often still the most polluted today.[14]

In this chapter, we provide an overview of the historic and current burdens that the US energy system has imposed on millions of Americans, most often people of color and with low income. In this way, we demonstrate how energy decisions adversely affect some people more than others. Understanding this history is critical. Centering people in discussions of the energy transition first requires acknowledging past harms, an idea that environmental-justice advocates and scholars refer to as "recognition justice." History has a tendency to repeat itself, and attention to the past can help us better understand how the clean-energy transition may recreate similar challenges.

The discussion in this chapter is intended not to be exhaustive but rather to be illustrative of the many different ways that our use of fossil fuels has negatively affected disadvantaged communities throughout the United States. The evidence is overwhelming. In so many ways, current energy systems have created enormous burdens on the most vulnerable Americans, sacrificing their health and well-being as the country benefits from the use of fossil fuels. Climate change affects us all, but many of the burdens of fossil fuel use are localized, and our laws, policies, and practices have too often neglected those on the front lines.

Our attention will focus on the prevalence and patterns of the environmental and health effects associated with fossil fuels and other dimensions of the historical and current energy systems. However, the underlying political dynamics are unmistakable. Energy companies in the United States, especially those in the fossil fuel and utility industries, have exercised enormous political influence for decades.[15] This influence has created obstacles for making more progress on protecting human health and the environment and for taking on the challenge of climate change,[16] and it has contributed to the ongoing disproportionate impacts of their activities on historically marginalized communities.

Our discussion in this chapter focuses on four areas: extraction, distribution, processing, and consumption and disposal. One can think of these areas as stages of the fossil fuel life cycle, starting with how and where energy is produced, proceeding to how it moves and is converted for use in the economy, and ending with how it is consumed and disposed. The purpose of the discussion is to remind us that decisions about the energy sources we use have consequences for people's well-being. The transition to new technologies will come with similar challenges.

Pulling Energy from the Ground

The US economy still relies primarily on fossil fuels. Although renewable sources of energy like wind and solar power comprise an increasing share of the nation's energy portfolio, the fact remains that Americans still largely power and heat their homes and businesses with coal and natural gas and fuel our cars, buses, trucks, and planes with fuels derived from crude oil. In 2023, more than 80 percent of US primary energy consumption came from fossil fuels (see figure 2.1).[17]

The overall dominance of fossil fuels in the economy has existed for more than a century, and for most of the past sixty years, the United States has been a net importer of energy. Contentious relationships with many oil-rich nations in the Middle East, periodically resulting in violent conflict, often leave an impression that the United States is especially reliant on these countries for energy, especially oil. On this point, we are reminded of the popular bumper sticker in the aftermath of 9/11, *When you ride ALONE, you ride with bin Laden.* US foreign-policy focus on the Middle East, however, obscures two important realities. First, the United States actually imports more oil from its North American neighbors, Canada and Mexico, than from Middle Eastern nations. Second, and more importantly, the United States is also a major

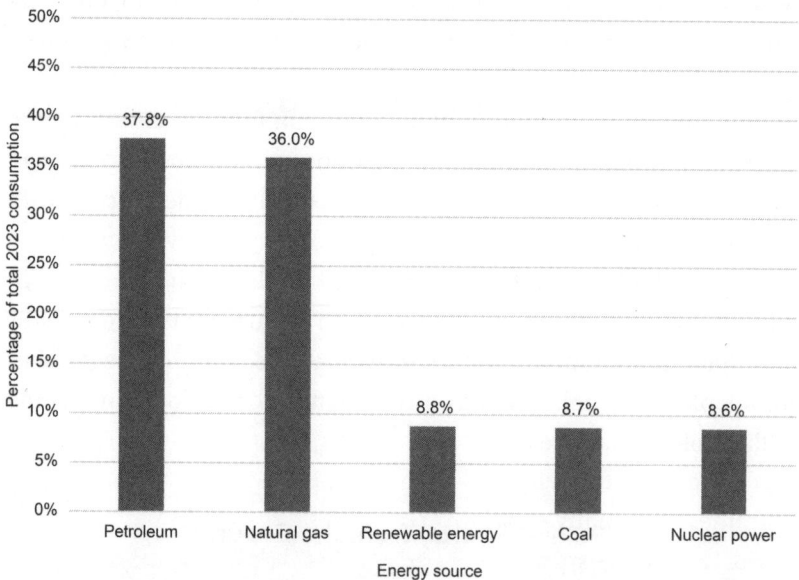

Figure 2.1 American primary energy consumption by source, 2023
Source: US Energy Information Administration, *Monthly Energy Review.*

producer of its own fossil energy. In fact, as a result of the "fracking revolution" that accelerated in the middle of the first decade of the 2000s, the United States became a net exporter of energy in 2019.[18]

Extracting fossil fuels is a dirty process. Whether drilling for oil, blowing off the top of mountains to expose coal seams, or fracking for natural gas from shale rock, fossil fuel extraction generates pollution, alters the landscape, and often exposes workers and nearby communities to hazardous substances that adversely affect their health. Fossil fuel production occurs throughout the country. Coal production takes place in twenty-three states, natural gas production in thirty-four states, and oil production in thirty-two states.[19] However, much of the production of these energy resources is geographically concentrated. So, while the country as a whole benefits from domestic production in terms of a reliable supply and lower energy prices, many of the environmental and health effects are clustered in specific communities.

Consider coal mining, which has a long and checkered history in the United States that dates back to at least 1775.[20] Coal mining brings to mind images of miners, often white, middle-aged men, emerging from dark shafts, faces blackened with dust. In 1950, there were about 380,000 coal mining jobs in the United States,[21] with a large majority in just a handful of Appalachian states—Kentucky, Pennsylvania, and West Virginia. As of 2023, there were just forty-three thousand active coal miners.[22] These substantial job losses are the result of many factors. Over many decades, mechanization and changing approaches to coal mining that involve less labor-intensive underground mining and more large-scale strip mining and mountaintop removal have reduced labor needs.[23] In more recent years, the massive decline in demand for coal to generate electricity, driven by declining prices for substitutes like natural gas and renewable sources, as well as by environmental regulations, has resulted in mine closures and company bankruptcies.

Coal mining has taken an enormous toll on miners and their communities, and many of its adverse consequences will persist for decades or more despite the industry's overall decline. In terms of impacts to coal workers, tens of thousands of miners have lost their lives while on the job over the past century.[24] In most cases, these deaths are due to accidents in underground mines from methane or coal dust explosions or roof collapses. The largest accident in recent years was the 2010 tragedy at the Upper Big Branch Mine in Raleigh County, West Virginia, which killed twenty-nine miners. The US Mine Safety and Health Administration's investigation of the disaster determined that the mine owner, Massey Energy, contributed to the accident by committing hundreds of safety violations. The CEO of Massey Energy, Don Blankenship, was later convicted of a misdemeanor for conspiring to willfully violate coal-mine safety standards, a crime for which he served a one-year federal prison sentence.[25] Fatalities due to on-the-job accidents are relatively uncommon, but mining for coal remains a dangerous job. On average, twenty-three Americans lose their lives each year working in coal mines.[26]

Coal mining is also associated with a number of respiratory illnesses, such as black lung disease, silicosis, and chronic obstructive pulmonary disease, all caused by the inhalation of coal dust.[27] Public-health researchers have routinely documented that coal miners suffer from these diseases at much higher rates than workers in other professions. Not only are they chronic, debilitating diseases that make breathing difficult, they can also all lead to premature death. A 2018 study from the US Centers for Disease Control and Prevention found that, on average, coal miners who die from black lung disease lose from eight to thirteen years of life.[28] Moreover, after a couple of decades of declining incidence, epidemiologists have identified a resurgence of black lung disease among coal miners, especially in Central Appalachia,[29] due perhaps to increased exposure to silica dust, the use of modern machinery, insufficient worker training, and longer work hours.[30]

The harsh health realities of coal mining are so well documented that, more than sixty years ago, Congress passed the Federal Coal Mine Health and Safety Act (1969) to partially address them. After years of lobbying by the United Mine Workers of America, this law, among other features, provided dedicated health and survivor benefits to coal workers and their families. The law created a worker's compensation program that included monthly benefits for coal miners dis-

abled because of black lung disease, as well as benefits for their de-
pendents and survivors in cases where the disease led to their death.[31]

The process of extracting coal from the ground has also permanently scarred the landscape, especially in heavily mined areas of
Appalachia and the Powder River Basin of Montana and Wyoming.
In some cases the damage is so severe that it is permanent, and the
land cannot be reclaimed or returned to productive use. In fact, one
of the earliest uses of the term "sacrifice area" was in the context of
strip coal mining, a practice where vegetation, rock, and soil are removed to expose coal seams below the surface. In a 1973 report from
the prestigious National Academy of Sciences and National Academy
of Engineering, a committee of scientists referred to some coal mining
areas in the American West as "national sacrifice areas."[32] Other common ecological effects of coal mining include acid mine drainage and
damages to nearby aquatic wildlife.

Coal mining, buoyed by a strong labor union, has provided countless Americans with well-paying jobs and, for many families, an income enabling them to move from poverty to the middle class. During
boom periods, coal mining has also provided important benefits to
communities in Appalachia and beyond through tax revenue and
complementary economic activity. Yet it is also impossible to ignore
the physical toll that coal production has had on people and the environment. And it is also the case that in many areas where coal mining
has been the only real source of economic opportunity, communities
still experience widespread and persistent poverty.[33] In the next chapter, we expand on the conditions facing coal communities and how
they reinforce poverty and adversely affect public health.

While coal mining has a long history in the United States, modern
hydraulic fracturing, or "fracking," is a much more recent phenomenon. The basic process of fracking involves creating fissures in shale
rock to release natural gas and oil. The process entails injecting high
volumes of water, mixed with sand and various chemicals, under extremely high pressure. The key innovations that sparked the massive
expansion of fracking were the introduction of a water-based fluid—
previous efforts relied upon gel-like substances, and in some cases

more flammable fluids such as gasoline and napalm—and the ability to turn drills horizontally, even when a mile or more underground. These innovations unlocked access to vast amounts of natural gas and oil in Pennsylvania, North Dakota, Texas, Ohio, Louisiana, Colorado, and many other states around the country.

The emergence of fracking in the United States resulted in a more than doubling of domestic oil production in just a decade. In 2010, energy companies produced about 5.5 million barrels of oil per day; in 2018, this figure had increased to over 12 million barrels per day, and it reached nearly 13 million barrels per day in 2023.[34] To put these numbers in perspective, this increase alone is larger than the current annual production of any other oil-producing country, except for Russia and Saudi Arabia. The increased domestic production of natural gas is just as remarkable. In 2007, the US produced less than 2 trillion cubic feet of natural gas from shale resources (about 8 percent of total US gas production); in 2022, the number was over 32 trillion cubic feet (nearly 75 percent of total US gas production).[35]

These developments had huge effects on domestic and global energy markets. The United States became a net exporter of oil, ending decades of a trade deficit in this area. Closer to home, the growing and inexpensive supply of natural gas is a major reason for the sharp decline in coal demand, in part leading many electric utility companies to shutter their coal plants because they can no longer compete with plentiful and inexpensive gas.[36] Also due in part to these trends, electric utility companies have canceled nearly all plans to construct new coal plants. And this shift from higher-polluting coal plants to lower-polluting natural-gas facilities contributed to large declines in power-sector emissions of carbon dioxide, smog-forming pollutants, and toxic chemicals like mercury.[37]

The fracking revolution also reshaped local communities. One place in America that has experienced the transformative consequences of fracking is western Pennsylvania, which is on top of a productive area of the Marcellus Shale region.[38] The once largely rural landscape of this part of Pennsylvania saw a modern-day "gold rush," as companies large and small feverishly dashed to sign leases with property owners to frack wells. Many people became suddenly rich in the process, leasing their land to energy companies for development.

For these and other reasons, there was considerable optimism around the fracking revolution in the early years, particularly with respect to natural gas. Both Democratic and Republican politicians praised it. Republican officials, buoyed by the oil and gas industry, have long supported the domestic production of fossil fuels, often decrying opposition from "radical, Democratic environmentalists" who they claim would rather save polar bears than provide Americans with a cheap, safe, and reliable domestic supply of energy. Many Democratic leaders have often voiced opposition, or at least concern, about fossil fuel production. In the case of fracking, however, many Democratic politicians were quite enthusiastic about the prospects of natural gas, often promoting its environmental benefits relative to coal. President Obama celebrated the fracking revolution in his 2013 State of the Union address, declaring: "The natural gas boom has led to cleaner power and greater energy independence. We need to encourage that. And that's why my administration will keep cutting red tape and speeding up new oil and gas permits."[39] This belief in the benefits of natural gas found its way to policy too. When President Obama proposed a clean-energy mandate, natural gas was included as a "clean" fuel.

Some people in the environmental community also celebrated fracking, referring to it as a "bridge" fuel in the fight against climate change. The Sierra Club, for example, supported the expansion of natural gas as part of a strategy to move the United States "beyond coal," even accepting money from the natural-gas industry to partially fund its anti-coal campaign.[40] The national organization's support for natural gas, however, was opposed by many of its members, which in some parts of the country were leading local efforts to oppose fracking.

The "clean" luster of the fracking revolution quickly wore off. Many controversies soon emerged, first among a small number of detractors and over time by many in the environmental- and public-health community. Initially, most of the worries pertained to contamination of drinking water. People in communities such as Dimock, Pennsylvania, complained that nearby fracking operations were poisoning their water wells. Nearby residents complained of being misled by energy companies who had assured them of the safety of fracking. Many people directed their frustration and outrage to state and federal regulators who they believed were complicit in allowing energy companies to come to town with lofty promises only to ruin their water and despoil their land. As fracking operations became more geographically widespread, opposition got louder, but it was often met with derision. Early detractors, such as Josh Fox, director of the 2010 Oscar-nominated anti-fracking film *Gasland*, were dismissed as alarmists.

Over time, however, scientific studies began to confirm many of these initial worries and document additional environmental damage from fracking. Problems such as chemical spills, methane leakage, air pollution, and contamination of groundwater and surface water turned up near fracking operations.[41] Studies also began to attribute a variety of adverse health effects related to fracking operations, due to poorly controlled emissions from well sites and inadequate management of wastewater. For example, research has found higher rates of preterm births, low birth weight, and birth defects for babies born closer to fracking sites.[42] Some studies suggest that the underlying cause of these poor birth outcomes are contaminated public drinking-water sources around fracking operations, affecting pregnant women nearby.[43] Importantly, these types of negative effects have been shown to be highly localized. One recent evaluation found that the largest health effects occur for pregnant women living within one kilometer of fracking sites, with few effects for pregnant women living at least three kilometers away.[44] Studies have also shown that fracking can reduce fertility; increase the incidence and severity of asthma, especially in kids; and lead to other respiratory and cardiovascular diseases.[45]

Uranium mining provides another illustration of the geographically concentrated, adverse effects of historical energy extraction in the United States. Uranium is a naturally occurring substance that, when enriched, can be used in a variety of ways, including for nuclear weapons and as a fuel source for generating electricity or powering naval submarines. Nuclear power remains an important part of the US electricity portfolio, accounting for about 20 percent of annual generation.

Since the late 1980s, US nuclear power plant operators have mainly imported the uranium they need. According to the most recent statistics from the US Department of Energy, about 95 percent of uranium utilized in nuclear reactors comes from abroad, largely from Kazakhstan, Canada, Australia, Russia, and Namibia.[46] This has not always been the case.[47] Historically, the US federal government pushed for domestic production of uranium, first to provide nuclear material for weapons and later as a fuel source for commercial power plants.[48] From 1948 to 1971, in fact, the US government was the sole purchaser of US-produced uranium, as mandated by law.[49] US reserves of uranium are mostly concentrated in Texas and the western part of the country, especially on Native American lands.

Although uranium mining is now uncommon in the United States— despite recent efforts to restart it—its troubled history is well documented, especially for people of the Navajo Nation.[50] According to one estimate from the EPA, there are over five hundred abandoned uranium mines on Navajo land, though other estimates place this number at over a thousand. Often, these abandoned mines have piles of tailings left on site by companies, tailings that often contain high concentrations of uranium and other toxic metals.[51]

One of the most well-known examples of energy companies leaving behind a toxic legacy of their operations on Navajo land are the abandoned uranium mines operated by the former nuclear giant Kerr-McGee (bought by Anadarko Petroleum Corp in 2006, which itself was later acquired by Occidental Petroleum). Kerr-McGee mined over seven million tons of ore on or near Navajo land from the late 1940s through the 1980s, including uranium mines, many of which the company hastily closed when demand diminished, and without any remediation. In 2015, Anadarko reached a $5 billion settlement with the US EPA to resolve claims that the company had fraudulently transferred assets to partially evade liability for these contaminated sites, with most of the money to be directed to remediation efforts at these abandoned mines and at other locations.[52]

The uncontrolled waste from these mine sites is not just an eyesore but a health hazard. The Navajo people have been exposed to radio-

active waste from abandoned mines through airborne dust and contaminated drinking water, and many of their homes were constructed
using contaminated materials.[53] The Navajo people also worked in the
mines when they were operational, often exposing themselves and
their families, unknowingly, to long-term health risks.

Public-health scholars Doug Brugge and Rob Goble have written
a detailed history of uranium mining and Navajo people, finding that
not only were Navajo miners exposed to dangerous levels of pollution,
but the facts were often hidden from them. Oral histories done with
former Navajo miners indicate that most were neither provided information about the potential hazards of working in the mines nor given
any protective equipment or ventilation to reduce their exposure to
radioactive materials.[54] The risks of uranium mining, and specifically
the potential for radon exposure to cause lung cancer, were known by
US government officials as early as the early 1940s, but records show
that little of this knowledge was communicated publicly. The US Public Health Service conducted a study of uranium miners working in
Colorado in 1950, finding higher rates of lung cancer among miners,
but focused their published findings on white miners only and ex-
cluded findings for minority miners. Later studies demonstrated some
of the devastating health effects for Navajo miners, in which about
two-thirds of lung cancer deaths among former miners came from
radon exposure.[55]

Historian Tracy Voyles characterizes the Navajo experience with
uranium mining as "wastelanding," whereby certain places are considered pollutable,[56] an idea similar to the notion of "sacrifice zones."
The toxic legacy of uranium mining on Native American lands extends
to other parts of the country as well. Open-pit uranium mines in Washington State have been blamed for high cancer rates for the Spokane
living on the Spokane Indian Reservation, and similarly for Lakota
living on the Pine Ridge Reservation in South Dakota and for Eastern
Shoshone and Northern Arapaho in the Wind River Reservation in
Wyoming.[57]

Communities on the front lines of energy extraction often resist,
fighting back against the activities that adversely affect their health

and the environment where they live. Examples include the Navajo people and other Indigenous groups for whom energy extraction is perceived as another form of land dispossession;[58] local towns near fracking operations worried about pollution and the industrialization of their communities; and people in Appalachia concerned about permanently scarred landscapes and the short and long-term effects of coal mining.[59] This resistance usually faces strong countervailing pressure from industry, often from government agencies, and from others who economically benefit from the extraction.[60] Of course, perspectives on what is right and wrong when it comes to energy extraction vary, but often there is a feeling of unrequited sacrifice.

A common theme with energy extraction, whether in the context of coal, natural gas, or uranium, is that most of the environmental and health effects are concentrated in the places where it occurs. And, quite often, these effects are far away from where the fuels are ultimately used by consumers or companies. This geographic mismatch—places where the benefits are enjoyed being different from the places where the costs are incurred—is present across all energy sources, including newer energy sources that are a part of the clean-energy transition, as we will discuss in future chapters.

Moving Energy Across the Country

Often, when one thinks about energy infrastructure, the images that come to mind are soaring smokestacks at power plants, massive cooling towers at nuclear facilities, or sky-touching wind turbines. These large industrial facilities leave an impression. For people who live nearby, they are an inescapable part of the local landscape, whether perceived as attractive or not. Most Americans, however, rarely see these energy behemoths. For most, the sight of a large power plant is rare, perhaps occurring only when driving along an interstate on a long-distance trip.

Much of the energy system is hidden from plain sight. Pipelines provide perhaps the best example. Few Americans think about, let alone see, the mostly invisible network of pipelines that criss-cross the country. Yet there are about three million miles of pipelines in the United States, transporting oil, natural gas, and finished fossil fuel products (e.g., jet fuel and propane) from coast to coast.[61] By comparison, the entire US interstate system consists of about forty-seven thousand miles of highways.[62] These pipelines serve a critical role, providing a mostly safe and efficient way to distribute the energy that powers our economy.

However, pipelines can leak, sometimes catastrophically so. On average over the past decade, there are 644 total spills a year, about half of which are considered significant,[63] and ruptures to natural-gas lines can lead to explosions causing loss of life and property.[64] Natural-gas explosions in residential neighborhoods are often deadly. In 2010, a natural-gas pipeline owned by Pacific Gas and Electric burst into flames in San Bruno, California, leveling dozens of homes and killing several people.[65] At a national scale, one recent study found that natural-gas pipelines are more concentrated in US counties with more socially vulnerable populations, as measured by indicators such as poverty levels, unemployment, educational attainment, and race and ethnicity.[66]

In general, we do not think about the vast network of pipelines because there is no reason to—"out of sight, out of mind," as the saying goes. Yet, many in the environmental community have made a concerted effort to block new pipelines, as part of their fight to address climate change. This effort has brought considerable attention to pipelines, to the point where projects like Keystone XL and Dakota Access PL have become household names for many Americans.

Much of the recent controversy over pipelines began with the Keystone XL project, a proposed pipeline backed by the Canadian company TC Energy to transport oil from the tar sands fields of Alberta, Canada, to Steele City, Nebraska. From here, the oil would be sent through other pipelines to oil refineries in places like Whiting, Indiana, and Port Arthur, Texas, for processing. Proponents of Keystone XL emphasized the thousands of construction jobs that would be created to complete the project, as well as the potential to enhance the reliability of the oil supply for the United States.

Much of the opposition to the Keystone XL project came from environmental activists, such as Bill McKibben, the founder of 350.org. For McKibben, Keystone XL represented a step backward in the fight against climate change. In his view, the pipeline would only enable the production of more oil, and with it, more carbon pollution contributing to climate change. Moreover, in McKibben's view, President Obama was uniquely positioned to stop it. McKibben wrote: "Keystone XL is such a huge deal because the president can actually stop it himself, without consulting our inane Congress. That's why we'll be surrounding the White House on Nov. 6 [2011], circling it with people simply holding signs with quotes from his campaign. Like, 'it's time to end the tyranny of oil.'"[67] The opposition was ultimately successful, leading the Obama administration to withdraw a key permit for the project. Although the Trump administration provided TC Energy a lifeline, the company officially abandoned the project several months after President Biden denied a necessary permit on his first day in office.

Most environmental activists fought Keystone XL because of climate change, and this part of the story captured most of the national headlines. However, local opposition to Keystone XL was also fierce. Many landowners in places like rural Nebraska—not typically thought of as a bastion of environmental activism—had deep concerns about the project's direct effects on them, their families, and their land. In Nebraska, many of the concerns centered on the potential for spills that could contaminate already-threatened groundwater aquifers, sources of water critical for drinking water, irrigation, and the raising of livestock.[68] The types of political and ideological concerns that often explain attitudes toward energy projects nationally are often less prominent for those living near a proposed pipeline route.[69] For local neighbors to pipelines, the concerns typically are more about spills, loss of property value, and the unfair use of eminent domain by government officials to seize their property. As a result, the coalitions of opposition that emerge in some cases include strange bedfellows.

At the same time that the debate about Keystone XL raged on, controversy erupted over another pipeline project, the Dakota Access Pipeline (DAPL). DAPL, a project developed by Energy Transfer Partners, began construction in 2016 and was designed to carry crude oil extracted from the fracking fields of northwestern North Dakota through South Dakota and Iowa, ending at a distribution terminal in Patoka, Illinois. By the time DAPL became a national story, much of the 1,168-mile pipeline had been completed. However, a small segment of the pipeline intended to cross underneath the Missouri River first required a Clean Water Act permit from the US Army Corps of Engineers. In July 2016, the Army Corps issued the permit, with the Omaha district commander, Col. John Henderson, writing, "I have evaluated the anticipated environmental, economic, cultural, and social effects, and any cumulative effects" of the river crossing and found it is "not injurious to the public interest."[70]

"The public interest" is a curious phrase. It implies that the US Army Corps made an objective evaluation that weighed the aggregate costs and benefits of the pipeline. In so doing, however, the Army Corps seems to have willfully ignored the vocal concerns expressed by the Standing Rock Sioux tribe that the pipeline would harm culturally important sites and threaten the quality of their source of drinking water. Moreover, the choice to direct the pipeline under the Missouri River came only after another route was determined to be inappropriate. The original route called for the DAPL project to run north of Bismarck, North Dakota, the predominantly white capital of the state. The Army Corps, however, determined that the site was a "high-consequence area" that might threaten the water supply. Ironically, or perhaps not, the Army Corps did not make a similar judgment regarding the water supply for the Standing Rock Sioux tribe.[71]

What ensued was a legal and political standoff. The Standing Rock Sioux tribe both pursued a legal injunction in federal court to stop the pipeline and simultaneously staged a demonstration that lasted for six months to prevent completion of the pipeline. The camp at Standing Rock at its peak consisted of hundreds of protestors, living outdoors through a bitter winter to stand their ground and to fight for their cultural heritage. The demonstrations were also violently repressed, with at one point, police responding with water cannons on a cold night to break up the gathering of protestors.[72]

The standoff at Standing Rock was about more than a pipeline, water quality, and risks of accidents. At the heart of the controversy were questions about Indigenous rights, tribal sovereignty, and disputed claims to land. The route for the DAPL is on land that Sioux believe is theirs, granted to them as part of the 1868 Fort Laramie Treaty. This claim is not recognized by federal and state authorities, who instead maintain that Energy Transfer Partners owns the land. In this way, the fight over DAPL was about land appropriation and tribal self-determination. This is a common story. Many of those who carry the burdens of our energy system are those who have been historically repressed, diminished, or enslaved, and these same people are now fighting for representation and equality as part of the transition to a cleaner energy system.

In different ways, the controversies over Keystone XL and DAPL illustrate the major disruptions that fossil fuel infrastructure can have on local communities. Even if one sets aside questions such as whether pipelines are necessary and whether this mode of transporting energy resources across long distances is safe (or, at least safer than the alternatives), they illustrate that certain communities feel the burdens more than others. These burdens come in the form of loss of control—either through eminent domain or assertions of sovereignty over disputed lands—and potential environmental and health burdens if accidents later occur. In many cases, these burdens are not new but a continuation of past exploitation. As is the case with many parts of the fossil fuel economy, it is often the most vulnerable populations who experience the impositions on their lives and communities.

Processing Fossil Fuels for Use

Pipelines are just one small piece of an immense fossil fuel infrastructure that traverses the United States. DAPL and projects like Keystone XL are intended to move oil and natural gas from their points of extraction to where they are processed or delivered to markets. And, just as the impacts of pipelines are concentrated on those living nearby, the same can be said for the impacts of the facilities that process fossil fuels.

Over the past several decades, countless studies have documented that people of color, and especially Black Americans, are more likely to live in close proximity to major industrial sources of pollution and contaminated sites. The evidence is overwhelming.[73] Whether one is considering hazardous-waste facilities, garbage dumps, incinerators, power plants, oil refineries, or other large polluters, study after study has shown that these types of facilities are disproportionately located in disadvantaged communities. Companies often site these facilities, with government approval, despite vocal opposition from local communities. This type of opposition in places like Kettleman City, California; Warren County, North Carolina; and Chester, Pennsylvania gave rise to the environmental-justice movement in the 1980s.

Certain places appear targeted for these types of facilities, and it is not uncommon for some communities to "host" numerous major sources of pollution. These overburdened areas have been characterized as sacrifice zones, and neighborhoods adjacent to facilities are often called "fenceline communities"—places where people live next door to large sources of pollution often separated only by a chain-link fence. For many Americans, the view from their front porch is not a white picket fence or a friendly neighbor, but an oil refinery, a petrochemical storage tank, or a concrete batch plant.

The tendency for large sources of pollution to be located in communities of color is especially prominent in the case of facilities that process fossil fuels. Oil refineries provide a good example. In 2022, there were 130 oil refineries operating in the United States, varying in size from small operations that process a few thousand barrels per day to incredibly large ones that process half a million barrels per day.[74] For context, processing one forty-two-gallon barrel of crude oil produces about twenty gallons of gasoline.[75] The refineries are owned and operated by some of the largest companies in the world like Exxon,

Valero, BP, and Marathon, as well as state-owned enterprises such as Saudi Aramco and Petroleum of Venezuela. Many of these companies feature prominently on lists of companies that release the most toxic and climate pollution in the United States.[76] An analysis of the population characteristics of these facilities shows that, within one mile, on average 21 percent of the residents are Black, 25 percent are Hispanic, and 15 percent are people living below the federal poverty line.[77] Unlike the location of coal mines or oil wells, which are mostly dictated by natural factors, companies select the location of oil refineries for other reasons, which may be related to where historically marginalized populations reside.

Oil refineries are important cogs in the modern economy. Refineries not only process crude oil into gasoline for cars and trucks, they also produce a variety of other products such as jet fuel, heating oil, and asphalt. The refining of petroleum comes with a substantial environmental footprint. Oil refineries are major sources of hazardous air pollutants, including benzene, toluene, and xylene(s), as well as conventional air pollutants such as particulate matter, sulfur dioxide, nitrogen oxides, carbon monoxide, and hydrogen sulfide. A typical oil refinery releases thousands of pounds of these pollutants each year. And they do not only pollute the air. Oil refineries can also pollute groundwater and surface-water resources. The wastewater created in refining may include high concentrations of sulfides, ammonia, and suspended solids, all of which can pollute water if not first treated.[78]

As large industrial sources of pollution, oil refineries are highly regulated. Legally operating an oil refinery requires numerous government permits that specify limits on the releases of pollution to the air, water, and land. However, the EPA and state agencies routinely find that oil refineries violate pollution control laws. Since 2000, the EPA has entered into thirty-seven separate Clean Air Act enforcement settlements with US companies that comprise over 95 percent of the country's petroleum refining capacity. These settlements, which resolve violations of national air-quality regulations, cover 112 refineries in thirty-two states and territories, including BP, Chevron, ConocoPhillips, Exxon,

Koch Industries, Shell, and Valero, among others.[79] Moreover, even when these types of facilities fully comply with their permits, they still may release dangerous amounts of pollution.

Port Arthur, Texas, could be considered "refinery central" in the United States, and its circumstances highlight the burden these facilities place on local communities. Located at the southeast corner of Texas, along the border of Louisiana, Port Arthur is situated just north of the Gulf of Mexico, enabling easy access to both vast oil fields and shipping channels. The city is bordered on its east side by Sabine Lake, a ninety-thousand-acre saltwater estuary. The Port Arthur side of Sabine Lake is composed of a heavily developed industrial landscape, mostly made up of petrochemical facilities. Sabine Lake is connected to the Gulf of Mexico by the Sabine-Neches Canal, a dredged shipping lane. In stark contrast, the Louisiana side of Sabine Lake is largely undeveloped, and the area is mostly preserved as part of the Sabine National Wildlife Refuge.[80]

Port Arthur is a city of fifty-five thousand people, with about two-thirds being people of color (over 40 percent Black and over 30 percent Hispanic). Median household income in Port Arthur is $43,000, which is about $25,000 less than the country as whole, and a quarter of its residents have income below the federal poverty line.[81] It is perhaps no surprise that a community with these characteristics is the home to some of the largest refineries in the country. In fact, Port Arthur is home to the second largest oil refinery in the United States, now fully owned by Saudi Aramco, the Saudi government's oil company.[82] Saudi Aramco's Port Arthur refinery has a capacity of 626,000 barrels per day. Two other large refineries are also located in Port Arthur, a Valero refinery that has daily capacity of 335,000 barrels and a Total refinery that has a daily capacity of 238,000 barrels. Altogether, the capacity of these three Port Arthur refineries represents about 7 percent of the total refining capacity in the United States.[83]

When environmental-justice advocates talk about fenceline communities, they have places like Port Arthur in mind. The Saudi Aramco and Valero refineries abut each other, separated only by Texas State Highway 82 (also known as W. Levee Road). Tucked in between is a neighborhood of modest homes on tree-lined streets, separated on the Saudi Aramco side by a combination of wooden and chain-linked fences and railroad tracks and on the Valero side by fencing and Highway 82. In both cases, homes are within just a few hundred feet of the refineries, and not surprisingly, residents complain about the constant sulfur smell and the toxic chemicals spewing into the air. The refineries are located in parts of the city where segregated Black Americans were compelled to live and in areas that were redlined for much of the twentieth century.[84] Residents who live nearby feel disposable and unseen. Local community activist Hilton Kelley summarizes this sentiment: "People look at us as expendable. They say, 'Statistically speaking, only a small minority of people are going to be impacted.' But the people they're talking about are Port Arthurians. My town is *not* expendable. We're living, breathing human beings who deserve a better quality of life than what we're getting."[85]

The spotty environmental compliance record of the oil refining sector is on full display with the Port Arthur facilities. An examination of EPA data shows that the refineries operated by Motiva (owned by Saudi Aramco), Valero, and Total have all frequently violated their environmental permits. These facilities also routinely have large "excess emissions events," which are releases of air pollutants due to facility malfunctions, startups, and shutdowns. These pollution events are violations of the federal Clean Air Act, though they are not always treated as such as by the state of Texas, which often excuses facilities from any wrongdoing in these cases. Yet these events are frequent in magnitude, especially in the oil refining sector.[86] As a case in point, Valero's refinery in Port Arthur had nearly eight hundred such excess emission events between 2005 and 2022 (an average of forty-seven a year), and the Total refinery had over a thousand such events over a similar period.[87]

Directly attributing adverse health outcomes to specific pollution sources is difficult, but the data for Jefferson County, where Port Arthur is located, is certainly suggestive. Adult asthma rates in Jefferson County are 33 percent higher than in the rest of the state.[88] Other sources of pollution beyond the Port Arthur facilities certainly affect air quality in the county—for example, Beaumont is also located in Jefferson County, and it too hosts numerous petrochemical facilities, including an Exxon refinery which is the seventh largest in the country. The general point, however, remains the same.

Residents of Port Arthur are not passive observers of the pollution burdens they endure. There has been a tremendous amount of community activism to protest the high levels of pollution generated by the refineries and other industrial sites in the city. The Environmental Integrity Project, an environmental advocacy organization that works extensively in Texas, has documented the efforts by Port Arthur residents, which has included the formation of local organizations to push back against refinery expansions and the creation of community air-quality monitoring networks. In October 2021, the EPA agreed to investigate a claim by local residents and environmental groups that the state of Texas is violating the civil rights of Port Arthur residents through its lax oversight of the Oxbow Calcining plant, which produces petroleum coke.[89]

Port Arthur may be an exceptional case, but oil refineries and other petrochemical plants are located throughout the country, quite often quite close to where historically marginalized groups live. From south Los Angeles, California, to Joliet, Illinois, to Paulsboro, New Jersey, these refineries impact the quality of life for local residents, providing one more example of the concentrated burdens of the US fossil fuel economy.

Pollution and Waste from Using Fossil Fuels

Public attention to fossil fuels in recent decades has focused mostly on climate change. More so than any other activity, the burning of fossil fuels to generate electricity for homes and businesses and to power cars and trucks has resulted in the warming of the planet. The effects of carbon dioxide pollution from burning coal, oil, and natural gas are not geographically concentrated. It does not matter if a ton of carbon dioxide comes from a coal plant in Petersburg, Indiana, or Beijing, China. While the geography of climate change impacts will vary—for example, some places will experience drought, while others will see extreme rain and flooding—the location of the impacts is incidental to the location of the source of the carbon pollution.

The same cannot be said for most other pollutants that are emitted from the burning of fossil fuels. The combustion of coal, oil, and natural gas releases sulfur dioxide and nitrogen oxides that contribute to smog and can lead to the formation of particulate matter. In addition, burning fossil fuels results in the emissions of toxic substances, including heavy metals like mercury and arsenic. These same hazardous chemicals may also find their way into water discharges from fossil fuel plants and eventually into rivers and lakes. Importantly, this pollution tends to have more localized impacts. Due to these factors, the populations that live near places where fossil fuels are used—power plants, highways, and ports—often feel the brunt of the pollution.

Local air and water pollution, of course, are not new problems, and one might assume that an advanced, wealthy nation like the United States has already addressed them. After all, it has been more than fifty years since Congress enacted the Clean Air Act, the Clean Water Act, and other pollution control laws. While air and water quality have unquestionably improved as a result of these laws, pollution has by no means been eliminated or even reduced to a point where it does not create health and ecological damage.

Take the example of air pollution, according to a 2022 report from the American Lung Association, four in ten Americans—about 130 million people total—live in an area of the country that does not meet EPA's national air-quality standards.[90] Poor air quality is not just a nuisance or inconvenience. Exposure to polluted air causes and exacerbates respiratory and cardiovascular disease, and it is also a leading cause of death. One estimate from the World Health Organization found that outdoor ambient air pollution caused 4.2 million premature deaths worldwide in 2016, mostly from exposure to fine particulate matter (i.e., very small solid or liquid particles that are inhaled and get lodged in our lungs).[91] While most of these deaths occur in low- and middle-income countries, studies have estimated that more than a hundred thousand Americans die prematurely each year because of exposure to airborne particulate matter.[92] This tally is more than double the number of people who die from gun violence,[93] suicide,[94] or car accidents.[95] Environmental scholars characterize the harm and death toll caused by pollution as "slow violence," though it is rarely talked about in this way.[96]

Disguised in these numbers, however, is the recurring pattern that we have highlighted throughout this chapter—the adverse impacts of burning fossil fuels are not shared equally.[97] The same American Lung Association report found that people of color in the United States are much more likely to live in places with poor air quality, a fact supported by considerable research.[98] The main reason is that people of color are more likely to live in close proximity to the places where we burn these fuels.

Let's consider some examples. First, take the case of power plants. EPA recently conducted an analysis of the demographic composition of people living in close proximity—within three miles—to power plants compared to the rest of the country, finding that 53 percent of the people living near power plants are people of color and 34 percent are low-income, both of which figures exceed the national averages of 40 and 30 percent, respectively.[99] Another recent study shows that electric utility companies have often sited coal plants in neighborhoods in "redlined" sections of cities,[100] thereby exposing already-marginalized groups to further burdens.[101] These decisions did not rest with utility companies alone, and they also reflect local zoning practices that often placed industrial zones in close proximity to redlined areas. The study of power plants in redlined areas, moreover, showed that current air pollution is highest in these areas. That is, decisions by electric utility providers to site power plants many decades ago, are responsible for today's disproportionate air pollution burdens. This fact does not absolve current policy and practice, however, which help perpetuate these burdens through ongoing environmental permitting and sometimes lax enforcement.[102]

Pollution from power plants, like that from oil refineries and other large petrochemical facilities, places burdens on local residents. In many cases, however, these kinds of pollution are not the only environmental burden facing these residents. Quite often, residents also find themselves living in heavily industrialized areas where there are other sources of potential harm to their well-being. In many of these same places, there are major highways, railroads, ports, and other types of transportation infrastructure near their neighborhoods. The transportation sector is not only the largest contributor of greenhouse gas emissions in the United States, it is also the largest contributor to several other major air pollutants that are emitted from the exhaust pipes of cars and trucks.[103] For instance, EPA data indicates that more than 50 percent of all nitrogen oxide (NOx) emissions and 15 percent of all volatile organic compound emissions come from the transportation sector.[104] Traffic on highways is a particularly important source of these emissions. Although tailpipe exhaust from all gas-powered cars contributes emissions, emissions from diesel-powered heavy commercial trucks—such as the big rigs that you see on highways—are the largest emitters of pollutants like NOx.

A variety of studies have documented high levels of PM2.5, NOx, and CO emissions near major roadways. Moreover, researchers have found higher levels of adverse health outcomes for people living near highways, including higher rates of asthma and lung cancer, heart attacks, cardiopulmonary and stroke mortality, preeclampsia for pregnant women, and preterm births. And, in a now all-too-familiar pattern, these health burdens are concentrated on some segments of US society. People of color and those with low income disproportionately live near highways, and consequently they are more likely to experience the pollution burdens from the associated traffic.[105]

The co-location of highways and low-income communities of color is often no coincidence. In many cases, in cities across the country, federal, state, and local transportation agencies decided to route federal highways through Black and Hispanic areas. In the process, historical neighborhoods and business districts were carved up, families and long-term neighbors were physically separated, and property values were decimated. Just some of the examples include Interstate 40 in Houston, Interstates 65 and 70 in Indianapolis, Interstate 75 in Cincinnati and Detroit, Interstate 81 in Syracuse, Interstate 10 in Los Angeles, and Interstate 95 in Miami.[106] As was the case with redlining and the siting of coal power plants, the cases of placing highways in communities of color illustrates how land-use decisions, often of generations past, have persistent and long-run effects on those who live nearby. And, as in other instances of environmental injustice, grassroots organizations and local communities have often fought back.[107]

The most harmful pollution from the use of fossil fuels is air emissions, but air emissions are not the only byproduct of fossil fuel consumption. Waste comes in other forms too, such as the different types of water pollution that we have highlighted in this chapter—coal mining can destroy streams; fracking-related operations can contaminate groundwater aquifers; coal plants and oil refineries directly discharge heavy metals and chemicals into rivers and lakes; and pipelines leak, sometimes catastrophically, affecting nearby waterways.

Beyond these forms of pollution, there are other byproducts of fossil fuel use that may result in contamination of land. As an example, of the more than 1,300 contaminated areas around the country that have been designated as Superfund sites, many are related to the processing or disposal of fossil-fuel-related waste, including the Diamond Head Oil Refinery in Kearny, New Jersey; Wilcox Oil Company in Bristow, Oklahoma; and the Sinclair Refinery in Allegany County, New York.[108]

Another type of waste byproduct that has received considerable attention in recent years is coal ash. Coal ash, also known as coal combustion residuals, includes many byproducts of burning coal (e.g., fine ash, boiler slag), and it is usually either disposed of by placing it in on-site surface impoundments or landfills, or recycled as fill material in products like concrete or wallboard (about half the coal ash produced in the US is reused in this way).[109] Coal ash often contains dangerous chemicals that remain after combustion, including mercury, arsenic, and cadmium. These heavy metals are highly regulated when released into the air or water, and the EPA also has strict standards for cleaning up soils contaminated with these metals, for example when they are found at Superfund sites.

Coal ash is not new. After all, the United States has relied heavily on burning coal for electricity for nearly 150 years. In fact, coal ash is the second-largest waste material in the United States, ranking only behind household trash. Nonetheless, it did not emerge as a major policy issue until 2008, and only after a major accident. In December of that year, a coal-waste storage pond in Harriman, Tennessee, at the Tennessee Valley Authority's Kingston power plant caved in, spilling nearly four billion liters of a coal ash slurry.[110] As NPR reported at the time, the toxic sludge "gushed out like a tidal wave" and in the process destroyed several homes and covered nearly three hundred acres with as much as nine feet of "grayish muck."[111] The cleanup took until 2015 at a cost of $1.1 billion.[112]

The coal ash spill in Harriman spurred the federal government to take action. The EPA in 2015 put in place new regulations that included standards for coal ash ponds as well as new groundwater-monitoring requirements. However, in deciding how to handle coal ash, EPA decided to classify it as a "solid" rather than "hazardous" waste, a decision derided by many in the environmental community as inadequate given that coal ash is well understood to contain toxic substances.[113] One implication of this decision was that states, not the EPA, would take the lead role in managing coal ash, which also troubled many given states' uneven commitment to protecting the environment.

Coal ash is an environmental-justice issue for a couple of reasons. First is the fact that coal plants themselves tend to be located in communities of color, as we discussed earlier. The on-site disposal of coal ash thus creates an ongoing risk for these communities, especially in places where the impoundments are located near drinking-water sources.

An example is AES Indiana's Harding Street Station, which is located in south Indianapolis. Harding Street Station is a power plant that has provided electricity for Indianapolis since 1931. Previously owned and operated by Indianapolis Power and Light (now by AES Indiana), the former coal power plant was converted to a natural-gas power plant in 2016.[114] Harding Street Station is located just feet from the banks of the White River, one of the main sources of drinking water for the more than eight hundred thousand residents of Indianapolis. There are four coal ash ponds at Harding Street Station, and recent groundwater monitoring reveals that each has leaked heavy metals, including arsenic, at levels above current standards.[115] In fact, the groundwater monitoring that has occurred across the country as part of the 2015 EPA rule shows that nearly all coal ash ponds, especially those constructed without the use of liners, have produced similar contamination. A 2022 report from the Environmental Integrity Project that reviewed these monitoring results nationwide found that 91 percent of coal power plants, across forty-three states, have coal ash ponds that are contaminating groundwater.[116] The report further found that at nearly half of the plants, the owner has not accepted responsibility for the contamination and is not currently planning remediation.

A second way that coal ash may concentrate risks on vulnerable populations is through off-site disposal. While most coal ash waste is managed on-site at the same location where it is generated, in some cases it is transported elsewhere for disposal. This was the case after the Kingston accident in Tennessee. As part of the cleanup of the toxic sludge generated by the coal ash spill, the Tennessee Valley Authority (TVA) decided to send the contaminated soil three hundred miles away. The location was the Arrowhead Landfill in Uniontown, Alabama. Uniontown is a small, rural town in the central part of Alabama. It has a population of about two thousand and is composed mostly of poor African American residents—97 percent of the residents of Uniontown are Black, the median household income is $22,159 (compared to $54,943 for all of Alabama), and more than 60 percent of the population are living below the poverty line.[117] Arrowhead Environmental Partners, the operator of the landfill, proudly advertises the facility as an ideal location for disposing of waste, and notes that it serves thirty-three states as an Alabama Department of Environmental Management (ADEM)- and EPA-permitted facility that accepts waste of numerous kinds.[118] In the case of the coal ash, ADEM modified the permit for the Arrowhead Landfill to license it to take the coal ash.[119]

Setting aside debates about the safety and performance of the facility, the disposing of the TVA coal ash waste at the Arrowhead Landfill represents a classic environmental injustice. The waste from one community—in this case, a mostly white, middle-class community in

Tennessee, is shipped a hundred miles away to a poor community of color for disposal. And, to be clear, most Uniontown residents did not accept the coal ash waste with open arms. Quite the contrary. Community residents have tried to fight back, buoyed by local churches and civil rights organizations and supported by environmental groups such as Earthjustice. Uniontown residents filed a Civil Rights Act complaint against ADEM for changing the Arrowhead permit without, in their view, appropriate attention to safety and public involvement.[120]

Uniontown, Alabama, has emerged as a contemporary battleground for environmental justice. Yet, as we have made clear throughout this chapter, it is just one more place on the front lines of the environmental and health burdens resulting from a general reliance on fossil fuels. Throughout the energy system, from extraction and distribution to processing and consumption, the cases of concentrated benefits on disadvantaged communities are ubiquitous.

The focus of this chapter has been on the past and present. We have demonstrated the various ways in which our historical energy system has had disproportionate effects on some people and some communities. By and large, the impacts have been more severe for people of color and those with low income; the same Americans who have been historically repressed and marginalized and the same Americans who often lack equal access to health care, high-quality public schools, and fair opportunities in housing and employment. These social vulnerabilities are exacerbated by the environmental and health issues associated with the production and consumption of fossil fuels.

We turn our attention in the chapters to come on the burdens that may come with the transition away from these fossil fuels. In important ways the burdens will be different, but they may also fall disproportionately on some of the same populations that have borne the burdens of the past.

CHAPTER 3

Beaten, Broken, Forgotten

In a church basement in Williamson, West Virginia, in the summer of 2016, a group of former and current coal miners and community members sat in a circle. They were a part of a focus group series that we were convening in West Virginia, including residents from surrounding areas, on the impacts of the energy transition on local communities. Sean, Tom, and Carla sat scattered around the circle, with rough familiarity of each other before the focus group but so in sync during the session that they were finishing each other's sentences.

SEAN: My biggest issue is that I don't have a problem with the transition. I really don't as long as we can bring something here that's gonna replace the jobs and the money.

TOM: That's the way I feel too.

TOM: I don't care what the job is as long as there's a job to do.

SEAN: Exactly.

TOM: I could care less what it is.

SEAN: Exactly, if our people can have good careers and good healthy lives and work.

CARLA: And that's the thing, good, healthy lives, I mean our miners here they dedicate their life and they you know they . . .

SEAN: Beaten and broken.

An hour and half north, in Huntington, we also interviewed Roger, who was working for a nonprofit focused on coal and the energy transition. Roger, a lifelong native of the region, and someone who is deeply enmeshed in both the politics and the human side of the energy transition in the region, was very direct. He explained that those around him feel forgotten: "I feel people are really bitter and nobody has their interest at heart. They feel like the country used them for energy while it bothered them and now that it's done with them, they kind of, like, forget."

Such sentiments of sacrifice and abandonment are connected to a long history of US dependence on coal to power its economy and a reliance on certain regions to extract their local resources for the sake of this broader mission. In exchange, energy communities were offered jobs, in some places well-paying jobs, and local revenue to support public services. But these jobs also came with a set of costs—or sacrifices—in the form of public-health, environmental, and economic challenges for host communities. This chapter summarizes these challenges, contextualized in history, and discusses the ramifications for modern energy communities as the energy transition unfolds. We begin with a discussion about coal, and specifically coal consumption and production, before transitioning into other fossil fuel resources and the impacts of the energy transition on legacy fossil-fuel communities.

Coal Addictions

In March 2021, the United Nations Secretary-General, António Guterres, gave an impassioned speech to the Powering Past Coal Alliance Summit in which he urged countries to end their "addiction to coal." He explained: "Phasing out coal from the electricity sector is the single most important step to get in line with the 1.5-degree Celsius goal. This means that global coal use in electricity generation must fall by 80% below 2010 levels by 2030. Once upon a time, coal brought cheap electricity to entire regions and vital jobs to communities. Those days are gone."[1]

The United States and many other countries around the world unquestionably have been "addicted" to coal. US reliance on coal started as early as the late 1800s, and even earlier in other parts of the world such as Europe, when coal replaced wood as the dominant fuel source and also produced the electricity that replaced other leading lighting sources such as whale oil, kerosene, and gas. From that point onward, up through the early years of the first decade of the 2000s, coal accounted for the vast majority of electricity generation in the United States. Over the same time period, there was also a rapid growth in electricity demand and thus significant coal production expansion. Coal brought inexpensive, ample, and reliable electricity.

Coal also brought relatively high-paying jobs for generations of coal miners. In many coal mining communities across the United States, multiple generations of family members have worked in the industry, with sons following in the footsteps of their fathers and fathers following in the footsteps of *their* fathers. These jobs often required relatively little formal education (e.g., a high school diploma), and for many they have been an attractive alternative to seeking a college degree. In the words of Daniel Alexander Hawkins, a coal miner in Appalachia, being a coal miner requires "common sense, a strong work ethic, and a clean drug test,"[2] which, in turn, provides steady employment and a good salary. These benefits, however, do not come without personal health sacrifices. As just one important example, discussed in the last chapter, coal miners have historically been exposed to coal dust while operating in the mines. Over time, coal dust builds up in one's lungs and generates scar tissue, and ultimately contributes to coal workers' pneumoconiosis, or black lung disease, among other pulmonary complications.

In the words of UN Secretary-General Guterres, "those days are gone"; and in the words of energy economist Susan Tierney, "the coal industry is troubled."[3] US coal production has declined from a peak in 2008 of 1.17 billion short tons to 0.54 billion short tons in 2020 (one short ton is two thousand pounds), for a total loss of 54 percent. The decline has many causes: the low prices of natural gas, the rapidly declining price of wind and solar power, environmental regulations, and anticipation of climate mandates.[4] Coal mining jobs have followed a similar pattern. In 2011, nearly ninety thousand people worked as coal miners in the United States; this number had fallen by more than half to about forty-three thousand in 2024.[5]

The decline of coal in the United States started sooner in Appalachia and other parts of the country, where the mined coal has a high sulfur content. With amendments to the Clean Air Act in 1990 that set sulfur dioxide emission limits for coal-fired power plants and deregulated railroads—thereby reducing the cost of importing lower-sulfur-content coal from the western part of the country—the demand for eastern coal fell precipitously.[6] The retirement of US coal-fired power plants has further diminished the demand for coal. Between 2015 and 2020, energy companies retired 414 coal power plants,[7] and they plan to retire almost two hundred more plants before 2030.[8] Not a single utility company has announced plans to build a new coal plant in the foreseeable future,[9] and several major coal companies have sought bankruptcy filings in recent years.

The United States is not alone. Despite 2023 setting the highest record for coal demand across the world, many countries are also decreasing their reliance on coal, although with notable exceptions such as China, India, Indonesia, South Africa, and Vietnam.[10] International trade disruptions such as wars and economic sanctions (e.g., the Russian-Ukraine war) can lead to spikes in short-term demand, but most experts predict general downward demand for coal over the long term.[11]

Burning less coal will result in fewer greenhouse gas emissions, which is promising and necessary from a climate change mitigation perspective. However, the diminished need for coal also introduces a distinct set of challenges for traditional coal communities. With coal jobs no longer ample, miners will need to seek new employment opportunities, and residents, local officials, and businesses in coal communities will have to navigate the economic, social, and cultural upheaval that accompanies the reduced demand for the industry that has historically sustained their local economies.

Other Fossil Fuel Communities

Coal mining communities have captured the most media attention and political rhetoric over the past decade in the discussion about our changing energy systems. During then-candidate Donald Trump's campaign rallies in 2019, the future president disparaged then-President Barack Obama's "war on coal," a war allegedly fought through environmental regulation. This same slogan emerged in the national scene many times thereafter, including in Trump's declaration that he was ending the "war on coal" when reversing several of the Obama administration's environmental and energy regulations and executive orders and when initiating a withdrawal of the United States from the Paris Climate Agreement.[12]

But there are several other types of fossil fuel industries that have become central to the economic, social, and cultural fabric of communities. The clean-energy transition will bring massive changes to these industries too, as well as to the operations along their supply chains and to the communities that host them.

Oil and natural-gas extraction, distribution, and refining provide well-paying jobs too, particularly within fields of engineering. These industries also provide physical-labor positions, such as in the operation of oil and gas rigs and in the construction and maintenance of the pipelines, storage tanks, and other types of infrastructure required to move these products to markets. Fossil fuel power-production industries—including coal, gas, and oil—have also historically provided steady employment for local residents. The exceptions here are some oil and gas drilling, as well as hydraulic fracturing operations, that are more temporary in nature due to fluctuations in market demand. Oil and gas production operations are sometimes characterized as "boom and bust," producing transience in the workforce, especially in the early years of operations in a given location, with more steady local employment if a boom lasts for several years.

US coal production is concentrated primarily in five states— Kentucky, Montana, Pennsylvania, West Virginia, and Wyoming—

and even further concentrated in just a handful of counties within these states. Oil and gas operations are somewhat more widespread, and they employ many more people, with the most significant oil and gas production in Texas, North Dakota, California, Pennsylvania, Louisiana, Oklahoma, Colorado, and New Mexico. In addition, oil and gas operations are common on many tribal lands, including those of the Navajo Nation in Arizona, the Crow Tribe in Montana, and the Northern Cheyenne Tribe in Montana.[13] Impacts from the transition to clean energy will most severely be experienced in these states, and especially within counties such as McKenzie County, North Dakota, and Midland County, Texas, for oil and Susquehanna County and Washington County in Pennsylvania for gas.[14]

The current predicament for the oil and gas industries, at least in the near or medium term, is not as bleak as it is for coal mining and coal power production. The United States is the world's leading producer and exporter of gas and oil, and extraction for both resources has continued to rise in recent years[15] and is projected to continue to rise through 2050 given the current regulatory and policy environment.[16] The return of Donald Trump to the presidency in 2025 may provide further boosts, at least in the short term.

Yet energy and climate experts warn that these trends must change if we want to address climate change, and the United States, as well as the rest of the world, will eventually need to move to curb all fossil fuel production. In 2015, 196 countries adopted the Paris Agreement, an international treaty on climate change, with an established goal to limit global warming to within at least 2 degrees Celsius, but preferably 1.5 degrees compared to pre-industrial levels. To achieve something close to the 1.5-degree target, countries must seek net-zero emissions by the year 2050. According to the International Energy Agency (IEA), achieving this goal in turn requires heavy investment in low-emissions technologies, on the order of a doubling capacity between now and 2030 and fading out all coal production in advanced economies by 2030 and in emerging and developing economies by 2040. To reach net-zero by 2050, both oil and gas production must decline precipitously by 2030, and there must be a phaseout of internal-combustion-engine cars beginning around 2030.[17]

Of course, the reality of the energy outlook may not align with what the IEA and other energy experts predict as necessary investments to mitigate climate change and stay within the targeted temperature thresholds. Indeed, there is plenty of evidence that countries' leaders talk out of one side of their mouths when making climate pledges and the other when making firm policy. Many companies also celebrate their ambitious climate goals, while simultaneously falling short of taking the steps necessary to meet them.

Yet there is still reason to believe that many countries around the world will reduce their fossil fuel usage, and dramatically so, in the coming decades. Consider the case of gasoline: 20 countries around the world have either set future bans on the sale of new internal combustion engines cars or imposed upcoming mandates for 100 percent zero-emissions vehicles.[18] Twelve US states have similar zero-emissions goals. Countries have made other commitments as well. As of 2024, 89 percent of the world's population are covered by net-zero targets.[19] At the 2023 international climate negotiations in Dubai, countries approved a plan for a fossil fuel phaseout, even if the road map to actual commitments remain unspecified.

With such drastic reduction of fossil fuels, oil and gas communities—including places of extraction, refining, and combustion—will soon be on the front lines of the energy transition. Much as coal communities are already experiencing, these communities will face changes to their economic and social conditions, including job losses, reductions in tax revenue, and social change.

Just how significant will these changes be? While many who currently work in the fossil fuel industries will retire between now and when deep decarbonization occurs, there is still a large contingent of younger people working in these sectors who will need to find new employment. One study of the US employment effects associated with the carbon targets set by the IPCC predicts a loss of around two thousand coal mining jobs per year for twenty years, six hundred per year of which would be lost by younger workers who would not have retired otherwise. This same study found a loss of about seventy-seven thousand oil and gas jobs, or about 3,800 jobs per year over twenty years, and nine hundred per year lost by workers who would not have retired over this time frame. If one additionally includes related fossil fuel industries that provide ancillary services to coal, oil, and gas markets, these totals come to about 16,475 jobs lost per year.[20]

The whole world benefits from moving away from fossil fuels. Burning less coal, oil, and gas means cleaner air and water, and, if done fast enough, the potential for a more stable climate. Some people may benefit from these changes more than others, such as people who live near power plants or major highways and those likely to feel the brunt of hot temperatures and extreme weather exacerbated by climate change. Yet the costs of this transition—such as through the loss of jobs—will be concentrated in certain areas. These labor market effects illustrate the uneven geographic distribution of benefits and costs that the clean-energy transition may create or perpetuate. To add on, many of those facing these costs are from the same communities that often bear the environmental and health costs of our current energy systems.

Economic and Environmental Conditions Faced by Fossil Fuel Communities

Although no two locations are the same, communities that largely rely on fossil fuel industries tend to face a similar set of economic and environmental conditions. In most communities, revenue generated from fossil fuel activities is an important source of income for the local government, community, and residents. In some locations, a coal or natural-gas power plant can account for the majority of local tax revenue, which in turn is used to fund local public services and schools.

Colstrip, a rural community founded around coal in Montana, presents a stark case of the economic impacts power plant closures can have on a town. In 1975, the first Colstrip Station unit began operating, with three more units opening by 1985. As of 2017, the Colstrip power plant employed 80 percent of the state of Montana's coal workforce, accounted for $76 million of the $95 million in taxable value of properties in Rosebud County, and made up a large share of the local school's tax base. When two of the four units closed in 2021 in response to expensive air pollution regulations and increasing competition from other fuel sources, $22 million, or 24 percent, of the county's property tax base disappeared.

To provide another example, a team of researchers recently conducted a study of the economic impact of the closure of four coal-fired power plants in the state of Indiana. The power plants—Schahfer, Michigan City, Petersburg, and Rockport—collectively had an average annual output of over thirty-three million MWh of electricity generation.[21] The researchers found that these closures would result in a loss of 652 jobs, $77.5 million in employee compensation, and $354 million in gross state product. These closures also cause ripple effects: When a large number of individuals in a local area lose their jobs and have less disposable income to spend, local restaurants will sell fewer meals, gas stations will pump less gas, and construction companies will sign fewer contracts. In the case of the Indiana coal power plant study, these ripple effects added up to another 1,732 jobs, $98.4 million in employee compensation, and $184.7 million in gross state product.

The most tangible and immediate impact of a declining fossil fuel industry on local residents is economic hardship. When a coal mine closes, the miners lose their jobs and the local economy loses the tax revenue. Miners may also find themselves, however, without pension funds. As of 2015, the five largest coal producer companies in the US—Peabody, Arch, Cloud Peak, Alpha, and Alliance—carried $1.9 billion in unfunded pension liabilities. Given that several of these companies are now bankrupt or carry significant debt, the chances of them allocating resources to these pensions appear small.[22] In the oil and gas sector, unfunded pensions are even higher: The five largest US-based oil and gas companies—Exxon Mobil, Chevron, ConocoPhillips, Anadarko, and Devon—carry $14.2 billion in unfunded pensions.[23] Unemployment benefits, when available, may provide temporary relief, but one has only so much time before one must find new sources of income. Unemployment benefits vary by state, but the majority provide only twenty-six weeks of support.

Many US coal mining communities, including places in both Appalachia and the American West, are extractive mono-economies. The conditions that lead to a single industry growing in such strength and influence, and to overall economic specialization, tend to play out over a long period of time,[24] stifling alternative industrial development and entrepreneurial efforts.[25] In some cases, particularly in Appalachia, the roots of economic and social isolation grew in the late 1800s and early 1900s, when mining companies recruited workers from across the country to work in the booming coal industry in company controlled towns and camps. In the words of Shannon Elizabeth Bell, a sociologist who has conducted extensive fieldwork in regions of Appalachia, the mining companies owned everything—"the houses, the streets, the schools, the water systems, the churches, any recreational facilities, the doctor's office, and the company store"—and paid their employees entirely in a script redeemable at their own stores, where they set the price for all commodities and services—but not elsewhere.[26]

The structure of local economies in many coal mining regions leave them particularly vulnerable to price fluctuations and boom-and-bust cycles,[27] conditions to which adaptation is difficult.[28] These conditions then reinforce each other over time: The boom-bust cycles limit other economic opportunities, thereby leading the local economy into further dependence on the extractive industry for jobs and reinforcing fragmented decision-making that locks in further economic dependence.[29]

Many energy extraction communities are also marked by high rates of poverty and overall economic decline.[30] In the case of coal mining communities, there are some situations where people living in the very places where the fuel powering electricity generation is mined are unable to access or afford the resulting electricity. In the case of coal generating communities, there are situations where the community cannot access the actual electricity from the local power plant. The irony is not lost on people. One such example is the case of the Navajo

Nation, home of the now-retired (in 2019) Navajo Generation Station, a 2,250-megawatt coal generating power plant, where electricity access is not available to all. Marie Gladue of Black Mesa Water Coalition in the Navajo Nation, for instance, told reporters that "power lines were built over communities on Navajo land, taking electricity and water to cities across the Southwest, while thousands in the Navajo Nation remained with no electricity access or water."[31]

In some cases, energy communities may actually be worse off than comparable communities that are not connected to the extractive economy. A study by the Ohio River Valley Institute in 2021 revealed that fracking communities in West Virginia, Pennsylvania, and Ohio disproportionately contribute to their respective states' gross domestic product but have lower rates of income and jobs than other counties within the state. These trends were only made worse during years of significant natural-gas market growth, from approximately 2008 through 2019, despite claims by the gas industry that fracking would deliver these regions significant economic benefits.[32]

These conditions are further reinforced by low levels of educational attainment and household mobility. Studies have found Appalachian and other eastern coal communities to be below the national average in high school and college completion rates.[33] These conditions are connected to the allure of well-paying coal mining jobs that often do not require more than a high school diploma. These circumstances, however, create a cycle: Lower educational attainment begets higher rates of poverty and a lack of economic opportunity, and repeat. These communities are also marked by low levels of household mobility,[34] whereby residents have a strong connection to the land and landscape and are rooted by generations of family members who resided there before them.

Oil and gas communities face similar conditions, particularly communities with drilling operations that experience "boom-and-bust" cycles. In "good" times when oil and gas prices are high, jobs are plentiful and revenues stream in, but in "bad" times when prices are low, unemployment rises and economic activities dry up. Oil and gas communities are also often situated in regions that specialize in resource extraction, with fewer alternative industries to help support the local economy during the bust periods.[35] Even locations with long-term and steady production of oil and gas extraction and refining over time have economies that are heavily reliant on the oil and gas industries for employment and public revenues, as well as significant capital investments in oil and gas infrastructure, such as pipelines and distribution terminals.[36]

Recent experiences with hydraulic fracturing or fracking provide evidence of additional types of challenges. Fracking operations are common in rural places that have high rates of poverty.[37] People in these communities, and especially those who live in close proximity to fracking operations, tend to suffer adverse health outcomes due to exposure to air pollution as well as contaminants in the ground, as discussed in the last chapter.[38] These communities also experience increased traffic,[39] crime, and social disruption.[40] These conditions, in turn, have the potential to affect quality of life and mental health.[41] Fracking operations have also been found in some regions to encroach on the traditional way of life and conception of place, one marked by agricultural operations, rural landscapes, and peacefulness, leading one scholar to characterize fracking operations for host communities as a "destroyer of human flourishing."[42]

Clearly, the siting and operations of fossil fuel activities are accompanied by a challenging set of trade-offs, such as those between having a large local employer and the potential for economic overdependency or between a few hundred well-paid jobs and lower overall economic prosperity. Similarly, there are many trade-offs that these communities face that relate to their local environment; when one accepts—or is forced to accept—fossil fuel production within a close geographic proximity, they may gain access to jobs in that industry, but they must also face the risk of energy-production-related environmental pollution.[43]

Energy extraction and production activities contribute to many localized sources of air, water, and land pollution. These pollution burdens, however, as faced by local communities, are not restricted to the time in which fossil fuel operations are locally active. The environmental impacts of power plants, mines, and wells often persist even after operations cease. Such lingering impacts include coal ash disposal sites, acid mine drainage from coal mines, and methane leaks from abandoned oil and gas wells, among others. Several abandoned mines in southeast Ohio provide an example. Coal mines that operated in the late 1800s and early 1900s have been found to have contaminated local water sources through the entire twentieth century. By the early 2000s, over 340 miles of local streams in southeast Ohio were still contaminated and in need of costly remediation.[44] Not only will, or have, some fossil fuel communities lost their jobs and tax revenues, but they are also left with the pollution and without the resources for cleanup.

As featured on BBC in 2019, Jason Walker lived in southern West Virginia near coal mining operations. When the mining company switched from underground mining to mountaintop removal, Jason observed a dramatic change in the water quality coming out of his taps. It smelled like rotten eggs and turned a deep orange rust color, a color that quickly stained his faucet, bathtub, and clothes. When he had his water tested, he was told that it was so toxic that "there was a risk that direct sunlight could set [his clothes] on fire."[45]

This orange pigment and distinct smell is due to local water flowing through mining waste of mountaintop removal debris in waterways or due to underground water flowing through mines. This "acid mine drainage" contains acidic ions and metals such as nickel and arsenic, which are picked up by the flowing water and eventually make it into residential water systems, and into people's homes. The contaminated water supply forced Walker to use bottled water for cooking. For other purposes, he hauled in water from a stream near his house and used pool cleaning chemicals to prepare it for use.

When Jason and several of his neighbors joined a lawsuit against the mining company, the community became deeply divided between those who wanted to protect the safety of their water and those who wanted to protect their jobs. As one resident put it, "It's how people make their living and support their families around here. If you don't work in the coal mines you either flip burgers or you have to move out of state and do something else."[46] This comment reveals a

harsh reality—many coal-dependent communities must make difficult trade-offs between their health and the environment on the one
hand and employment opportunities on the other hand, and in which
the next best employment opportunity may be significantly less lu-
crative than mining. These circumstances also produce a dynamic in
which residents overlook or discount the downsides of energy production[47] and, in some cases, either contest that the energy transition
is occurring or oppose a transition to another way of life.[48] Of course,
not everyone will have the same views of these trade-offs. Community
members may be divided depending on how they personally weigh
the costs and benefits of health, the environment, and employment,
and due to their own personal history with the energy industry.

Shannon Elizabeth Bell offers another reason for the lack of resistance or protest against coal companies in Appalachia despite the
environmental and other socioeconomic conditions. In her book
*Fighting King Coal: The Challenges to Micromobilization in Central
Appalachia*, she argues that a lack of public mobilization against coal
companies can be largely explained by the high levels of isolation experienced in these regions and by coal companies' efforts to reinforce
the idea that "coal is both the economic backbone of the state and the
cultural identity of the citizenry."[49]

Another illustrative case is the town of St. Paul, in Wise County,
Virginia, which was once the site of coal mining operations. When the
mining industry in the region began to falter, the county supported the
construction of a new power plant to fill the economic gap, despite
protests from local residents who voiced concerns about its potential
environmental consequences. The Virginia City Hybrid Energy Center, a large 668-megawatt-capacity power station, was commissioned
in 2012 and burns coal, biomass, and coal mining waste, the latter of
which includes ash and other waste products that are collectively referred to as "gob." Since beginning operations, the plant has been a
heavy polluter and has frequently exceeded legal limits of particulate
matter, carbon monoxide, and sulfur dioxide emissions.[50]

Although Wise County has a low median income, the power plant
has been a major contributor to the local economy. As of 2019, the

county's median income is just below $39,000, relative to the national median of above $74,000, with about 22 percent of respondents below poverty and with significant unemployment. The power plant provides about $8.5 million in tax revenues, which is approximately 15 percent of the county's total revenues.[51]

Due to a Virginia policy to close most of its coal plants by the year 2025, this plant will close. When it closes, the county will need to somehow account for the loss of income in an already economically depressed region, while maintaining their schools and public services. The surrounding coal mines will no longer have a local source of demand, and thus may suffer economically and possibly close as well. And the county will need to figure out what to do with the remaining ponds of unused "gob" stored near the plant, which are at risk of overflow if disturbed by extreme weather events such as flash flooding.

These sources of local environmental toxins and pollution often linger for decades if not longer, and they are very costly to maintain. In a recent case with a utility in Indiana, Duke Energy has spent about $212 million since 2010 cleaning up a coal ash pond at one of its power plant sites, which was found through sampling to be leaking toxins into the local groundwater. Duke Energy asked the Indiana Utility Regulatory Commission to recover these costs through a rate increase for all its paying customers. Although the request was initially approved by the commission and supported by the Court of Appeals, the Indiana Supreme Court denied it on grounds that Duke Energy did not follow the right procedures to first ask permission for accruing those expenses.[52]

Although consumers were spared from paying for the cleanup in this Duke Energy case, they still bear the burden of the local contaminants in their groundwater. Similar burdens with coal ash are experienced elsewhere. As just one example, consider people living in close proximity to coal ash ponds in Mobile County, Alabama, where there is threat of the ponds spilling into the Mobile-Tensaw Delta, one of the most biodiverse regions in the entire United States. As one local resident referred to the coal ash pond: "We've got an A-bomb up the river."[53]

Similar stories emerge from oil and gas communities as well. Again citing the group of researchers working in the Gulf Coast region of

Louisiana and Texas, we highlight the trade-offs as well as the sacrifices that local communities make to keep their largest employers present and aligned in a mission of economic prosperity for the region. As the authors explain:

> The political and economic dominance of the energy sector is well known by individuals within communities and is most often manifested through expressions of fear. Leaders of nonprofit organizations, for example, often fear being labeled as unpatriotic if they advocate for an energy transition . . . Moreover, residents and local government officials often fear criticizing nearby petrochemical plants due to their political and economic power within the community. As a result, crucial matters facing residents often go unaddressed. In St. John Parish, many residents are dependent upon such plants for their income. Yet the same residents are also reluctant to drink their tap water because they fear that it's unsafe due to pollution from the plant. Nonetheless, because of their reliance upon the plant as a vital source of income, residents do not press for clean water as they fear that doing so would cause the plant to relocate.[54]

When one considers the fate of fossil fuel workers, it may be tempting to diminish this set of issues to one of "jobs, jobs, jobs," a slogan that Hillary Clinton coined when running for president in 2016. What the public wants are jobs, and good-paying, steady jobs. And what politicians then campaign on are promises of more jobs and, in the case of presidential candidate Donald Trump in 2020, specifically the return of coal mining jobs and the end of the "war on coal."

We use the word "diminish" because the fixation on jobs neglects to account for the complex history that fossil fuel communities have had, and continue to confront, with employment, health, and the environment. The single-minded focus on jobs obscures the difficult trade-offs that these communities face, and it often also conceals the personal and communal sacrifices they have made and continue to make—sacrifices to their environmental and economic well-being that can linger well beyond the time past which fossil fuel operations expire. Reducing the debate to a question of merely jobs lost or jobs gained not only masks the complicated layers of what fossil fuel communities experience, it conceals their deep-seated feelings of loss, which condition their views about the future.

Add on top of these challenges the threat of climate change to specific regions, such as the Gulf Coast. The Intergovernmental Panel on Climate Change and the US National Climate Assessment both predict that coastal areas in Louisiana and Texas will face increased incidence of flooding, hurricanes, land loss, and rain.[55] Such weather and related events will stress these areas' infrastructure, financial stability, and economic activities—while they are simultaneously transitioning their energy systems and economic foundation.[56]

Confronting Loss

Beyond the numbers, what happens to a community when their local fossil-fuel industry shuts down, and after generations of economic isolationism at the expense of human health and the local environment? This question led us to conduct interviews in 2016 in Appalachia in regions of West Virginia, Kentucky, and Virginia, and focus groups specifically in Williamson and Ghent, West Virginia. We interviewed individuals who worked with local communities on issues related to the energy transition, including those within economic development units, workforce training programs, community and religious organizations, research groups, foundations, and the media. We also conducted focus groups with former coal miners and other community members, many of whom had family and friends with close ties to the coal industry.[57]

Although participants in these conversations did not agree on everything, we heard a common narrative of economic decline, limited opportunities, personal grief, and a cultural reckoning. We also found conflicting elements of despair and optimism, all shrouded in a recognition that coal is a fading industry and that life had become very difficult for those living within these transitioning coal communities.

For those who entered the mining industry out of high school, or in some cases without graduating from high school, alternative employment opportunities may be limited. In our focus groups, Greg, a coal miner in his fifties, explained this to us:

> Because to some extent it's everybody who is vulnerable here, that doesn't know how to adjust, but I think the 18–25 kind of ranges because they're just graduating high school, and that's their point to make a decision whether they want to go to college or whether they want to get a job, ah, and it used to, if you want to go to college, great, but if not, you can get a job, a very, very good job, working in the coal mines, support your family for the next fifty years you're alive. Now that's almost not an option, and that takes almost all the options off the table, unless you want to work in fast food or retail, which we don't even have a lot of that.

Without money to spend at restaurants, the grocery store, or the barber shop, other local businesses then suffer. Since many of these communities have economies that are heavily reliant on the coal industry for jobs, when the miners cannot support these other businesses, they shut down as well. When residents leave to seek employment and economic opportunity elsewhere, but without new residents moving in, property values go down.

To manage their financial strain, many former coal miners with whom we spoke told us that their families had to renegotiate their plans, often painfully, including plans for who works and how much, as well as financial planning of debt and expenditures. In some cases, other members of the family had to seek new work or additional hours to make ends meet. In other cases, the former coal employee needed to travel long distances for employment to keep their families afloat. These sudden changes had cascading effects for some families, who then also needed to shift childcare and other commitments to accommodate their new circumstances.

Although investments in the clean-energy sector generally yield more jobs on a per-dollar basis than similar investments in fossil fuel industry jobs,[58] alternative job opportunities do not always exist in the same place as where the jobs were lost. In 2021, just seven out of every thousand workers who left a fossil fuel job were able to find a job in clean energy in the same place. By comparison, two hundred out of every thousand workers who left a fossil fuel job found another one in the same industry. Older or less educated workers stood an even lower chance of finding a clean-energy job. States with stronger energy transition policies, such as California, had higher rates of transfer from fossil fuel to clean-energy jobs, while states without such policies, like West Virginia, had higher rates of workers staying in the fossil fuel industry.[59]

Indeed, renewable-energy potential infrequently aligns with displaced fossil-fuel jobs. One study estimated that the cost of building new renewable capacity to locally replace fossil fuel capacity will raise the cost of development by at least one-quarter.[60] We heard from some former coal miners that they now have to travel for weeks at a time to work truck driving jobs, and from others that they leave their families during the work week to take distant electrical linesman jobs, sometimes a hundred or so miles away.

There may be other lifestyle changes as well, notably in the form of income. Sean, the same focus-group participant that we mentioned above, shared: "There's absolutely nothing here to do that you can get paid more than minimum wage." Andrew, who works at an economic redevelopment organization in Central Appalachia, told us:

> It's a tough transition, it's almost because the area in the past 100, 150 years has been nothing but coal. That's all they know, that's all generations used to know. They've all worked in the mines, they never had to go to college, they'd make just as much as any person in the area. So if somebody tells them to stop doing that, and they're used to making $120,000 a year, you can live off $40,000 a year job . . . but they don't know how to. And never had to. It's basically taken away everything they have ever known.

These coal miners were lured from high school with six-figure-paying jobs and a promise of a stable income and lifestyle for the duration of their lives, as their fathers, grandfathers, and great-grandfathers had experienced. They started their families accordingly, bought new houses and cars or trucks, and never had to think about a backup plan.

They never had to take financial administration courses or consider their debt-to-income ratio or how to deal with losing their promised pension. This story is no different for oil and gas communities, such as in the Gulf Coast region where many oil and gas jobs do not require an advanced degree and offer salaries that are roughly 150 percent of the average salary for a region.[61]

To transition to a new career path, several coal miners with whom we spoke had enrolled in technical-training programs, often to learn the skills necessary to work as an electrical lineman or to install solar panels. While this opportunity is great for many, not everyone is guaranteed these positions. Training programs are not always available in regions that are losing fossil fuel employment; they may train individuals to perform jobs that are not locally available; and they often last longer than unemployment benefits do, thereby putting trainees at risk of further economic hardship.

Without the major source of tax revenue, the local government struggles to provide basic services, including support for the local schools. As two focus-group participants in Williamston shared, the affected parties include schoolteachers. The first explained that "a lot of friends have moved away . . . to keep jobs, new jobs, they've moved onto other states," while a second confirmed, "My son lost nine teachers at his school cause they lost three hundred students." A lack of revenue—and overall local economic depression—also affects the ability of economic development officials to attract new businesses and launch new efforts. In some cases, the energy industry also formerly supported the community financially through, for example, the sponsoring of local sports leagues or community events.

In addition to stories about financial and economic hardship, we also heard deep and personal accounts of social and cultural loss experienced by individuals and their communities. Here again we see evidence of an uneven geographic distribution of benefits and costs related to the clean-energy transition, where the benefits of cleaner air are widely enjoyed but the costs, in the form of economic decline and personal sacrifice, fall disproportionately on certain individuals and communities. Emotions of loss and grief were intense, and not just due to their personal economic predicaments but also due to the difficulty of the conditions facing them, their changing sense of identity, and their complex feelings of sacrifice for and betrayal and abandonment by both the US government and the rest of the country.

The vast majority of those with whom we spoke during our field work recognized that coal was a dying industry, as Jack, a longtime resident of Ghent, West Virginia, conveyed to us by likening the coal industry to a horse: "I can tell you what my granddaddy always said: No matter how many times you beat and kick that dead horse, it's not getting up to plow again." Even with short-term production spikes due to, for example, temporary increases in export demand, most of our interviewees recognized that coal would fade in the long run and their coal mining jobs would become obsolete.

This acknowledgment in and of itself, and its implications for households and the broader community, is worthy of note. But the emotions run deeper and connect to the personal trade-offs borne by residents within these communities, trade-offs that they have made between earning a living and sacrificing their personal health, or between economic development of the community and environmental health of their land and resources. As Carla, Tom, and Sean conveyed in the opening passage of this chapter, coal miners in their West Virginian community gave their bodies and careers to the coal industry and, as the energy transition unfolded, were left feeling "beaten and broken," and, as Roger further articulated, "forgotten."

Another source of grief comes from the changes that one must make to one's own conception of self and identity ("If I am no longer a coal miner, what am I?") as well as to family and community identity ("If my family is no longer a coal mining family, what are we?" "If this community is no longer a coal mining community, what are we?"). Research has shown that a strong sense of identity connected to extractive industries is common.[62] In our work in Appalachia, that connection was apparent, due not just to generations of workers within the industry but also to economic and social interdependence between the residents and the fossil fuel industry. As an economic development official, Susan, explained to us: "There is also a sense of grief that comes along with it, you know, coal mining is really a part of the culture here and it's interwoven into the way people feel about themselves and their own identity and their identity as a community."

With an identity crisis also comes introspection, and the opportunity for reimagining oneself and their surroundings. Several people with whom we spoke expressed optimism that they would take on a new identity and seek out new personal and economic meaning. Lance, a former coal miner, expressed such an outlook:

> I think slowly but surely, you kind of go through all these stages of mourning. So there's still those that think coal will come back. But there's more than ever in my lifetime many that say it's not coming back, at least not how it was and so there is, definitely more among the younger than among the older, but there is kind of this excitement and possibility, that now coal is gone and we can rebuild our economy into what we want.

Lance's reflections on the conditions facing his community reveal a deep personal sadness about the changes that his community has been forced to endure, yet they also harbor an optimism about the future. And although this optimism is not universally shared across US coal and other fossil fuel communities, it is noteworthy nonetheless. These sentiments are underpinned by hope for a different future, a future in which his community does not need to face difficult trade-offs between jobs and the preservation of the surrounding environment, or jobs and their residents' personal health. They are also underpinned by hope for a future world in which community members are not powerless, in which they are not born into the decision to work for the local energy company or flip burgers. It also underscores a hope of recreating an identity that is present and not forgotten, contributing to a cleaner energy future for all.

•••

These sentiments—and broader ideas about an inclusive and fair energy transition—are beginning to find their way into the public discussions, both in the United States and in other parts of the world. For example, coal communities and their transitions are a growing focus of an academic literature on "just transitions." The notion of a just transition grew out of labor movements in the United States in the 1970s and 1980s and eventually expanded to other regions and other social movements. The topic was eventually built into the preamble of the Paris Agreement in 2015 in mention, though it made its big debut in terms of substantive discussions and negotiations on the international scene at the international climate negotiations in Katowice, Poland, in 2018. Just weeks prior were the formative "Yellow Jacket" protests in Paris, France, and the introduction of the "Green New Deal" in Congress, where advocates argued for a just transition toward cleaner energy for all of society, both across and within countries, that also included other objectives—such as labor and healthcare protections for those who had historically worked in energy industries or done environmental cleanup that included exposure to toxic chemicals.[63]

Over time, the notion of a just transition has evolved to include not only labor concerns but also a range of energy, environmental, and climate justice issues. At root, modern conceptions of a just transition, in the words of scholars Dimitris Stevis, Edouard Morena, and Dunja Krause, "reflect a growing awareness of and concern about deepening inequalities between the world's rich and poor, and how the climate and environmental crises, and efforts to address them, are accentuating them."[64] And in the words of Manuel Pastor and Mijin Cha, leading just-transition scholars in the US context, such a shift is a "transition away from poorly funded educational systems, away from depriving communities of resources, and away from a racially unjust system to a system that is built around the needs of communities, particularly those that have borne the burdens of the carbon-intensive economy."[65] Legacy fossil-fuel communities, especially those based on extraction, have long borne such burdens.

It is not just academic and political pleas for just transitions that have grown; so too has policy action, particularly at the state and local level. The state of Colorado, for example, opened a Just Transitions Office in 2019 with primary responsibilities of ensuring investments in Colorado coal communities and coordinating state and local pol-icies.[66] Also in 2019, New Mexico passed the Energy Transition Act, which establishes a priority of investing millions of dollars into coal communities.[67] These states and others that have also prioritized just transition–oriented policies. Michigan and Illinois, for example, have introduced apprenticeship programs, required energy developers to adopt project labor plans, created scholarship programs for children of displaced workers, and mandated severance pay and job training for workers who lose their energy industry jobs.[68] The US Depart-ment of Energy also runs the Community Local Energy Action Pro-gram (LEAP), designed to help transitioning energy communities build community action plans for energy-based economic develop-ment.[69] These emerging policy developments signal the growing com-mitment to just transitions across the country, as well as a general understanding that the challenges facing transitioning energy com-munities are multifaceted and, if they are to be addressed, will require procedural justice in the form of substantive and genuine community engagement.

These recent policy developments signal a recognition that the cur-rent plight of and difficult long-term prospects for fossil fuel commu-nities require attention and deliberate action. It is too early to judge the effectiveness of any given policy or program, and effective solu-tions will need to be tailored to the specific needs of communities and developed in close coordination with communities themselves. Nev-ertheless, their emergence reflects a realization that the clean-energy transition will adversely affect many whose economic livelihoods and, in some cases, social identities are deeply interwoven with our fossil fuel system.

CHAPTER 4

Life Without Energy

A lifelong native of West Virginia, Darrin Johnson lives in an austere home in a rural community with his eight-year-old son. Darrin[1] is forty-two years old and suffers from several medical conditions that have, among other complications, impaired his vision. He can no longer drive a car and must rely on walking or carpooling with neighbors to gather groceries and other necessities from the nearest store, which is twelve miles away.

As of 2021, Darrin has never missed a payment on his energy bill. His home energy needs are modest, with working AC and heating systems; he occasionally watches a small TV, mainly because he likes to hear some company in the house, and he uses basic but small appliances. He keeps his home fairly cool, with minimal air conditioning in the summer and heat in the winter, in an effort to keep his bills down. His electricity bill typically hovers around $40 per month. The average utility bill in West Virginia is $117 per month.[2]

In February 2021, after Darrin's first missed payment, Darrin's power went out. He thought it was a local blackout, and even called the utility tip line to let them know. This was in the early afternoon, and he and his son continued through the evening making do with their powerless situation. It was not until it was fully dark outside and he could see streetlights nearby that he realized that there was no power outage. Rather, his utility had shut off his electricity due to a $13 outstanding charge.

Darrin is among millions of Americans who are disconnected from their energy service each year because of a missed bill payment or underpaying a bill. Thirty-four million households, or one out of every four, struggle to afford the energy they need in their homes, and twenty-five million households, or one out of every five, report having to make difficult trade-offs between paying their energy bills and paying for other essential household needs such as food or healthcare.[3]

For the last several years, we have compiled and tallied utility disconnection reports from across the United States. We find that, each year, energy utility companies report disconnecting nearly three million households from their service due to nonpayment. Three million is actually a drastic underestimate, because only some utility providers across the country—on the order of about 10 percent of utility companies, serving about half of the US population—publicly disclose information about their disconnections. Twenty-three states mandate that utility providers routinely gather and post publicly this type of disconnection information, and, even in these states, typically only the investor-owned utility companies are required to report.[4]

We study the incidence of energy insecurity, also often referred to as energy poverty. A household or individual is energy insecure when they have difficulty securing their basic energy needs, like heating, air conditioning, water heating, and lighting. Among the indicators of energy insecurity are a reported difficulty paying an energy bill (e.g., electricity, natural gas, heating oil), receipt of a shutoff notice due to nonpayment, and service termination (or "disconnection") by the utility company. People in energy insecure households may forgo expenses on other essential items, carry utility debt, or maintain an uncomfortable temperature in the home to reduce energy costs.

In our research on low-income households,[5] we find that families with young children—similar to Darrin's—are more than twice as likely than families without young children to struggle to pay their energy bill, and they are about three times more likely to be disconnected. Yet young children, as well as the elderly, are more at risk of medical complications due to exposure to extreme heat or cold—complications like dehydration, heat exhaustion, or hypothermia.[6] Similar disparities exist with other vulnerable groups. Households where someone relies on an electronic medical device—such as a respirator, oxygen concentrator, or dialysis machine—are also more likely to be disconnected. When doctors prescribe a medical device, they often do not consider how much it costs their patients to operate the device, and whether this additional expense might compromise their ability to pay their bills. Nor do they consider what their patient would do to maintain their health in the event of a power shutoff. The electricity costs of operating these medically necessary machines, moreover, are generally not covered by public insurance programs such as Medicare and Medicaid.[7]

People of color are particularly susceptible to utility disconnections. Black and Hispanic households, for example, are much more likely to have their power shut off by their utility companies in response to nonpayment. A Black household is three times more likely than a white household to face disconnection, and a Hispanic household is four times more likely.[8] These stark disparities occur even after accounting for income, household size, and the type of home where people live.

An important contributor to energy insecurity, especially for low-income households, is the amount of money that they spend on energy relative to their income. Energy researchers refer to this figure as a household's energy burden, a metric measured as energy expenditures divided by a household's total income. Low-income households, on average, have an energy burden of 8.1 percent compared to the national median of 3.1 percent, meaning that they spend nearly three times as much of their income on energy than the median household.[9] Low-income homes are typically older and less energy-efficient than homes owned by higher-income households, which contributes to their energy burden. As an example, an inefficient or single-pane window or door can cost a household an additional $43 per month in heating bills, averaged across all seasons of the year.[10] For comparison, the average energy bill in the US is $141.41.[11] By one

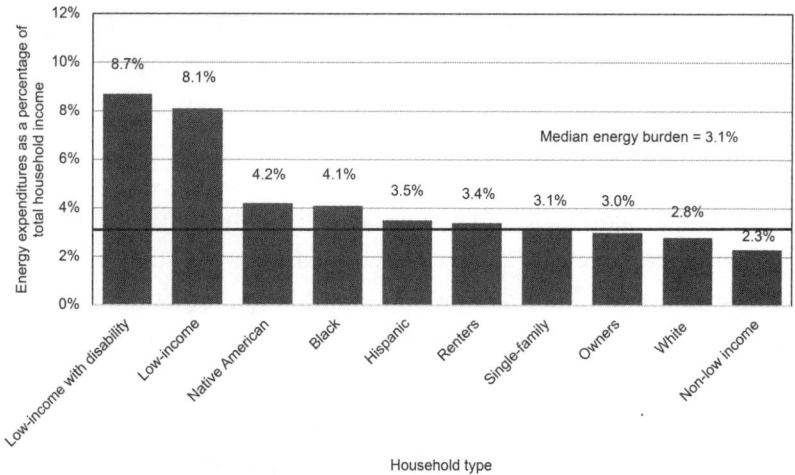

Figure 4.1 Energy burden by household type subgroup, 2017
Notes: Figure shows average energy burden by subgroup, where energy burden is money spent on energy divided by total household income.
Source: Adapted from Drehobl, Ross, and Ayala, *How High Are Household Energy Burdens?*

estimate, improving the energy efficiency of an average low-income home could cut energy costs for those consumers by about one-third.[12] According to data compiled by researchers at the American Council for an Energy-Efficient Economy, shown in figure 4.1, energy burden also varies across other population groups. On average, households of color also spend much more of their income on energy—this finding also holds true for households in multifamily dwelling units and manufactured homes.

The US government tracks energy insecurity rates in its Residential Energy Consumption Survey, which the Energy Information Administration administers every five years. When comparing results from the 2015 and 2020 surveys, energy insecurity declined slightly on average, but the rate of energy insecurity among households of color remained the same, and it was only white households that saw an overall improvement.[13]

Housing conditions also matter. When a family has cracks or holes in their walls; broken or inefficient refrigerators, HVAC systems, or air conditioning units; or mold or exposed sockets, they are much more likely to be energy insecure. The recommended Energy Star heating and cooling temperatures, under 70 degrees for heat and over 78 degrees for cool, are also harder to maintain with subpar housing conditions.

In figure 4.2, we show the rate of disconnection by various self-reported household characteristics from a nationally representative survey of low-income households we conducted in 2020–2021. The data illustrate the importance of household demographics and living conditions. The average rate of disconnection is about 4 percent across low-income households, but this rate goes up significantly for certain groups, including Black households (6 percent), those with young children (12 percent), those with residents who rely on electronic medical devices (12 percent), and those living in poor housing conditions (9 percent).

Darrin was $13 under payment when his power was disconnected. When the utility company turned off his power, they did it remotely,

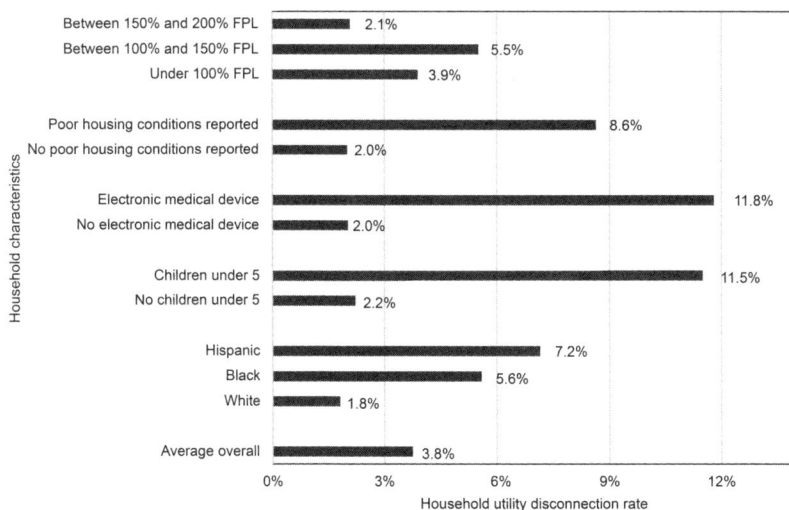

Figure 4.2 Rates of household utility disconnection, 2020–2021
Notes: "FPL" refers to the federal poverty line. The estimates are averages, with the application of survey weights.
Source: Carley and Konisky, *Survey of Household Energy Insecurity in Time of COVID*. The data presented in this graph were gathered via surveys to a representative sample of households within 200 percent of the federal poverty line. We administered surveys at four points in time between May 2020 and June 2021.

via a new smart meter that the company had installed in his home the year before. The cost of the smart meter was passed on to Darrin and his neighbors through an extra charge on their monthly energy bills, a charge that likely contributed to the extra money that he owed within a matter of a few months. Darrin's bills had also increased recently for other reasons that were unclear to him; whereas he had formerly paid about $40 per month, his bill had since unexpectedly doubled to $80.

Ultimately, individual utility providers make decisions about whether and when to disconnect a customer due to late payments. Some utility companies may provide a grace period, offer to help the customer with a payment plan, forgive some or all of the debt through an arrearage management plan, or at the very least provide ample notice before ordering a disconnection. According to Darrin, in this case the utility company did not make an effort to notify him in advance—at least not that he knows of—and instead it made the decision to shut off his power due to his modest outstanding balance.

When Darrin was finally able to get someone on the phone from the utility company the next day—he spent considerable time navigating his way through their automated phone system—the customer service representative was unable to locate his account to give him details about his disconnection. He was passed along to another representative, and then to another. These representatives were compassionate and eager to help, though as Darrin explains, "they were no different from myself . . . they couldn't help." Darrin was eventually transferred to a manager, who was able to identify that when Darrin was disconnected, the company had canceled his account. Darrin would need to open a new account, and pay a new customer fee, in addition to paying off the balance on his old, discontinued account plus an additional disconnection fee. The amount that he owed was now in the hundreds of dollars, which he reported to well exceed his monthly income.

These fees added up quickly and Darrin had difficulty pulling them all together. He paid everything that he had available but still came up short on the bill. He was out of options, and he could not apply for the government's Low Income Home Energy Assistance Program

(LIHEAP) funds because he had been disconnected and was therefore no longer eligible. If Darrin had instead received a notice of disconnection first, he could have applied for LIHEAP funding to pay off his utility bill.

Within about a thirty-hour period, Darrin lost the perishable food in his fridge and his cell phone died. To charge his phone, he walked to his neighbor's house and asked to use their power and sat there while his phone charged. He did this multiple times over the month that he was disconnected. Back in his home, he camped out. He took cold showers, even though his home sat at what he reported as an unbearably low temperature. And he continually worried about the possibility of his pipes freezing.

In our discussions with child advocates, we have learned that in some places across the United States, if a child is living in a home that has its heat shut off, they are removed from the home by child services on account of negligence and held in foster care until the home is habitable again.[14] Darrin was thankful to be near his son's grandparents—his former parents-in-law—who could take his child until the power was restored. Darrin feared that the house was too cold for his son and that he might not be able to call for help in the event of an emergency due to the unreliability of his phone charge.

Darrin was also lucky to have a friend who learned of his circumstances and went to his church to see if they could help. Even though Darrin was not one of their parishioners, they used their emergency relief fund to pay off the remaining balance on Darrin's account. They paid the utility company directly and let him know that they did so.

But four days later, Darrin was still without power. He again called the utility company and made it through the labyrinth of pre-recorded directories to finally get a real person on the phone, one who could again track down his closed account. Only then did Darrin learn that the account had, in fact, been paid off four days before, with a balance drawn to zero, but the utility company would not turn the power back on without a phone request to do so from Darrin. He made the request, and the utility company was able to switch his power back instantaneously via the remote smart meter.

The whole experience was traumatic for Darrin. Never before had he had to live under such conditions, with such stress, and with such fear about his own personal health and well-being. As he relayed his story to us, he explained, "I hope that there are not that many people who experience something like that . . . my circumstances, it was one of the most difficult things that I ever went through." The situation also set him back on many other facets of his daily life: he was now behind on other bills like his water bill; he had had to replace all the food in his refrigerator; and his health was suffering due to eating less and the overall stress of the events. He also faced the shame of not being able to provide a warm and stable home for his son.

In our work, we have found that it is more common for households to struggle paying their energy bills repeatedly over time than it is for such struggle to be a one-time occurrence. About half the households that have been disconnected face a repeat disconnection within a year's time. In fact, one of the leading predictors of someone being energy insecure—either struggling to pay an energy bill, receiving a notice for disconnection, or being disconnected—is previous incidence of energy insecurity.[15]

When Darrin was disconnected, he could have entered a financial deficit spiral had his friend's church not paid his bill in full. It will take him time to rebalance his finances. He will also now have an account with the utility company that shows that he has been disconnected, increasing the potential risk for disconnection in the future. If he falls behind again on a payment, even just another $13 worth of missed payments, will the utility representative who makes disconnection decisions observe that he is a risk and make the same call as before?

What Happens When Households Are Energy Insecure?

Darrin's disconnection experience was both upsetting and disruptive, and his unexpected loss of basic energy services illustrates the burden of this most severe form of energy insecurity. The energy insecurity problem, however, runs much deeper for many Americans. And, often, the true nature of the problem is obscured when one thinks only

in terms of people's ability to pay their utility bills. Another important manifestation is something that energy scholars Shuchen Cong,
Destenie Nock, Lucy Qiu, and Bo Xing refer to as the "hidden form of
energy poverty," in which households keep their houses at uncomfortable temperatures to keep costs down.[16]

When the weather outside gets hot, households with air conditioning usually turn down the temperature on their thermostat to
stay cool. The difference between when an average household turns
on their AC and when a disadvantaged household turns on theirs is
what these researchers refer to as the "energy equity gap," and it also
pertains to people's use of heat during cold temperatures.[17] Not surprisingly, this gap is very large for low-income households. People on
tighter budgets are by necessity more frugal with their energy use. In
the summer, lower income groups typically wait longer to first turn
on their AC and, when they do turn it on, they keep their homes less
cool than higher-income groups do. This gap reveals that the problem of energy insecurity is actually much worse than researchers and
government officials have estimated to date using the more standard
measure of energy burden. Using just the energy burden measure un-
derestimates the total number of households that are struggling with
their basic energy needs by approximately 150 percent.[18]

When families have to keep their homes at uncomfortably cold or
hot temperatures, they will seek out short-term relief. In particularly
hot conditions, a family may use fans or swamp coolers (i.e., a circu-
lating fan that runs over a damp pad to circulate cooler, wet air), drink
or apply ice to their skin, close off a single room that has a window AC
unit, or leave the house for a location that has air conditioning, such as
a local shopping mall. When exposed to particularly cold conditions,
families use several similarly common techniques: Wear or curl up
under blankets or heavy clothing; drink hot water; take a hot shower
or bath; or heat a single room in the house. To stay warm, families
may use several other, much riskier coping strategies. For example,
families may use space heaters, burn trash or other items within their
homes, turn on and leave open an oven, or sit in a running car with
the heat on, all for warmth. Each of these strategies can be damaging

or deadly, and they force people to weigh risks to their health from exposure to cold temperatures with the risk of adverse consequences from seeking warmth.

During Winter Storm Uri, the February 2021 extreme weather event that hit Texas and caused rolling blackouts, families were desperate to stay warm as the temperatures dipped into the low digits. Many piled under as many layers as they could, as well as piling on top of each other to share body heat. Some additionally brought their grills inside and burned charcoal; others ran their gas stove burners. In Sugar Land, Texas, the Nguyen family used their fireplace for heat. The mother put the kids and her mom to bed—her mom was staying overnight because her own power had gone out and she wanted to be near family for safety—but woke to a raging fire that had already consumed her three young children and her mom.[19]

In another setting, in a nineteen-story Bronx apartment building in January 2022, a family was running space heaters to stay warm. In the middle of the night, one sparked and started a fire. The smoke poured into other apartments and killed a total of seventeen residents, including eight children. Space heaters have proven deadly elsewhere: Eighty deaths and 1,700 fires are caused each year by space heaters.[20]

In our own research, people have shared with us details on how their families cope when facing particularly cold or hot temperatures. One explained that he disconnects his dryer vent, puts his daughter behind the dryer and closes the door to the laundry room, then runs the dryer cycle to warm her up.[21] Others share stories about using wood stoves, open ovens, gas burners, and burning trash to keep their home indoor temperature livable.

These isolated stories are tragic and in some cases deadly, but they are neither exceptional nor mere anecdotes. In a nationally representative sample of low-income Americans (i.e., households within 200 percent of the federal poverty line, or a four-person household with an annual income of $60,000), we found that 55 percent of respondents reported engaging in at least one coping strategy at some point during the year, as shown in figure 4.3.[22]

Temperature and space heating strategies are not the only ways that families try to reduce their energy costs and to avoid being disconnected. Families use a range of different coping strategies, many of which are either risky or difficult to achieve. These strategies reveal just how desperately families want and need to avoid high energy bills and shutoffs and how they must weigh trade-offs between the risk of being unable to pay their bills or having their power shut off, and their own health and well-being.

Families most commonly deal with an inability to pay their energy bills by accruing debt. The National Energy Assistance Directors' Association estimated that in 2021, households owed about $22.3 billion in arrearages, at an average of about $1,060 per household.[23] Debt accumulation can make a household especially susceptible to a future disconnection.

Nearly one in five low-income households copes with energy insecurity by forgoing expenses on other essential items, such as food or health care.[24] In some of these cases, the parents are the first to skip meals in order to leave enough for their children. In more desperate situations, the children must also skip meals. Researchers refer to this general phenomenon as "heat or eat," and it is often seasonal; in the cold weather months, when heating bills rise, the amount of discretionary money left for food or other essential needs declines, and families then spend less on food.[25] Yet it is during the cold months when one's body most needs nutrients and calories to stay warm. A lack of food and nutrition can lead to other complications as well, like diabetes and heart disease. Studies have found that when a family is energy insecure, the affected children are more likely to experience developmental delays and have lower health ratings.[26]

Families also engage in bill balancing, where they pay down one bill one month and another bill the next month, or pay all of them down just to the level that is needed to avoid punitive consequences such as late fees, accruing interest, or disconnection. This practice underscores just how interwoven various forms of material hardship are for millions of families, in which they must make strategic decisions about which bills they can pay in any given month. In order to make their monthly payments, families may have to pay off some bills that might otherwise trigger an immediate negative outcome, such as an eviction or disconnection, and leave other basic needs such as food and medical care unmet.

When a household faces the risk of being disconnected, they can apply for LIHEAP. LIHEAP is a program funded by the federal government and administered by states. LIHEAP typically provides annual assistance of about $350 to qualified households.[27] LIHEAP is not an entitlement program like the Supplemental Nutrition Assistance Program or the Medicaid program; demonstrating eligibility does not guarantee access to program benefits. Rather, eligible households receive funds until funds are exhausted. Funds are allocated to the states in early spring and winter[28] and are often fully spent by the late summer and late winter; thus, they are unavailable when households are struggling to cool their homes in the summer or warm their homes in the early winter. Studies show that LIHEAP only reaches about one in five eligible households,[29] though these numbers may be a dramatic underestimate of the actual need.[30]

LIHEAP, the federal Weatherization Assistance Program that we discuss below, and similar bill assistance programs offered by local governments require that a family know that these programs are available, know how to access them, and meet the criteria for eligibility. These criteria vary by location but typically include income or asset thresholds, a limited number of uses per year, provision of documentation, and, as in Darrin's case, that the customer is not already shut off from their energy utility service. Families may also borrow money from friends, neighbors, or family, ask for assistance from charitable or faith-based organizations, or seek payment assistance from their utility service directly.

Many also start a payment plan with the utility company, by which they pay a set amount each month or allow for more flexible payment periods to pay down their bill. Although such plans may seem like an obvious first step, many of the people we have interviewed who face disconnections report significant challenges communicating with their utility companies when seeking explanations and assistance. Three of them told us, respectively:

It was one of those situations where you get a hold of one service agent who tells you they're going to do something, and then you think that they are making note of it. And I always have the lesson of please email me right after we get off the phone because I want proof that this was communicated to me. But then you never do. And then it gets lost, I guess, in transmission, and then that's how the communications happen in these situations.

I was very upset that day. I mean, very upset. I even argued with these people and told them, how could they do something like this when I thought they weren't doing that anymore? They know what's going on with COVID, with everybody getting fired. And you and I—I mean, everybody knows what kind of situation anybody was in. So they just don't care. Companies like that just don't care. They got a lot of money. But anyhow.

I feel bad for calling them or you know what I mean? Call them, be like, "Hey, can I get an extension?" Or whatever. It makes me feel kind of down, you know what I mean? I don't want them to be like, "Well, you should pay the bill." You know what I mean? That's why I definitely just prefer to talk to them online, because I feel like I can't hear any emotion or anything nervous. I feel like if they're upset or if they're rude, I won't be able to hear it.

These experiences add to the distrust that people often already have toward their utility providers—as well as feelings of being "locked in" when they have no choice but to be served by a single utility company in regulated service areas—and it likely helps explain why many turn to other strategies to cope with the difficult circumstances.

While various bill assistance approaches may seem like good options when one needs financial support, we have found in our surveys with low-income households that people more often use riskier financial strategies. This type of behavioral response suggests that many people would rather avoid the administrative and social burdens, as well as the shame that can come with asking for assistance, than taking on the financial burdens associated with carrying debt.

Families also seek out loans, either from banks or from other lenders that can provide cash immediately without strict credit requirements, such as payday lenders. In our work, we find that taking out a loan is less common than other techniques and is only practiced by about 2 percent of all low-income households. But, for these families, taking on a high-interest loan only provides temporary relief and increases their overall debt.

In figure 4.3, we show how often low-income families engage in these various coping strategies. Note that the more common strategies are also the riskiest from a financial and physical health perspective.

Families often engage in many of these different coping strategies, and particularly vulnerable households use them all, often simultaneously.[31] Households with young children and those with individuals who rely on electronic medical devices are both more likely to be energy insecure and more likely to use a combination of behavioral and financial coping strategies, such as using the oven for space heat and bill balancing. And, for some, the choices require accepting difficult trade-offs.

Consider the case of an individual with an electronic medical device such as a home oxygen concentrator. Their doctor prescribed the

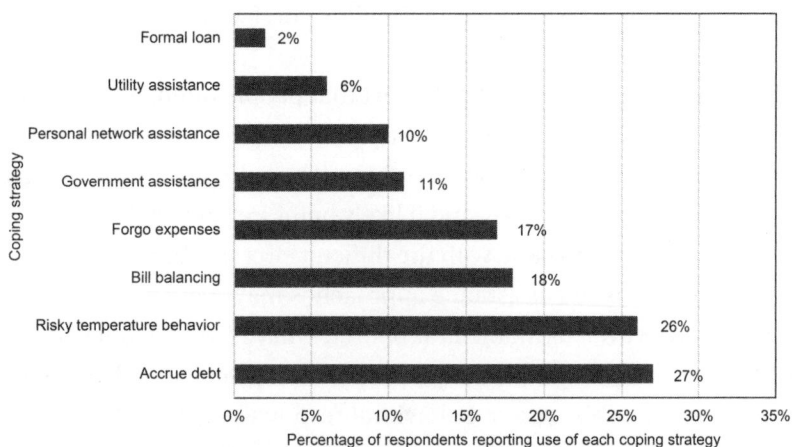

Figure 4.3 Household coping strategies for energy insecurity, 2020–2021
Notes: Survey weights are applied to all values.
Source: Carley and Konisky, *Survey of Household Energy Insecurity in Time of COVID*. The data presented in this graph were gathered via surveys to a representative sample of households within 200 percent of the federal poverty line. We administered surveys at four points in time between May 2020 and June 2021.

machine because they were having trouble breathing due to a worsening lung condition. At the time of prescription, the doctor likely did not think about the cost to the patient to rent or purchase the machine, nor about the ongoing cost of its operations. And while insurance, depending on one's plan, may cover a portion of renting or purchasing the machine, it typically does not cover the cost of its operations. Constant use of the machine in a state with median electricity prices would cost this individual about $750 more for their electricity bill each month.[32]

With a constrained budget, this additional expense may require sacrifices of other essential items. These additional expenses will also mean that in months with higher bills, such as during times of cold spells or heat waves when the heat or AC must run more often, people will find it particularly challenging to make ends meet. At no point can this individual decide to just not use the device, since their life depends on it. To get by, they will need to take other measures, such as cutting costs on food, or trying other measures to keep their heat or AC costs low.

Health Consequences of Living Without Power

Most Americans take for granted the ability to flip a switch and have instant electricity. We are heavily reliant on electricity and other forms of energy from the moment that we wake up each day. We use household appliances to preserve and prepare food, do our laundry, charge our devices, and keep our homes at comfortable temperatures. A household without power is one that is isolated from modern life, disconnected from family, friends, and coworkers, and unable to carry out everyday activities.

A lack of power can also produce severe health consequences such as an increase in the incidence of asthma, respiratory infections, and death,[33] and mental health issues in the form of sleep loss, stress, and depression.[34] One study has found that winter-related deaths from exposure to cold conditions and lack of adequate energy access have killed more people in Vermont than die in car crashes.[35] The Center for

Disease Control and Prevention (CDC) estimates that each year about sixty-five thousand people visit the emergency room for acute heat illnesses, and over six hundred people experience premature mortality from heat exposure.[36] Another study puts the mortality estimate closer to 5,600 deaths annually due to heat exposure.[37] Climate change will make matters worse.

Consider Darrin's case: His circumstances produced extreme stress, which, combined with very cold temperatures in the home, made it difficult for him to sleep and function. This type of mental and physical stress could have exacerbated preexisting medical conditions, for example, if Darrin had had a heart condition or diabetes.

Energy insecurity also affects children's ability to learn and to fully participate in their schooling. If a child cannot do their homework at home because there is no power, or if they are living in a home with uncomfortably cold or hot temperatures, it may affect their cognitive functioning and their academic performance on assessments and other graded material. Similarly, if a child is removed from the home and put in foster care while their parents are left to deal with a power shutoff situation, the child's sense of security and stability will invariably be affected.

What we know less about are the long-term health consequences for households that face chronic energy insecurity. How do traumatic experiences such as being disconnected affect a family's ability to function, to stay together, to achieve long-term educational and professional success? How traumatic is being disconnected as a child, and do the effects linger as the child becomes a teenager and young adult? How can families exit the cycle of energy insecurity once it starts, and, if they cannot exit, how does the cycle perpetuate other forms of material hardship?

Avoiding Energy Insecurity

An obvious way for someone to reduce their energy costs is to reduce the amount of energy that they use. To do so effectively, however, requires that people have a good sense of how much energy they consume while carrying out their daily activities and of what is needed to achieve a certain quality of life.

Several years ago, behavioral scholar Shahzeen Attari and her team of researchers asked people about the different ways they could save energy. The most common answer they received was to turn off the lights. Yet, while turning off the lights helps, this action is much less impactful than using efficient models of air conditioning units, dishwashers, clothes dryers, and other appliances.[38] For example, an efficient washing machine can save users around $37 per year, an efficient dryer can save up to $25 per year,[39] and an Energy Star–certified water heater can save over $450 per year.[40] In addition to the efficiency of appliances themselves, people can also save money with how they use them; for instance, running a washing machine with cold rather than hot water can save 64 cents per load.[41] While more efficient appliances can save households money on their energy bills, so too can energy conservation efforts, such as hanging clothes to dry or turning the thermostat down in the winter or up in the summer, still keeping within the recommended safe indoor temperature range.[42] Yet, these techniques may not be common practice for most Americans. There is little public information about how to save energy, just as there is little public information about weatherization programs, bill assistance, and disconnection protection practices that a utility must follow. When one is disconnected, it is often too late to learn about these options.

Purchasing energy-efficient appliances requires spending more money up front, since high-end dishwashers, HVAC systems, and washers and dryers with energy saving features typically cost much more than standard models. The returns, in terms of lower energy bills, can be substantial, but they will be enjoyed only gradually over the lifetime of the appliance. Moreover, most renters—about one-third of all US households—do not have the ability to even make such purchases, regardless of their affordability. Instead, landlords make these decisions, and they generally do not have much reason to be concerned about energy efficiency, since typically it is their renters who are paying the utility bills (this is referred to as the "split-incentive" problem). The nature of this problem is much worse for renters and owners in older housing, which are more likely to have less energy-efficient appliances and a lack of sufficient insulation, sealing, and high-quality building materials.

Beyond changes to personal behaviors in the home, how else might one address the energy insecurity problem, particularly for those populations most likely to experience challenges paying utility bills and therefore the potential for disconnections from service? Private and public investment in energy efficiency is one approach.

For their part, utility companies do not invest heavily in their energy efficiency programs directed to low-income households where the benefits would be especially impactful. And this lack of investment contributes to the higher energy burden experienced by low-income households. By one estimate, utility companies only devote about 8 percent of their electricity and 26 percent of their natural-gas energy efficiency expenditures to low-income households.[43] The American Council for an Energy-Efficient Economy provides a scorecard for utility companies and their energy efficiency efforts. In 2023, when the council first started tracking utility efforts aimed at lower-income customers, they found that only thirteen of the fifty-three utility companies they scored had a practice of directing customers struggling to pay their bills toward utility energy efficiency programs.[44]

If utility programs do not go far enough, what other supports are available? The federal Weatherization Assistance Program (WAP) and similar state and local programs could help vulnerable households improve the efficiency of their homes, which could help them lower their energy bills. These programs provide financial assistance to help income eligible individuals upgrade the energy efficiency of their homes. Weatherization techniques include installing insulation, replacing windows, sealing ducts, putting weather strips along external doorways, filling holes and gaps in the wall, and repairing heating and cooling units. Fully weatherizing older homes can be quite expensive. According to one report from the US Department of Energy, the minimum average weatherization expense per house is about $6,500.

However, over time, these investments have been demonstrated to reduce household costs—on average about $283 per year—as well as to reduce other costs, such as an average of $514 in avoided medical costs each year from ailments that are exacerbated by too much exposure to heat or cold.[45]

Given these demonstrated returns, WAP might seem like an obvious answer. However, historically, WAP has been woefully underfunded. The program only serves a fraction of the eligible population and those in need: Thirty-six million US households are eligible, and estimates range from 4.6 million[46] to 15.7 million[47] for households that severely need weatherization assistance, yet only seven million households have received such support in the past forty years.[48] At the current rate of WAP funding, it will take 360 years to weatherize all homes that have particularly high energy burdens.[49] One obvious way to expand the program is for the federal government to allocate more funding. Another way to improve the program is to funnel more of the current funds toward those households most in need of support; such funneling can take place through targeted information campaigns as well as through active coordination with local community members who can help communicate program benefits to eligible households.

Further complicating matters are changing energy markets and climate conditions that are leading to an increase in the price of electricity. A recent study found that these price increases can affect energy burden; in Phoenix, Arizona, for example, low-income customers can face energy burdens of up to 14 percent during the summertime.[50]

In some places around the country, the households most in need of WAP assistance are unable to access it. In Indiana, for example, there is a waiting list for WAP assistance that can extend as long as two years.[51] Of course, just a couple of months could make a huge difference for someone who is struggling to pay their energy bills and potentially facing the threat of disconnection. Two years could feel like a lifetime. In Rhode Island and plenty of other states across the country, a landlord must approve of a WAP application if a renter seeks assistance. In addition, households that have significant mold and other structural problems, such as foundation or roofing needs, are ineligible to access WAP until the conditions are fixed. In such cases, homes might first be deferred to other programs to address these conditions.[52]

In Darrin's state of West Virginia, WAP program eligibility includes income under a certain threshold ($25,760 for a one-person household and an additional $9,080 per person); proof of home ownership or rental agreement; and living space over 219 square feet (with an additional 100 feet per person). One is automatically eligible if they receive supplemental security income or temporary assistance to needy families, and those who spend over 20 percent of their income on energy or more than $2,100 per year on energy are given higher priority.[53]

Eligibility for WAP or LIHEAP—the federal energy-assistance program we discussed earlier in the chapter—varies across the country. In some states, income eligibility limits differ. In other states, some people are automatically enrolled, whereas elsewhere, the application process is completely separate and people must apply annually.[54]

Other challenges to accessing WAP and LIHEAP benefits include the sheer difficulty of the application process. Public policy scholars refer to these types of obstacles as administrative burdens, and they are ubiquitous across social welfare programs.[55] For example, assuming eligibility requirements are clear, applicants often have to provide multiple types of documentation, such as utility bills, bank statements, and proof of income. In addition, they may have to provide multiple years of tax documentation. In some states, people can apply online, whereas in others they must show up to an office, often in the middle of the workday.[56]

Another major barrier to accessing WAP and LIHEAP is simple awareness of these resources. Studies have found that many people do not apply because they wrongly think that they are ineligible due to their employment, immigration, or income status. Others express concern that they will have to pay back the money, or they have language barriers that make them hesitant to explore this option. In many cases, those who need the help simply do not know that the programs exist. Overall, only about 20 percent of all eligible households currently take advantage of them. New York is the state with the highest number of LIHEAP eligible households, yet less than 46 percent of eligible New Yorkers are served by the program. Overall, estimates state that federal funding would have to increase to ten to twenty times its current levels to provide enough funds for all eligible families.

A final, important barrier is that many people distrust the government officials that implement energy-assistance programs; that distrust may extend to the companies contracted to install energy efficiency upgrades in people's homes. This distrust is reflected strongly in figure 4.3, which shows that only 6 percent of all low-income households called their utility company when they were struggling to pay their energy bills, and only 11 percent sought government assistance.

In a study of the implementation of WAP in a primarily African American neighborhood in Kansas City, Missouri, energy justice scholar Tony Reames uncovered high levels of distrust among eligible households.[57] Reames's interviews with energy efficiency program staff revealed that homeowners were especially distrustful of people perceived to be "outsiders" and not from the neighborhood. Moreover, Reames's research found that *who* knocked on the door was critical, and that local neighborhood association leaders and residents who offered to share information about WAP were perceived as the most credible and trusted sources.[58]

Reames's finding of distrust is important and speaks to another, more general pattern with energy efficiency and clean-energy technologies—there is a lack of diversity of the workforce in this area, which has the effect of limiting awareness of ways to reduce energy use and raises skepticism among people when approached about making upgrades to their homes. We discuss this topic in more detail in the next chapter.

WAP and LIHEAP potentially address energy insecurity, by providing resources to eligible households at potentially different points in time. WAP as a program is designed to be a long-term remedy, whereby public support for energy efficiency upgrades can reduce a person's energy bill for the long haul. LIHEAP, by contrast, is mostly designed to provide immediate financial support to help people through acute situations where they cannot afford to pay their energy bills. In this sense, we can think about WAP as an early-stage intervention into the problem of energy insecurity, while LIHEAP comes toward the end. Another policy approach comes even later, and that is to place limits on when and under what circumstances utility companies are allowed to disconnect their customers from service, even in the case of nonpayment.

Let's revisit Darren's situation. Had he been properly notified of a pending disconnection, he could have tried to borrow money from a friend or neighbor or sought LIHEAP funds. In West Virginia, the policy is that a utility must notify its customers before disconnecting their service. It is possible that the utility company did so, but, as Darrin recounts the story, he does not recall any such notification—via email, mail, a door knock, or otherwise. Darrin is not alone here; many of the interviews that we have conducted with people who have been disconnected reveal that the household either did not receive notification at all, or it came in the mail on the day of the actual shutoff. In other words, they learned of their circumstances much too late to take action. Without notification, Darrin was literally in the dark about the situation. In West Virginia, there are no additional disconnection protections beyond those offered during extreme cold weather (days when the temperature is forecasted to fall below 32 degrees F). If he had lived in another state that offered special protections for those with medical impairments, Darrin might have had another way to avoid being shut off.

There are no uniform, federal protections that limit when utility companies can shut off customers' services. Rather, because energy utility companies are regulated at the state level, any such protections come from either state laws and regulations or rules imposed by state public utility commissions, who are responsible for regulating utilities in each state.

The most common state disconnection policies are seasonal moratoria designed to limit disconnections during potentially severe weather. These types of protections are either based on temperature or date thresholds. An example of a temperature threshold is in Arkansas's policy, which holds that no one can be disconnected on a day in which the temperature is above 95 degrees Fahrenheit or below 32 degrees

Fahrenheit, or Nevada's policy, which sets temperature thresholds at 105 degrees Fahrenheit and 15 degrees Fahrenheit. An example of a date threshold is in Indiana's policy, which prohibits disconnections between the months of December and mid-March, though this prohibition only applies to people who first apply for LIHEAP assistance. All told, most states across the country have some sort of protection in place during the winter months, whereas similar protections during the summer are less common (as of 2024, just twenty-three states and Washington, DC, have heat-related protections).[59] The other common disconnection limits that states impose are to protect specific vulnerable populations, such as people with medical conditions, the elderly, or families with young children. As an example, in Michigan, customers can have their disconnections delayed for up to twenty-one days if someone in their household has a documented medical condition.

The degree of public awareness about these protections is unclear. In most, but not all, states, there is some requirement that utility companies regularly share this information with their customers. The extent to which people pay attention to these notifications, however, is unknown.

We do know that state protections, where they exist, can effectively limit the number of utility shutoffs, thereby reducing energy insecurity and other adverse outcomes. During the COVID-19 pandemic, many states included shutoff moratoria as part of emergency orders to respond to the public health crisis. We conducted research on the effect of these temporary state moratoria on both incidence of disconnections and household "heat-or-eat" trade-offs. We found that the moratoria significantly reduced both forms of energy insecurity.[60] Other work also found additional benefits of avoiding disconnections.

• • •

Access to affordable, reliable energy is something most Americans take for granted. One may think about the loss of energy utility service as infrequent and temporary, often the result of outages caused by severe storms. For most Americans, this is the case. The loss of energy services is a rare inconvenience.

However, as we have explored in this chapter, for many households, the risk of losing access to essential energy is common and constant. Energy insecurity is especially prevalent among low-income Americans, people of color, those who are medically vulnerable, and those living in poor housing conditions. Not only do these vulnerable populations struggle to pay their monthly bills, they are more likely to face the risk of disconnection altogether. And the risk of a shutoff is not just a one-time thing but one that recurs month after month.

People cope with these circumstances in a variety of ways, sometimes using risky financial and behavioral strategies to get by. In doing so, they often rely on their social networks, in part because they have lost faith in their utility companies and in the government agencies that implement programs intended to provide at least some assistance.

The current challenge is immense, and, moving forward, the challenge is likely to become more pressing. Climate change will bring more extreme weather, including heat waves that will strain family budgets even more. Increased exposure and more erratic temperature swings introduce two challenges for energy insecure families, or families who are not yet energy insecure but may become so in the future. First, hotter and colder temperatures require households to use more energy in the form of air conditioning or heat. Thus, climate change has the potential to raise energy bills. Second, more erratic weather during periods that formerly were not covered under state disconnection protections will result in more overall disconnections, and these disconnections will occur during times when people are particularly vulnerable to heat and cold exposure.

Government will need to recalibrate its safety net by assuring that protections historically designed for the winter (e.g., LIHEAP funding, disconnection protections) are adjusted for the stresses that will come. In addition, transition to clean sources of energy, the need to modernize the electric grid and other energy systems, and transmission and distribution upgrades are also likely to result in higher energy costs, disproportionately affecting those who are already struggling to pay for energy.

One way to mitigate some of these future burdens is to be sure that those who are already experiencing the weight of the current energy system are able to access technologies that can alleviate these problems through lower energy bills. As we see in the next chapter, however, there are challenges here as well.

CHAPTER 5

Where New Technologies Don't Go

A decarbonized future, especially one of net-zero emissions, requires that energy and utility companies invest rapidly and with great vigor in clean-energy systems, such as solar and wind systems, as well as in technologies that limit and capture and store emissions.[1] Many argue that to do so efficiently also requires that governments price carbon in ways that lead companies that produce or use fossil fuels to internalize the costs of climate change impacts. Government subsidies for both early-stage research and development and later-stage technology deployment are also important pieces of the puzzle.[2]

However, achieving the emissions reductions necessary to avoid the worst effects of climate change requires more than government policy and changed practices from utility providers and other sectors of the economy. Most of us will also need to change our use of energy, including adopting clean technologies and altering our daily behavior.[3]

Human and household consumption accounts for a whopping 72 percent of all greenhouse gas emissions across the world.[4] The majority of these household emissions come from travel via personal cars and airplanes, meat and dairy consumption, and home heating and cooling.[5] By one estimate, people across the world will, on average, need to reduce their carbon footprints by a factor of thirty—or, alternatively phrased, they will need to use one-thirtieth of the amount of energy they use currently to meet global carbon mitigation targets,[6] with those in industrialized countries needing to reduce their footprints much more than those in industrializing countries. For the United States to meet its targets for the Paris Climate Agreement, it will need to reduce its carbon dioxide emissions from 4,954 million metric tons in 2022 to 3,630 million metric tons by 2030,[7] an amount that is equivalent to the total carbon emissions of Japan, France, and Monaco combined.[8] To reach net-zero, the United States will need to reduce carbon dioxide emissions to 558 million metric tons by 2050.[9]

To reduce emissions to anywhere close to these levels, households will need to replace their older, less efficient, and fossil-fuel powered cars and household appliances and pursue significant lifestyle changes that involve their food sources, overall building efficiency, and other material consumption as well. Households may also consider purchasing new energy devices that can monitor household energy consumption, such as smart thermostats, or technologies that help enable new forms of mobility, such as electronic scooters, to help shift consumption patterns at home and on the go.

The global debate about how to reduce greenhouse gas emissions tends to be macro in scale. Policymakers and academics tend to ask questions like: How many fossil fuel power plants must be decommissioned? How many new renewable-energy facilities need to be built? How to make these changes at the necessary speed and scale? And, how to best price carbon in a way that efficiently reduces greenhouse gas emissions? These debates often neglect a consideration of the necessary changes at the individual and household level.[10] Yet, to meaningfully and substantively reach decarbonization objectives, massive individual and household behavioral change is also necessary.

Changing behavior is hard. Not only are humans creatures of habit, but structural barriers often work against us, making behavioral changes difficult—and sometimes nearly impossible. In this chapter, we focus on the structural barriers to the widespread deployment and use of low-carbon and efficient residential-energy technologies.

The most common home in America has three bedrooms, a kitchen, a living room, and a dining room, coming in around 2,400 square feet.[11] Picture the set of *Breaking Bad*,[12] the five-season drama and thriller that aired between 2008 and 2013, in which Walter White, played by Bryan Cranston, is a high school chemistry teacher with terminal lung cancer who takes to selling methamphetamine to financially support his family. To decarbonize his home, excluding for the sake of argument his meth operations, Walter would need to switch his heating and cooking sources to electricity—specifically electricity sourced from low- or no-carbon energy resources—and trade in his internal-combustion-engine car—a Pontiac Aztek—for an electric vehicle, a bicycle, public transit, or a pair of sneakers. To achieve additional efficiency enhancements, Walter would need to swap out his appliances for more efficient models, installing an efficient HVAC unit and a heat pump or solar water heater. Walter could also install more energy-efficient windows and add insulation to his home. And, in order to offset some electricity from the grid or pay less for energy, Walter could install rooftop solar panels, an in-home battery storage unit, and a smart thermostat.

All these efforts amount to no less than ten new technology investments for Walter and his wife Skyler. Using their home to represent an average US household, multiplied by 124 million households in the United States, equates to about 1.24 billion new energy technology purchases for household decarbonization. Of course, some people have already begun to make these types of investments, and some homes will require fewer changes, but the general point remains—the sheer scale of required investments is enormous.

In short, deep decarbonization requires massive individual and household technology adoption. Yet people face significant barriers to transforming their homes and lifestyles. Not a single clean technology is currently accessible and affordable to everyone. Solar panels, electric vehicles, heat pumps, and many other clean and energy-efficient technologies still have prohibitively expensive up-front costs for many individuals and households, and some of the forty-four million renters across the US in particular may simply not be in a position to adopt some of these technologies at all, even if up-front costs were not an obstacle.[13] The challenge is also not just about money; it is often about geography, race, and other factors, exacerbated by market dynamics and the structure of government policies.

Choose any residential-energy technology; the story is the same.

We will start with solar. The profile of the average solar owner is someone who has high income, high credit scores, many years of education, lives in predominantly white, urban areas, and is middle-aged or older.[14] As of 2023, over three million US households had rooftop solar panels.[15] At the same median household income, studies have found that Black-majority census tracts had 69 percent less solar than any census tracts lacking a majority demographic; Hispanic-majority had 30 percent less; and white-majority had 21 percent more.[16] These disparities remain even when statistically controlling for income and whether homes in the area are predominantly rented or owner-occupied.

The good news for solar is that some of these trends are starting to change. While early deployment was more likely to occur in higher-income neighborhoods, beginning around 2011 groups other than the highest income began investing more in solar, albeit with smaller systems, which itself has implications for energy savings.[17] The median income for solar-adopting households declined from $140,000 to $117,000 between 2010 and 2022, but this level still remains well over the median income of US households.[18]

Is energy efficiency any better? Energy efficiency refers to the use of less energy input for the same amount of energy output. One could achieve energy efficiency through, for example, using an appliance that needs less electricity to perform the same task. Energy efficiency is typically considered one of the fastest and easiest ways to reduce one's energy bill. Efficient appliances can save a household between 13 and 32 percent off the cost of operation.[19] Yet this story is the same; lower-income households are significantly less likely to own efficient technologies. One study found that, of all spending on energy efficiency in the United States in 2018, only 10 percent came from low-income households,[20] which account for approximately 28.3 percent of total households.[21]

Higher-income households more commonly have higher-efficiency appliances such as central air conditioning, furnaces, and heat pumps.

Conversely, lower-income households tend to have less efficient appliances, including portable electric heaters and window-unit air conditioners.[22] Even appliances that are more universal, such as refrigerators and washing machines, vary in their efficiency across the incomes of households. Higher-income households are much more likely to have higher-efficiency models compared to lower-income households.[23] In fact, lower-income households often purchase used appliances to save money, but as a result those households often end up with older and less efficient models. These conditions explain, in part, why some groups have higher energy burdens than others, as discussed in the last chapter.

The story is the same for battery storage systems, smart thermostats, smart appliances, electric vehicles, and more. People with higher incomes are much more likely to invest in these technologies than those with low or modest incomes. For example, 75 percent of all battery capacity in California is in the state's most advantaged communities.[24] While this story is not unique to energy technologies, as any household or personal technology that requires a large up-front investment will display similar adoption patterns, clean and efficient energy technologies are prone to especially large up-front costs.

The fact that higher-income households are more likely to have efficient appliances and invest in new, cleaner, and smarter technologies is not surprising. In addition to the high price tag of these items, income-based patterns of adoption follow conventional wisdom about the diffusion of new technologies. Prices tend to be high at first as companies continue to build, modify, and tweak their product designs, and as customers learn about the technologies and experiment with adopting them and integrating them into their lifestyles. With "learning by doing" and "learning by using," terms that describe the process of both producers and consumers, respectively, trying out new products and continually improving them with experience, these technologies become more mainstream. Consumer demand rises, and supply does as well, eventually yielding product efficiencies and a lower price. The lower the price, the more accessible the technology is to a broader set of consumers, including those with less income.

Think of cell phones, which started off as a luxury item for the wealthy but are now common for everyone.

But the story of access to new energy technologies is about much more than income. The adoption of these technologies is also marked by racial, ethnic, geographic, and other disparities, and it is then reinforced by governmental procedures, institutions, and market dynamics. Here, we present seven interrelated conditions that both shape the current state of technology-access disparities and serve to reinforce and magnify them over time. These conditions not only raise concerning questions about equity, they also reveal a set of significant obstacles for the type of deep and widespread decarbonization of energy that is required to meet climate goals.

Access to Benefits Is Uneven Across Sociodemographic Groups

Low-carbon and efficient technologies provide significant benefits to their users such as reduced energy bills and cleaner air quality. From a normative perspective, not only should everyone be able to access these benefits,[25] but arguably these benefits will be more impactful for more overburdened populations since they have historically faced the most significant energy-related problems.

This innovativeness-needs paradox, in which those who are poised to benefit the most from a new innovation are the last to adopt it,[26] is not unique to clean-energy technologies. Other examples include vaccinations for highly transmissible diseases in overcrowded and medically underserved populations, or low-income Americans not taking advantage of child tax credits.[27] However, the paradox is especially pronounced with energy technologies because most require large up-front expenditures. Moreover, the use of a residential technology may be resisted by utility companies that control the local market. Such technologies often require significant behavioral adjustments and long-term planning by all consumers, including both individuals and businesses.

To put solar panels on a home's roof, a household must pay the up-front and installation costs, which can easily total $30,000 or more.

Federal investment tax credits help offset these costs, and leasing options are increasingly available, but, in general, there remain high upfront costs. Once the solar panels are installed, the household can save on its utility bills thereafter by producing its own electricity. Households typically pay off their solar systems within six to ten years,[28] and then save thereafter. These types of savings are most pronounced in locations with particularly large solar resources and favorable net metering policies that allow solar owners to sell back their unused electricity to their utility company at the retail cost of electricity (i.e., the levels that customers pay for incoming electricity).

Efficient light bulbs or appliances similarly save households money, once the buyers have recovered the initial purchase cost of these technologies. Consider the difference between an LED light bulb and an incandescent, the latter of which was prohibited from new sales by the Biden administration as of August 1, 2023. The LED costs about $5 per bulb and the incandescent costs $1. But the LED uses less energy and lasts significantly longer—25,000 hours relative to the incandescent's 1,200 hours (for reference, a single year contains 8,760 hours), which means that one would need to purchase twenty-one incandescents during the lifespan of a single LED bulb. At a cost of electricity of 10 cents per kilowatt hour, the cost of 25,000 hours of light for an LED is $26.25 and for an incandescent is $171.[29] The incandescent has been banned because of its energy inefficiency, but we can make similar comparisons to fluorescent light bulbs, which remain on the market. Like incandescents, fluorescent bulbs are cheaper than LEDs, at about $2.00 per bulb, but their lifespan is only 10,000 hours—so you would need five fluorescent bulbs for every two LED bulbs.

And this is just for one lightbulb. Most houses use many light bulbs, which enhances these savings by multiples. Walter's house, with three bedrooms and several main living spaces, likely requires about ten to twenty light bulbs, for a saving of $1,710 to $3,420 during the lifespan of a single generation of the bulbs.

Other household energy technologies can save additional money. Smart thermostats, for example, when used as intended,[30] can save a household up to 20 percent on their energy bills.[31] More efficient

HVAC units can save 20 percent on energy bills, an amount that may equal $1,000 per year,[32] more efficient washing machines can save $48 per year, and switching from a gas-powered furnace to a heat pump can save $815 per year.[33]

When one cannot access solar or these other clean technologies, due to high up-front costs and lack of access to dealers or maintenance service providers, one obviously cannot reap their financial benefits.[34] As discussed in the last chapter in the case of Darrin's disconnection experience, even small energy savings may make the difference between a household that is able to keep its lights on and heat operational and avoid disconnections, and one that cannot. Or it may constitute the difference between whether members of the household need to skip meals to cover their energy bills, or engage in other coping strategies.

Energy savings are not the only benefit, however, to clean-energy technologies. Another major benefit is resilience. Residential batteries, for example, can improve the resilience of one's home by providing a backup source of power which, when coupled with solar, can run even when the grid is down due to blackouts. Batteries can also serve as a form of what's called demand-side management by shifting one's energy load (i.e., one's demand) to different times of the day. When one lives in a region with time-varying electricity prices (meaning prices that might shift up and down over the course of a day), being able to shift demand to lower-cost times can both reduce costs and insulate a household from volatility during higher-price, more supply-constrained times of the day. This demand-side management can also help the electric grid by reducing the need for investment in more expensive energy sources that otherwise need to produce sufficient electricity to meet peak demand.

Replacing older, less efficient, and more carbon intensive sources of energy with clean energy can also reduce air pollution and improve health outcomes. However, to achieve these benefits, widespread adoption of these clean-energy technologies is necessary. Doing so will ultimately reduce the demand placed on fossil-fueled power plants—and the primary beneficiaries of that reduction will be those who live nearby who have historically been exposed to the most pollution from these facilities. And, as discussed in chapter 2, a vast number of empirical studies show that low-income households and households of color, are significantly more likely to face such exposure and to suffer mental and physical health consequences as a result. Thus, over time, one should observe positive external benefits from these technologies for those who have historically been the most burdened by our energy systems.

Barriers to Adoption Are More Pronounced for Certain Populations

Disadvantaged populations historically face more extreme versions of barriers to the adoption of clean-energy technologies, as well as unique and, in some cases, compounding barriers.

The most pronounced barrier to adoption for most clean-energy technologies is the up-front cost of the technology, particularly in the early years of the technology's market presence. These cost barriers are more pronounced for disadvantaged populations, particularly for low- to moderate-income individuals or households. In order to pay a large up-front cost, one needs to have the cash on hand or acquire the product through credit. Lower-income individuals are less likely to have the ability to save up for such a purchase or to have discretionary income readily available. Similarly, lower-income individuals are commonly more credit-constrained, facing lower credit limits, higher interest rates, and less access to loans from lending agencies.

In order to take advantage of a tax credit that the government may offer to help reduce the cost of the technology—an incentive that could prove particularly valuable for those with limited resources— one must also have a tax liability. If someone only pays a few hundred dollars in taxes, for example, they will only be able to recover a few hundred dollars in tax credits.

As an example, a household with a tax liability of $30,000 could take advantage of the $7,500 electric vehicle tax credit offered by the federal government, whereas someone with a tax liability of $3,000 could only receive up to $3,000, assuming that both purchases qualify for the full tax credit. In other words, both individuals—the one with the higher tax liability and the one with the smaller one—could buy the exact same car, but only one of them can get the full credit. Phrased differently, the lower-income individual pays more for the exact same car. For a $60,000 electric vehicle, the former would pay $52,500 and the latter would pay $57,000. For a $30,000 electric vehicle, that would be $22,500 versus $27,000 for the same car.

What is more likely in this situation, rather than the low-income individual purchasing the electric vehicle, is that the higher-income individual will sell their older, less efficient car (i.e., it consumes more energy, and thus costs more per mile to run) in the used-car market. The lower-income individual, who was in need of a car, may then buy it.

Almost all clean-energy technologies come with a higher up-front price tag than an alternative conventional technology. Compare a smart or Energy Star dishwasher versus a regular dishwasher, for example, or an electric vehicle versus a conventional internal-combustion-engine vehicle. But some clean and efficient energy technologies also come with monthly or annual maintenance or subscription services, even though they may not actually cost more to maintain. A smart thermostat, for example, may require a subscription. So one must have not only the financial means for the initial purchase but also funds to pay a monthly subscription, and potentially regular access to Wi-Fi to operate it. Yet if the smart thermostat produces savings against what the customer would have paid using older technology, then the initial purchase and monthly subscription may make the up-front investment worthwhile.

Information barriers to energy efficiency upgrades or programs can also be worse for disadvantaged populations. A decision to participate in energy efficiency subsidy programs can be influenced by uncertainty about eligibility, the application process, and program value, along with general distrust of the program or institution. Reducing the uncertainty through clearer communication, like redesigning program websites to increase transparency about eligibility and program benefits, could help reduce the informational barrier.[35] Additionally, people often make the decision to purchase a new appliance when their previous one breaks and during a time of emergency. It is difficult, on the spur of the moment, to access enough information about which appliances are most efficient and why one might work better than another.

Barriers extend beyond costs and information for certain populations. Those who do not own their homes, for example, are typically limited in what they can install. Renters generally do not choose their appliances, thermostats, or air conditioners, so they do not get to choose an efficient or "smart" anything. They also would not be able to install solar panels without the consent of their landlord, nor would they be particularly inclined to do so given that the panels would then stay with the home and the owner.

This is the classic "split-incentive" problem, whereby the owner is the one who needs to make the decision about whether to invest in a new purchase, but the tenant is the one who would benefit the most from an investment in more efficient or money saving technologies. The landlord sees the up-front cost and the tenant sees the savings; therefore, in absence of a mandate by the government for landlords to invest in a baseline level of efficiency, no rational landlord would make the investment unless there is some compensation or situation in which those expenses are returned in the form of savings or revenue.[36] This is precisely why the government sets certain standards for appliances such as refrigerators, to ensure at least a base level of efficiency.[37]

These conditions face a large and growing percentage of the population. Approximately 36 percent of the US population rents their home. The proportion of renters is even higher among some sociodemographic groups. A disproportionate number of people of color who are low income rent their homes; over 58 percent and 51 percent of African Americans and Hispanics are renters, respectively, which is 22 and 15 percentage points above the national average. Compare these numbers to the figures for white Americans, only 30 percent of whom rent their home.[38] Homeowners are much wealthier than renters, as the median homeowner has a net worth of $255,000 while the median renter has a net worth of just $6,300.[39]

There are many other similar examples of unique or compounded barriers for disadvantaged populations. Those who do not have structurally sound roofs cannot install solar panels unless they first repair their roofs.[40] Those who do not speak English well may struggle to track their energy consumption via smart technologies, or learn about government incentives for household technologies, when such information is are only offered in English. And the 14.4 million Americans who do not have access to the internet cannot take advantage of many smart gadgets and appliances.[41]

It is not just households that may struggle to access these technologies but in some cases entire communities. Consider the case of renewable-energy development in Native American territories across the United States, which is often bound by a dual set of barriers: financial and infrastructural.[42]

The Indian Reorganization Act of 1934, among other institutional arrangements, makes it difficult for Native Americans to secure the necessary outside investment from public or private markets required for renewable-energy projects. Additionally, legal advisers discourage private-equity firms from selling securities to nonaccredited firms, a category that includes many Native American tribes, thanks to a 1982 Security and Exchange Commission Ruling. Accredited firms are exempt from state and federal registration requirements that an organization needs to participate in private-equity markets. Without access to these private-equity markets, Native American tribes have struggled to secure enough capital for development projects.[43]

Due to the remoteness of many tribes, the infrastructure needed to connect renewable-energy projects to the electric grid is nearly nonexistent. Jon Canis, general counsel for the Oceti Sakowin Power Authority, a renewable-energy development company owned by seven Sioux tribes in the Dakotas, pointed out, "You're dealing with a class of customer that's uniquely situated . . . They're [the Sioux are] confined to reservations they didn't choose and have historic underinvestment. We don't have the kind of facilities we need to support our development of utility-scale wind and solar farms."[44]

Even though Native groups are eligible for funding from the Indian

Reorganization Act, the underlying infrastructure problems remain. The CEO of Energy Keepers, Inc., an independent energy developer owned and operated by the Salish and Kootenai tribes in Montana, stated that "the financial picture . . . it's no longer the limiting factor. So what becomes the limiting factor is transmission and interconnection, and the interconnection queue."[45] Transmission and interconnection delays are plaguing clean-energy developments in many parts of the United States, but they are particularly acute for tribal nations.

The problem is compounded when groups like Energy Keepers or the Oceti Sakowin Power Authority then have to pay the interconnection fee, combining both the financing problem and the infrastructure problem into one issue. Two wind projects that the Oceti Sakowin Power Authority was developing had estimated interconnection costs of a whopping $48 million, causing the group to cancel the project due to a shortfall of funding.[46]

While these trends are beginning to change,[47] large obstacles still remain. In even wealthier tribes, up to one-third of households lack electricity access, and researchers estimate that it would cost $70,000 per household to connect these houses to the grid.[48] And even Native American tribes that are relatively energy secure are having trouble developing projects. To access Indian Reorganization Act funds, tribes must first fully develop a project and connect to the grid. But connecting to the grid is a slow, expensive process. Tribes are thus left doubly exposed: Each interconnection study can cost up to $10 million, only for a tribe to receive news that transmission costs would be exorbitant—and if a tribe decides to develop anyway, many Indian Reorganization Act tax credits will run out by 2026, causing tribes to miss out on millions of dollars.[49]

Access to Government and Other Incentives Is Also Uneven

Government incentives for energy efficiency and clean technologies generally are financed by taxpayers and ratepayers. Yet it is predominantly wealthier households who benefit from these types of policies, and they benefit more. Part of this story is predictable: Higher-income residents disproportionately benefit from clean-energy programs. Someone without the discretionary income to purchase an energy-efficient appliance cannot take advantage of the government subsidies for this purchase.

Consider an example. Let's say that the government offers a rebate of up to 10 percent off the purchase price of an efficient washing machine and you are looking at a $650 Energy Star appliance. Assuming that the rebate is mail-in, you will need to have $650 plus tax to pay today, with an eventual return from the government of $65, for a total expenditure of $585. If you only have at most $400 to spend, however, that rebate will not make a difference; you simply cannot afford the washing machine.

Studies confirm this problem. A Congressional Research Service report found that 78 percent of electric vehicle tax credits claimed before 2019 were by households making $100,000 or more.[50] Another study found that 90 percent of all electric-vehicle tax credits went to higher-income households, and 60 percent of other residential clean-energy tax credits for solar, energy efficiency, and weatherization—amounting to over $18 billion in credits in total between 2006 and 2012—went to higher-income households in the years predating 2012.[51] In a 2023 *Harvard Law Today* interview, Ashley Nunes states this problem succinctly: "Knocking $7,500 off a $60,000 car matters little to middle- and low-income households because $52,500 is far more than these households ever could—or would—spend on a car."[52]

These statistics are what led the US government to extend electric-vehicle tax credits in the 2022 Inflation Reduction Act (IRA) to used-vehicle markets as well. In the IRA, one who purchases a used electric car can get up to $4,000 back, or up to 30 percent of the sales price of the car.[53] The tax credit has historically been granted at the time of tax returns, however, and not at the point of sale, although this too has changed with the IRA, which now allows a consumer to capture a rebate at the time of sale if the dealership is prepared to offer it.

Other policy interventions are helping here as well. Low-interest loan programs, for example, can help finance rural and low-income energy efficiency or renewable-energy projects. The Energy Efficiency and Conservation Loan Program, through the US Department of Agriculture, helps finance energy efficiency projects for rural business and residential customers. Rural utility providers can borrow money at Treasury rates of interest and lend money to customers for energy efficiency upgrades, allowing borrowers to pay back the funding over time on their electric bills.[54] A program through the Connecticut Green Bank, Smart-E loans, offers low-interest loans that can be used on energy efficiency or renewable-energy projects, helping cover potentially high up-front costs.[55] Low-interest loan programs like these could help finance and increase opportunities for low-income and disadvantaged populations pursuing energy efficiency or renewable-energy projects, who might not otherwise be able to invest in these projects.

The other part of this story is less commonly discussed and acknowledged: In addition to the unevenness in the distribution of access to these programs, there is also unevenness in the distribution of societal costs for these programs. Specifically, those who pay disproportionately more toward the program receive disproportionately less in return.

Revenue for clean-energy programs such as tax credits or public-benefit funds comes through tax payments and utility bill payments. Most consumers pay into at least one of these sources. The revenue is distributed to consumers of clean-energy technologies through, for example, the Nonbusiness Energy Tax Credit program, the Residential Energy Efficiency Program, the Qualified Plug-In Electric Drive Motor Vehicle Credit, the Alternative Motor Vehicle Credit, net-energy metering programs, or other utility incentive programs.

A study of the California Electric Vehicle Program[56] between the years of 2010 and 2017, when electric vehicles were still quite expensive compared to conventional counterparts, found that higher-income zip codes in California received more rebate outlays than they paid into the system to fund the rebates, and lower-income zip codes received more of the costs than the benefits. The higher the income level, the greater the imbalance between what a person paid into the program and what they received in benefits. The highest 10 percent of households by income paid only 11 percent of the total costs of the rebate program and received 33 percent of the total rebates. In other words, poorer populations cross-subsidized wealthier populations.

Access Is Even Further Geographically Constrained

Opportunities to access clean-energy technologies are also geographically constrained. The geographic limitations often fall along rural-urban lines and industrializing-industrialized country lines, where technologies are less often available in rural areas and industrializing countries. There are also cases of geographic divides by income, race, and ethnicity.

Empirical evidence suggests, for instance, that solar installers target and work predominantly within higher-income regions. In such a case, those of lower economic status have a hard time locating and communicating with these installers.[57] And those providers that are accessible in lower-income or otherwise underserved areas have much less competition and are able to charge higher prices.

In the case of electric vehicles, studies have found that one of the biggest barriers for some is lack of electric-vehicle stock at their local dealerships. A 2023 Sierra Club study contacted over eight hundred auto dealerships from across the United States and found that 66 percent of them did not have any electric vehicles for sale.[58] This same study also found that the majority of dealers, located disproportionately in overburdened and rural areas, did not mention state or federal incentives for electric vehicles, and 10 percent of them had an electric vehicle available for purchase but claimed that it was insufficiently charged.[59]

The historical case of the LED light bulb provides another interesting example. Energy scholar Tony Reames led a study in which he and his team visited stores throughout Wayne County, Michigan, to track the presence and price of LED and compact fluorescent light bulbs.[60] They visited a total of 130 stores—from corner stores to big-box retailers—across neighborhoods of varying income strata. What they uncovered was the presence of LED deserts, akin to the idea of food deserts, where one lacks access to affordable and nutritious food. These LED deserts were located predominantly in low-income neighborhoods and communities of color. And, when LEDs were available in the locations, they were for sale at a premium, whereas higher-income neighborhoods tended to sell them at lower prices and in ample supply, primarily at big-box hardware stores. Of course, one can buy LED light bulbs online these days, and even in bulk, but the geographic disparities are notable nonetheless.

Return to Walter White's decisions about which light bulbs to buy for his house. He has a pretty easy decision: Swing by the local Home Depot or Walmart and stock up on LED light bulbs, each of which can save him $171 over a few years' time. Compare his situation to that of a less well-off homeowner, who may have to travel across town to find LED light bulbs or purchase them at a nearby corner grocery store at a higher price, thereby cutting significantly into the potential savings that this bulb could generate.

The discussion up to this point has made clear that vulnerable groups face many challenges in gaining access to clean energy and energy efficiency technologies. And, ironically, these populations might benefit the most from these technologies, as ways to reduce their utility bills and to improve the quality of their local residential and community environments.

Access to Clean-Energy Jobs

A related challenge is that individuals from these same communities—lower-income individuals, women, and people of color, in particular—are much less likely to hold jobs in the growing clean-energy economy, a state of affairs that has direct implications for technology access and compatibility. A report from the National Association of State Energy Officials (NASEO) found a lack of workforce diversity in the energy sector. With respect to gender, about a quarter of energy workers are women, compared to an overall national average of 47 percent. Similarly, representation of people of color is much lower across most sectors of the energy industry, as compared to other industries in the US economy. The differences are particularly prevalent for Black Americans, who are represented less than national averages across all five of the technologies analyzed (i.e., electric power generation; fuels; transmission, distribution, and storage; energy efficiency; and motor vehicles). The statistics do reveal some types of energy jobs where women and Hispanic Americans have higher levels of representation. For example, Hispanics are employed at higher than national rates in some fossil fuel sectors, such as coal production and oil refining, while women are better represented in electric power generation, though their representation is still below national averages in other parts of the US economy.[61]

A closer look at the data reveals that the gaps are even more pronounced when it comes to leadership roles. The survey of energy workers done for the NASEO report found that about 80 percent of white respondents reported that they were either an executive with their company or served in a supervisory role, with 35 percent indicating that they were a company executive. Only 17 percent of Black respondents and 19 percent of Hispanic respondents, by contrast, reported serving as a company executive, respectively. The same pattern held for Asians; only 22 percent of Asian energy workers indicated that they were company executives.[62]

The lack of diversity in the energy leadership positions is also evident at the very top of large energy companies. Consider CEOs of Fortune 500 energy companies. There are about twenty-five such companies on the list, including large utility companies such as Exelon, Duke, Southern, Pacific Gas and Electric, and large oil and gas companies like Exxon, Chevron, BP, and Marathon. About two-thirds of these CEOs are white men; only three women are on the list. Similar patterns characterize large energy companies' board of directors. According to an analysis by S&P Global, women occupy just 15 percent of the positions on the boards of energy companies on the S&P Global indices. The numbers are slightly higher among US companies, and there has been significant growth in recent years.[63]

A report from the Chisholm Legacy Project similarly documents a lack of diversity on state public utility commissions, which serve an important role in shaping policies on consumer access to clean technologies and utility energy efficiency programs. A main finding from their analysis is that there is minimal racial diversity on public utility commissions across the country. Of the 197 commissioners for which the organization was able to find information, they found that only 3 percent were Hispanic/Latino, 3 percent were Asian, and 1 percent were Native American (11 percent were Black). Moreover, they found that, in twenty-four states, commissions had no people of color, and that in eight more, there was just one person of color. With respect to gender, about two-thirds of public utility commissioners were men, with 35 percent identifying as women.[64]

Disparities also exist in the emerging clean-energy sector. Consider workforce diversity in the solar industry. A recent census of the solar workforce found growing representation there of women and people of color, but it is still an industry that lags behind other parts of the economy in terms of its diversity. In 2022, about 31 percent of the total solar workforce were women, while just 9 percent were people identifying as Black.[65] The wind industry has nearly identical statistics, as 32 percent of the wind energy workforce were women and 9 percent identify as Black.[66]

How might workforce diversity in the energy sector be related to access to clean energy and efficiency technologies? At one level, we posit that people are more likely to gain familiarity and comfort with new technologies when they, and those whom they know, are able to work within the companies that help develop, market, and distribute these technologies. Moreover, the presence of historically marginalized individuals in the workforces of these companies, especially in management and leadership roles, may increase their recognition of the unique challenges that exist in diffusing these new technologies to diverse communities as well as the design of new approaches to overcome the obstacles. The same can be said for government officials responsible, in part, for designing policies to facilitate access to these technologies, such as those who serve on public utility commissions.

Wealth Begets Wealth

These interrelated conditions that shape the current state of technology-access disparities—whether intentionally or unintentionally—further perpetuate wealth disparities. It is typically only higher-income households that can participate, and when they do, they acquire more financial and health benefits. These compounding advantages of the wealthy in the face of differentiated energy landscapes, both in terms of access to technologies and to the programs that incentivize these technologies, reinforce inequities that have long been the hallmark of energy systems.

Consider several of the examples we have given thus far. Clean technologies like residential rooftop solar are only affordable if one has significant disposable income, or buys it with credit, assuming one has access to credit. Incentives are often only accessible to people with tax liability. Public EV chargers are only useful if they are publicly available. Rooftop solar—or batteries, or charging stations—can only be installed by homeowners who, in the case of charging stations, have access to private, off-street parking. Electric vehicles are only accessible if a nearby dealer carries them, or if one has convenient transportation options to get to dealerships that are farther away.

Size and scale differentials further exacerbate these disparities. The larger the solar panels that someone installs on their home, the larger the solar tax credit amount they receive. In other words, the more one can pay, the more one can save. And the more one consumes energy, the more production-based subsidies one can receive. The bigger your house and the more appliances you have, the more you can benefit from government subsidies.

Let's return to the case of residential solar. Smaller solar arrays generate less electricity both overall and as a percentage of total household energy consumption, and they also displace electricity on a marginal basis from the grid.[67] Larger systems are able to produce more solar energy. When a household is charged for energy consumption using an increasing block tariff pricing structure, the households with larger systems save more on a kilowatt-hour basis than households with smaller systems.

A study by energy economist Severin Borenstein considered the role of electricity pricing and tariffs in solar energy compensation. His study looked at California's Pacific Gas & Electric territory, which has increasing block tariff pricing. This type of tariff has an increasing price for increasing blocks of electricity consumed. Of course, this tariff structure may serve to disincentivize higher levels of energy demand; but it also yields more compensation for residential solar when the utility company compensates all solar owners for their solar generation at the retail cost of electricity. In his study, Borenstein found that solar adopters in the highest income bracket could save about 27.2 cents per kilowatt-hour on average, while those in the lowest income bracket could save about 21.3 cents per kilowatt-hour under the average increasing block tariff.[68]

Another way in which wealth begets wealth is in how the use of these technologies can provide additive benefits. Solar energy can lower one's energy bills. Solar energy with batteries can save even more money, enhance resilience, and allow users to shift their energy to capture extra electricity savings throughout the day or week. Having access to a combination of solar energy, batteries, and electric vehicles allows users to charge their cars with the solar energy that they captured on their roofs and stored in their batteries for use at a convenient time. These combined technologies can yield further cost savings in the form of both electricity and gas savings, while enhancing resilience, personal convenience, and flexibility. The more technologies, the more savings. The more a user can spend on these technologies, the more they save. Wealth begets wealth.

Given Market Structures, These Inequities Could Become More Pronounced in the Future

Preexisting market structures—a legacy of very early electricity market and rate designs—reinforce the wealth accumulation phenomena and have the dangerous potential to cause bigger disparities in the future.

Let's say that you live in a neighborhood with lots of neighbors.

One day your neighbor across the street installs solar panels and then subsequently tells all the neighborhood about how amazing they are and how much they can now save on their monthly energy bill. Soon thereafter, another neighbor installs panels, then another, then another.[69] Eventually you suspect that at least one-quarter of your neighborhood has panels.

With enough solar adoption in a service territory, non-solar adopters' bills may rise. Because you have not purchased solar—either because you could not afford to or because you are uninterested—you may end up with a bigger monthly energy bill. Electricity rate structures include both fixed costs and volumetric costs. In a world with direct translation of costs, the fixed costs would cover infrastructure such as transmission and distribution lines and the cost of using and maintaining those lines, and the volumetric costs would cover the cost of the electricity that one consumes from the grid and any other variable costs. What has historically been more typical, however, is that some of the grid's fixed costs are actually transferred to the volumetric cost category on users' bills, with the provider's objective being to keep fixed costs low and to allow payments to vary by how much energy one consumes. As a result, volumetric charges cover both the energy that one consumes and some of the grid costs. When one has residential solar and consumes less electricity from the grid, then the utility company recovers less of their grid expenses. With enough solar owners in an area, the shortfall must be recovered elsewhere, which means that rates will rise and those customers without solar will pay more.[70]

This grid cross-shifting or cross-subsidization has the potential to further burden the have-nots—that is, those who cannot afford or access behind-the-meter clean-energy technologies where older net-energy metering policies remain in place. If higher-income households are more likely to purchase solar and take advantage of the subsidies at mass scale and lower-income households are not, then less-wealthy households may end up cross-subsidizing wealthier ones.[71]

This cycle is self-perpetuating. As more people adopt solar, more of the costs are shifted to the non-adopters, and the differential grows between electricity prices and solar investments. As this differential rises, still more households will adopt solar. Yet it will be the least-well-off households who adopt last.

This issue as it relates to utility companies has been referred to as a "utility death spiral" by some, whereby the utility company increasingly struggles to cover its costs, renewable-energy generation drives down the wholesale price of electricity, and the company can less easily deploy their fossil fuel assets. These fossil fuel assets, such as coal and natural gas power plants, are long-term investments that often carry loans with long payback periods. When the company can no longer use the energy from a power plant, they will decommission the plant and it will then become a stranded asset.[72] These cost increases accompany fixed-cost adjustments for those customers who remain customers of the utility company. But as prices rise, more households will buy solar instead, which will further drive up prices, and onward in a moribund cycle for the utility provider.

The utility death spiral concept has been criticized by some, who argue that utility companies are unlikely to stand by idly as they become financially burdened and eventually obsolete; they will instead seek changes in rate structure and institutional arrangements, as many have already done across the country.[73] Yet the incidence of grid cross-subsidization is a potential problem: Startup policies to promote deployment have not been replaced with a more sustainable tariff structure for customers who remain connected to and reliant on grid-supplied power for at least some of their needs. This problem has deep regressive implications.[74]

These trends are not just evident in electricity markets but also appear in other energy markets as more consumers switch to building electrification—in other words, as they entirely defect from one system and replace it with another. In research on the US natural-gas industry, economists Lucas Davis and Catherine Hausman[75] find that natural-gas companies expand gas lines for new customers but rarely remove them with the loss of customers. The costs to maintain the infrastructure, however, remain, and must be spread across the remaining customers—i.e., those who have not switched to electrification. The authors estimate that a 20 percent grid defection leads to

$40 increase per year on each remaining customer's bill, and a 40 percent loss leads to a $115 annual increase. Those regions in the United States with declining gas industries include those with higher rates of poverty and with higher numbers of African Americans. Geographically, regions in the Rust Belt and rural regions in Appalachia will also be disproportionately affected.

•••

Deep decarbonization requires large changes to the US energy system, including the adoption of cleaner and more efficient energy technologies by most people. The scale of the climate problem means that individual and household behavioral changes must be part of any serious attempt in the United States and in similar Global North countries to reduce greenhouse gas emissions. Yet, as cataloged in this chapter, the barriers to widespread adoption of these technologies are severe, and they are particularly pronounced for some Americans.

Many individuals—often low-income people, and in some cases people of color or other marginalized populations—are locked out of the clean-energy transition due to prohibitive costs, unique and compounding barriers, and a lack of access opportunities. These challenges all operate within a set of institutions and markets that prioritize and benefit those with wealth at the neglect of, and sometimes even at the expense of, those less well off. And, if left unaddressed, decarbonization will continue to reinforce disparities between the haves and the have-nots.

None of this is to suggest that the pathway to decarbonization rests solely on the shoulders of individuals and households. The clean-energy transition also requires massive changes in how we produce and distribute the energy powering the economy. Doing so means siting new energy infrastructure, which presents its own set of challenges. We turn to these challenges next.

CHAPTER 6

Backyards and Ballots

Driving through the heartland of the United States, it is increasingly common to see large wind farms, with towering turbines dotting the landscape. Spurred on by lower costs, federal tax incentives, and state policies mandating that electric utility companies provide more of their electricity from renewable-energy sources, wind has become an important source of power in the American Midwest. Yet not everyone is happy with these developments. The changes to the rural agricultural landscapes that industrial-scale wind farms bring are increasingly being met with staunch opposition.

The Honey Creek wind farm project in Crawford County, Ohio, provides an illustrative example. The Honey Creek project was to include about sixty turbines that its developer, Apex Clean Energy, planned to build in this rural part of north central Ohio. When completed, the 300-megawatt project would have generated enough electricity to power about eighty-five thousand homes, more than four times the number of households in Crawford County.[1] The company estimated that, over the expected thirty-year lifespan of the project, it would generate $81 million for Crawford County, including millions of dollars in payments to landowners, local schools, and townships, and bring about one hundred local jobs during construction.[2] Despite these potential financial and economic benefits, the Crawford County Commission put the kibosh on the project in 2022 when it voted to impose a ten-year ban on wind development in the county.

What happened? Why would a clean-energy project that would pay landowners substantial sums and generate significant revenue for local communities be upended by the local government?

The Crawford County commissioners' decision to impose a decade-long prohibition against new wind development was enabled by the Ohio State legislature, which passed a law in 2021 (S.B. 52) to empower county governments to restrict solar- and wind-power development. Proponents of the bill argued that the legislation would give county officials the opportunity to preserve the beauty and serenity of their rural landscapes and to maintain control over their communities. At the time the legislature passed the bill, Matt Huffman, the Ohio Senate president, a senator representing a rural western Ohio district, stated, "The people who live there don't want to look at them [wind farms]." Large turbines "ruin the character" of the area.[3] Before the change in state policy, decisions about such developments rested with a state commission called the Ohio Power Siting Board.[4] Under S.B. 52, county commissioners could now designate "restricted areas" where renewable-energy facilities would be prohibited, unless specifically allowed by a countywide referendum.

The Honey Creek project was initially proposed to be about twice the size, spanning fourteen thousand acres of leased land in both Crawford and neighboring Seneca County. When the original developer, NextEra Energy, abandoned the development in 2015 to prioritize other projects,[5] four years after its initial proposal, Apex Clean Energy bought the project and the already purchased leases. Local reaction to the new, smaller project was mixed. Many additional landowners agreed to lease their land or provide access for transmission, while other members of the affected communities organized in opposition.[6]

The Honey Creek project split people with different notions about rural landscapes and property rights. In 2023, several local residents spoke to journalists about the project, including Crawford resident Bob Sostakowski: "I had no opinion one way or another on wind until this. There's an obvious and very provable negative impact on property values and people's standard of living."[7] Kimberly Groth, a resident of adjacent Seneca County, said something similar: "People want quality of life and people move to rural areas because of the peacefulness of it. When you introduce industrial-scale wind over tens of thousands of acres, you're interrupting that quality of life." She continued, "I think we've heard for 20 to 30 years now about renewable energy and there's just this assumption that it's good and that it's going to save us. So I think for me personally the more I looked into it, the more I realized it does have downsides. . . . Every form of energy has these pros and cons."[8]

Other local landowners viewed the proposed wind farm as an opportunity to supplement their incomes, and they emphasized that the siting restrictions resulted in a loss of their autonomy to decide how to use their land. Ann Fry had leased her land in Seneca County to NextEra Energy as part of the initial Honey Creek project, but Seneca County commissioners also imposed a ban on wind and solar projects in unincorporated areas, which invalidated her lease agreement. Fry's comments exhibit the loss of autonomy that came with this outcome: "It's our land. And if we want to grow corn, soybeans, or put green energy on our land, why is it someone else's choice, who might live on a postage-stamp-size lot, to tell us what we can and cannot do with our land?"[9]

Following the Crawford County Commission's decision, wind farm supporters secured enough signatures to potentially undo the decision through a referendum vote. However, the citizens of the county voted resoundingly to maintain the ban, with 75 percent of the vote,[10] bringing the project to a permanent halt.

The demise of the Honey Creek project is just one project in one community, but it represents a growing backlash against renewable energy in some parts of the country. In Ohio, a dozen counties have imposed bans or other types of prohibitions to limit wind-power development in their unincorporated areas, and these types of prohibitions are increasingly common in other states as well.[11]

Opposition to the siting of renewable energy—and to the transmission infrastructure necessary to harness this power—poses an immense challenge for the clean-energy transition in the United States. And the backlash against wind and other forms of renewable energy comes at the same time as the increasing urgency to move away from fossil fuels to address the growing climate change crisis.

Local opposition to siting large-scale energy infrastructure is nothing new. The American public intensified its turn away from nuclear power in the 1970s, in the years that followed the partial meltdown of the Three Mile Island nuclear reactor outside Harrisburg, Pennsylvania. Public concerns about nuclear accidents, in part, made it very difficult to find locations for new nuclear power plants, virtually ending the expansion of commercial nuclear power in the US electricity sector.[12] There are other examples. Concerns about hydraulic fracturing have led to bans in several states—California, Maryland, New York, Vermont, and Washington—and calls for such bans elsewhere. Concerns about spills and the continued reliance on oil has led to protests of pipeline projects, such as the Dakota Access and Keystone XL pipelines. Energy companies often anticipate local opposition to their projects and respond by siting infrastructure projects in communities with fewer political resources to resist and where they anticipate that people want the tax revenues. Following this path of "least resistance" can generate or exacerbate the types of disproportionate burdens for marginalized communities that we have discussed throughout the book.

Energy siting challenges may not be new, but they are likely to intensify in the coming years as the United States continues to remake the energy system with the replacement of fossil fuels with alternative energy sources. In this chapter, we focus on three interrelated challenges that highlight the equity and justice-oriented dimensions of the clean-energy transition.

First, the siting of new energy technologies will generate geographically uneven costs and benefits. The wind farms, solar panel fields, and transmission lines needed to bring electricity to where people live have broad environmental and health benefits. The operations of these sources of energy produce neither the greenhouse gas emissions causing climate change nor conventional pollutants that adversely affect air and water quality. However, these sources of energy do have impacts, and these impacts are concentrated on those individuals and communities who will be asked to host this new energy infrastructure. Moreover, in many cases, the host community is not the main beneficiary of the project. In the case of the Honey Creek project, for example, the new, clean electricity would have provided power mostly for people outside the county.

Similarly, the wide-scale deployment of carbon capture and storage technologies as a climate mitigation strategy, an example that we discuss later in this chapter, is also likely to generate an uneven distribution of cost and benefits. Any effective strategy to reduce carbon emissions from fossil fuel power plants and other major industrial facilities benefits everyone. However, the broad use of carbon capture and storage technologies would mean keeping fossil fuel facilities afloat, where they will continue to adversely affect people who live in close proximity. Capturing carbon mitigates the emissions causing climate change, but it does not reduce emissions of other harmful pollutants associated with the mining and refining of the energy resources, and it may not decrease emissions of other pollutants when the fossil fuels are burned.[13] Moreover, carbon capture and storage technology will require the construction of a network of new pipelines to move the carbon to storage locations across the country, which will generate impacts to the landowners and communities asked to host them.

A second factor that complicates the siting of new energy infrastructure is that companies in the energy sector are among the least trusted organizations in the United States. Gallup regularly solicits Americans' opinions toward various business sectors, asking them to indicate whether they have positive, neutral, or negative views. In these public-opinion surveys, Americans routinely rate companies in the energy sector the most negatively. Figure 6.1 shows two decades of results from these surveys. The figure plots the annual net favorability—the difference between the positive and negative ratings—for electric and gas utility companies and the oil and gas industry, and then compares them to the net favorability for all other business sectors included in

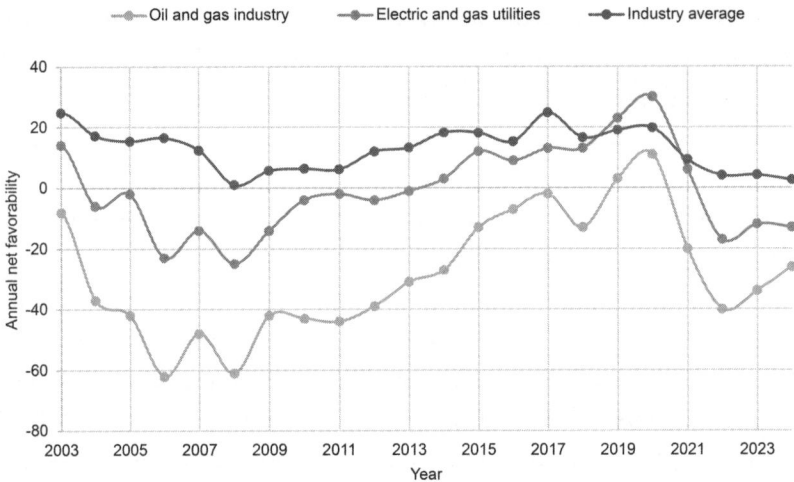

Figure 6.1 Net favorability toward energy sectors, 2003–2023
Notes: Annual net favorability is measured as the difference between the positive and negative ratings for each group.
Source: Gallup, Inc., "Business and Industry Sector Ratings."

the survey.[14] On average, the business sectors rated in the survey have net positive favorability. By comparison, the oil and gas industry has net negative favorability for almost the entire time period, with the exception of a short-lived positive rating in 2019 and 2020, which occurred before a sharp drop. Americans' views toward electric and gas utility companies were, on net, negative for most of the first decade of the 2000s, and then showed a gradual improvement for much of the 2010s, including eight years of net favorable ratings, before sharply declining in a similar way to oil and gas companies in the 2020s.

The mistrust that people have of energy companies, utility providers, and—often—the government agencies that regulate them means they are often not considered credible or reliable partners. At the project proposal stage, communities are often skeptical about the claims that companies make about the costs and benefits of their projects. This mistrust pertains to both "clean" and "dirty" projects, with project supporters and opponents both raising doubts. The lack of trust also results in the perpetuation of myths and misinformation about specific technologies, which means the debate about specific projects often takes place without a shared set of facts.

A third challenge is that the deep partisan divide that dominates so much of contemporary US politics also permeates discussion and decisions about energy infrastructure. Although Americans in general express overwhelming support for a clean-energy transition, and in the abstract enthusiastically support the wider deployment of clean-energy technologies, Democrats and Republicans often disagree on the form and timeline of this transition. Most importantly, they disagree deeply on the policies that could be employed to hasten the transition, and this disagreement increasingly shapes the conversation around energy infrastructure, both broadly and with respect to specific projects. These political differences are reinforced by our fragmented political system, which is quite important in the context of energy infrastructure, where decisions on particular projects often require the approval of state and local governments.

In the face of these three interrelated challenges—and considering the massive scale of new energy infrastructure needed—several important questions emerge. How can the United States bring about a clean-energy transition if it cannot site the necessary, new infrastructure? And, yet, if we do not hasten the transition, does the United States only risk perpetuating the inequities and injustices associated with its current energy systems?

Addressing these questions involves grappling with key energy-justice concerns. Not only is it necessary to recognize and address the geographically uneven costs and benefits of siting new energy infrastructure, but it is also necessary to do so in a manner that respects the preferences of those most directly affected. In other words, it is important to pay specific attention to the core principles of energy justice—distributive justice, procedural justice, and recognition justice—while still moving as expeditiously as possible to meet the climate challenge.

In this chapter, we illustrate this challenge in the context of three types of energy infrastructure—wind and solar farms, high-voltage transmission lines, and carbon capture and storage. To provide the needed context for this discussion, we first describe the scale of new energy infrastructure required.

Energy Infrastructure in the United States

The transformation of the US energy system from one built on fossil fuels to one that runs on clean energy is a huge undertaking. Historically, energy transitions are gradual, taking generations or often longer to materialize, spawned by technological innovations that then slowly make their way through economies and societies.[15] Yet, the current energy transition needs to happen on a much faster timeline. Any chance of achieving net-zero carbon emissions in the coming decades in the United States requires investments in the trillions of dollars to build and deploy new energy infrastructure throughout the country.

Consider the case of electricity generation. Despite the sharp declines in coal use over the past decade, as of 2023, fossil fuels still powered about 60 percent of the electricity system.[16] An additional 18 percent came from nuclear power. Although nuclear power does not generate carbon emissions, the United States has not added much new nuclear power to the grid over the past half century, and the aging power plants—mostly built in the 1970s and 1980s[17]—that still operate will eventually need to be retired when they are no longer safe or economically efficient to run. Taken together, US power providers will eventually need to replace nearly 80 percent of the electricity grid with renewable-energy sources or other zero-carbon technologies to reach net-zero targets (i.e., a point at which, on net, we are not adding more carbon emission to the atmosphere), not to mention additional capacity to meet rising demand—including, increasingly, demand from AI and data centers.

The demand for electricity will also increase as the United States electrifies other parts of the economy. In 2023, electric vehicles comprised just about 10 percent of new car sales. While this number reflects a huge increase over a short period of time—in 2020, the figure was only 2.5 percent[18]—the electric grid eventually will need to supply more electricity to support the growing demand from a future transportation sector where most people are driving electric-powered cars and trucks. Similarly, Americans have started to transition from natural-gas furnaces to electric heat pumps as a way to heat their homes. Heat pumps have the added benefit of serving as cooling units, producing air conditioning in a more efficient way than a standard system. National sales of heat pumps have been increasing steadily over the past decade, and in 2023 they reached a milestone of exceeding national sales of natural-gas furnaces.[19] The demand on the electricity system will only increase as more people adopt these technologies. Deep decarbonization at the scale necessary for avoiding the worst impacts of climate change will also require fully electrifying homes and other buildings, which eventually means turning away from natural-gas stoves and water heaters, heating oil, propane tanks, and other residential uses of fossil fuels.

Some of this electricity will be powered by distributed sources, such as rooftop solar power, but this will not be sufficient to meet the country's energy needs.[20] The National Renewable Energy Laboratory has estimated that rooftop solar could generate up to 1,432 terawatt-hours of electricity annually, which is an impressive amount, but it accounts for less than 40 percent of total annual electricity generation potential in the country.[21] In short, there are not enough rooftops, at least with current solar panel technologies, cross-subsidization concerns aside—meaning that most clean-energy generation will need to come from industrial-scale sites that can host large applications like wind and solar farms.[22]

Yet it is necessary to do more than *just* replace coal and natural-gas power plants with wind farms and utility-scale solar power. To reach decarbonization goals, it is also necessary that we move this new, clean electricity from the location of its generation to the places where people live and work. Wind and solar farms require large amounts of physical space, and they are most efficient when located in places with strong wind and frequent sunlight. As a result, many, if not most, renewable-energy facilities will need to be located hundreds of miles away from population centers, which will require building new transmission lines across the country. Related to the general challenge of distance is the problem of interconnection. The electric grid in many parts of the country is at capacity, and new projects must wait increasingly long periods to be added to the system. To illustrate the scale of the problem, at the end of 2021, there were more than eight thousand projects, mostly new wind and solar projects, that were waiting for permission from grid operators to connect. More recent data suggest that bottlenecks are only getting worse, and in 2023, they amounted to more than twice the total installed capacity of existing US power plants.[23] The wait is so long that it is not uncommon for project developers to abandon their projects altogether, not from a lack of community support but because the timeline to project completion is too long.[24]

More electricity use will also place more burden on the current electricity grid. Not only is it necessary to expand existing capacity, but the United States needs to better maintain and upgrade the infrastructure it already has. Most of the US electricity grid was built in the mid-twentieth century, and this old infrastructure is vulnerable to severe weather and general degradation from age.[25]

Deferred maintenance also increases the frequency and severity of power outages. In recent years, failure to properly maintain transmission infrastructure has sparked wildfires, including devastating fires that have ravaged parts of California, Hawaii, and the Pacific Northwest. These wildfires, made worse by climate change–induced drought, have resulted in hundreds of fatalities and untold millions, if not billions, of dollars in property damage. The scale of liability threatens the viability of some electric utility companies altogether. In 2019, Pacific Gas and Electric, the nation's eighth-largest electric utility provider in terms of customers, filed for bankruptcy due to its financial obligations associated with property damages from wildfires. A year later, the company pleaded guilty to eighty-four counts of involuntary manslaughter, acknowledging its responsibility for sparking the fire that destroyed the entire town of Paradise, California.[26] To manage wildfire risks, utility companies in California are now spending billions of dollars, and their customers' rates are going up accordingly.[27]

Estimates vary as to how much infrastructure investment is required to decarbonize the US economy. As part of its regular report card on American infrastructure, the American Society of Civil Engineers gave the electric grid a C- and projected a $200 billion shortfall in modernization requirements by the end of the decade. Energy system modelers have put the number in the trillions. An estimated $360 billion is required to put the nation on track for net-zero emissions by 2030, and, by 2050, the number jumps up to $2.4 trillion. Reaching these targets would require tripling the existing number of transmission lines within the next three decades. A Department of Energy study found that 47,300 miles will need to be built by 2035 to keep the country in line with net-zero goals.[28] For reference, only 386 miles of transmission line were built in 2021.[29]

Reaching net-zero carbon emissions by 2050 or sooner may also require the wide-scale deployment of carbon capture and storage technology, either through the direct capture of carbon from the upper atmosphere or from capture of emissions at their point of combustion. These technologies are not yet broadly in place in the United States or elsewhere, and countless questions remain about their technical and economic feasibility. The use of these technologies as a climate mitigation strategy, moreover, raises important moral-hazard concerns among many. Nevertheless, the United States is investing billions of dollars directly and indirectly through tax credits created by the Inflation Reduction Act to encourage development of these technologies, and many climate mitigation pathways toward net-zero by midcentury assume some reliance on them.[30]

In short, a massive amount of new energy infrastructure is needed to meet decarbonization goals and to bring about a clean-energy transition in the United States. Such a transformational change will require substantial investments, political will, and community buy-in, and not just in one place or a handful of places but in regions all across the country. This transition will also require private companies to lead the way, because most energy infrastructure in the United States is privately owned and operated.

That is one important way in which electricity and gas infrastructure differs from something like transportation infrastructure, which is mostly publicly owned and operated. Perhaps the best example is the US highway system, composed of federal interstates, state highways, county roads, and the like. When there is a need to expand interstates or make repairs to existing state highways, Congress and state legislatures appropriate money for these tasks and then delegate responsibility to transportation authorities to decide how to prioritize projects for public funding. As an example, the 2021 Infrastructure Investment and Jobs Act included more than a half-trillion dollars of federal investment in roads, bridges, public mass transit, water infrastructure, resilience, and broadband, where the money was funneled to state governments and then allocated to specific projects.[31] This public system is not without problems, but it does empower governments to make decisions, and it implicitly recognizes the need for the government to finance, build, and maintain this critical infrastructure.

Energy infrastructure is different. Power plants, pipelines, transmission lines, and most other energy systems are privately owned and operated (there are some notable exceptions, such as the federal Tennessee Valley Authority and municipal utility providers).[32] Decisions to build new infrastructure mostly rest with private utility companies, independent power producers, pipeline companies, and other private entities. These companies do not have carte blanche to act. New projects, particularly large ones, require government review and permits, and they are constrained by local zoning ordinances and other types of land-use restrictions. Recall in the Honey Creek wind farm case that it was the county commission that changed land-use rules to prohibit the project's development. In regulated markets, companies must also seek the approval of public utility commissions, who have the final say in determining whether companies can adjust their rates to pay for the costs and profit from new projects. Importantly, much of the decision-making is local, and in about half of US states, siting of wind projects rests with local governments.[33] But, the distinction to emphasize is that the infrastructure itself is typically privately, not publicly, owned.

An important implication of the largely private US energy system is that decision-making is decentralized. In this way, the energy system is the composite result of the decisions of countless companies operating in thousands of overlapping governmental jurisdictions. This is especially the case for the nation's electric grid. As energy economists Lucas Davis, Catherine Hausman, and Nancy Rose recently wrote, "The US electricity grid is a disorganized patchwork that is the result of over a century of mostly disconnected individual utility providers making independent decisions."[34] The United States has never had a cohesive national energy strategy,[35] which explains, in part, why the country over the past decade has both become a world leader in fossil fuel production and greatly enhanced its investment in renewable energy. The siting of energy infrastructure directly follows. There is no Grand Poobah or benign dictator of energy siting, making decisions on whether to green-light or halt projects, or to force energy companies to propose new projects or to make upgrades to existing infrastructure. It is not the job of a single agency, working with a coherent set of objectives.[36] Instead, the energy system is the result of hundreds, perhaps thousands, of agencies responding to the decisions made and projects proposed by numerous private companies.

This decentralized decision-making, coupled with the challenges of geographically uneven costs and benefits, mistrust of energy companies, and deep political disagreements about the direction of energy policy, further complicates progress in infrastructure development and meeting decarbonization goals. In the balance of this chapter, we explore these challenges with respect to the siting of three types of infrastructure that are central to the clean-energy transition: wind and utility-scale solar power, transmission lines, and carbon capture and storage.

The Renewable-Energy Backlash

Americans have expressed enthusiastic support for the wider use of renewable energy for many decades. Since at least the oil crises of the 1970s, Americans have regularly told pollsters that they want to see the country expand its reliance on alternative energy (i.e., non-fossil fuels).[37] In surveys we have conducted over the past two decades, we have regularly asked Americans which sources of energy they would

like to see the country use to generate electricity, and they have repeatedly indicated a strong preference for wind, solar, and other renewable
sources over traditional sources like fossil fuels and nuclear power.[38]

What explains these preferences? Studies show that Americans'
attitudes toward energy are quite logical—they want energy that is
cheap and clean. Specifically, people evaluate energy sources on the
basis of their attributes, such as how expensive it will make the electricity they consume and what the local environmental and health
impacts of its use will be. Energy sources that people see as inexpensive are more desirable, as are those that do not adversely affect the
quality of the environment of their local communities. And, it turns
out, people give much more weight, about three times as much, to
local environmental impacts than they do to costs. Concerns about
climate change play a modest role in understanding people's energy
preferences, but generally people who are most concerned about climate change want more renewables and less fossil fuel.[39]

People's preferences about energy sources are further reflected in
the public policies they support. Large majorities of Americans support federal investment into wind and solar power development, mandates that utility companies use renewable energy in their electricity
generation, and subsidies to increase the use of these technologies.[40]
The favorable support that people express for wind, solar, and other
renewable sources (e.g., hydropower) is further reflected in their support for power projects that use such renewables to generate electricity.[41] For these reasons, the types of policies that the federal government and states have put in place to advance renewable sources have
proven popular in most parts of the country. The federal government,
for example, has used the tax code to encourage the development of
both wind and solar power, using a production tax credit for wind
and an investment tax credit for solar. For their part, more than thirty
states have mandatory renewable portfolio standards that require covered utility providers to produce some percentage of their electricity
from renewable sources. Collectively, these policies—along with technological innovations that have reduced the costs of wind and solar

power—have resulted in a near fifty-fold increase in non-hydro, renewable energy from 2003 to 2022.[42]

These polling results seem to suggest that Americans are ready to embrace the clean-energy transition and even more widespread deployment of wind and solar power. Yet, at the same time, there are cases like the Honey Creek wind farm, where a local community objected to a new project and then pushed the local county government to outlaw the local use of the technology altogether.

And Honey Creek is no aberration. According to a report published by the Sabin Center for Climate Change Law at the Columbia Law School, as of May 2023, nineteen states and 395 local jurisdictions (in forty-one separate states) had imposed restrictions to block renewable-energy development.[43] The Sabin Center has been tracking these restrictions for several years, and its researchers have shown that there is an increasing trend toward more-restrictive local ordinances across the country. These ordinances take many forms. In some cases, they are outright bans like we saw in Crawford County, Ohio. In other cases, prohibitions take the form of temporary moratoria on wind or solar development, policies that put in place such restrictive requirements (e.g., setback rules or turbine or solar-panel spacing requirements) that they serve as de facto bans on wind or solar power development, and amendments to zoning policy that are designed to block specific projects.[44]

To illustrate, consider policies in Kansas and Oklahoma, two Great Plains states with considerable capacity for wind-power generation. In 2004, Kansas put in place a restriction on developing wind turbines in Flint Hills as a way to protect prairies. The protected area has been increased twice since then and now covers parts of twelve counties. Oklahoma also has restrictions on where wind turbines can be placed. As opposed to a specific region being protected, all wind turbines in Oklahoma must be located at least 1.5 miles from schools, airports, or hospitals, which severely restricts the places turbines can be built.[45]

These restrictive state and local zoning ordinances create a large potential impediment to the future deployment of renewable energy.

A study from researchers at the Department of Energy's National Renewable Energy Laboratory estimated that, if local governments across the country imposed restrictive setback requirements, it would limit wind generation to just 15 percent of what would otherwise be possible.[46] While some setback requirements are reasonable (i.e., few people want a turbine or solar array directly in their backyard), the key point is that restrictive setback requirements curtail the potential scale of renewable-energy deployment.

Of course, wind and solar power have been deployed in many parts of the country, and their development will certainly continue. Nonetheless, the growing policy retrenchment is real, and inarguably it delays progress toward a cleaner energy economy in the United States. The pushback against renewable energy, moreover, can be understood, at least in part, through a lens of energy justice.

First, consider the geography of the costs and benefits of wind and solar farms. In general, all Americans—and, really, people across the world—benefit from every project. After all, every megawatt of electricity generated from a non-carbon energy source that replaces a carbon source helps to address climate change. Moreover, generating electricity from wind or solar energy has the added benefit of not contributing to traditional air and water pollution; at the point of generation, there are no emissions of criteria or toxic air pollutants and no discharges of chemicals into nearby water sources.

While the communities that are asked to host wind and solar power projects also enjoy these benefits, they also are asked to absorb all the costs associated with the land-use impacts. This situation is no different from the historical siting of most energy infrastructure, where local communities were asked—or often not asked—to accept large fossil fuel power plants, even though the benefits of lower electricity costs or reliable supply were enjoyed by communities much farther away.

The amount of land required for a wind and solar farm is also considerable, exceeding that of a coal, natural-gas, or nuclear power plant in terms of power generated per acre.[47] Crucially at issue, however, is not just the *amount* of land that renewable facilities require but the type of land use itself. A typical case is like that of the Honey Creek wind project, where rural, agricultural communities are targeted for a wind project. A field of wind turbines is largely compatible with a field of corn or soybeans or with livestock grazing, but for people accustomed to farmland or prairie grasslands, the sight of towering wind turbines or flashing red lights in the sky can be unsettling. The same can be said in the case of a large, utility-scale solar project, which is generally less compatible with other land uses.

These types of costs, in the form of changed landscapes and disruptions to people's sense of surrounding place, are difficult to quantify, but they are real nonetheless, and they often evoke emotional responses from people in affected communities. These local costs may be offset by local benefits. The local jobs produced by renewable-energy infrastructure projects are modest and mostly temporary, but landowners who lease their land to energy companies may be paid handsomely,[48] and many communities enter into benefit agreements with the developer who pays annual fees that can be put toward local needs, such as roads, schools, and social services.

But the local environmental benefits are less clear, because in nearly all cases the decision is not whether to replace or build a new fossil fuel power plant or to instead install a wind or solar facility. Rather, it is to transform a current land use, one that is most likely not industrial at all, to a renewable-energy-infrastructure use.

Does local opposition to renewable-energy projects reflect selfish, knee-jerk, irrational "Not-in-My-Backyard" attitudes, or NIMBYism? That is, do people generally support the use of wind and solar power, as long as they are not put in their backyards? Although the media and renewable-energy advocates often portray opposition in this way, the evidence suggests otherwise.

The research we have done on Americans' attitudes toward energy infrastructure suggests a much more nuanced story. In a study of fourteen energy projects across the country, and with over sixteen thousand respondents, we compared levels of public support among those living in close proximity to a proposed project with those living farther away but in the same state, and we found very little difference. People living in close proximity to a proposed wind and solar project—and many other types of energy infrastructure projects as well, including transmission and distribution lines—were no different in their support for the project than those living farther away, and when we further analyzed people's opinions, we found little evidence that attitudes toward projects are explained by NIMBYism. The findings from this specific study suggest that proximity alone is not a very good indicator of how people evaluate projects.[49] In fact, although NIMBYism is often suggested to be the prime explanation for local opposition to energy projects, when we reviewed decades of empirical studies on the subject, we found little actual evidence supporting this conventional wisdom.[50]

What, then, explains opposition? Research suggests that whether people living near a proposed project object to it depends in large measure on whether they view the project as a disruption to their perception of place, or what human geographers and environmental psychologists sometimes call "place attachment." Place attachment refers to the emotional bonds and social attachments that people have with familiar locations where they live or regularly visit.[51] These attachments can be due to the landscape, the environment, or the full milieu of one's surroundings. Disruptions occur when people feel a threat to the positive feelings that they have attached to these places.[52] In the context of wind energy, the threat is the change of a landscape, for example from a rural place with farmland to an industrial land used for energy development. Recall Kimberly Groth, who—when reflecting on the Honey Creek project—noted that rural areas provide a special quality of life, one marked by peacefulness and serenity.[53]

Disruptions to place attachment due to renewable-energy development are not unique to rural landscapes in "red states" or to the Midwest. Proposed wind projects along mountain ridges and proposed utility-scale solar farm projects in open desert landscapes have also created these types of disruption for local residents.

One of the most famous examples of local opposition to a wind farm can, in part, be understood this way. In 2001, developers proposed the Cape Wind project off Nantucket Island, Massachusetts. Nantucket is a small island located thirty miles east of Cape Cod, best known as a tourist destination with beautiful beaches and views of the Atlantic Ocean. When Energy Management proposed the project, it was to include 130 wind turbines spaced over twenty-five square miles in Nantucket Sound, which, when completed, would generate about 468 megawatts of electricity. The Cape Wind project was to be the first offshore wind farm in the United States. Over the course of the next decade, Energy Management secured all the federal, state, and local approvals and permits it needed to start construction and lined up long-term power purchasing agreements with Massachusetts utility companies to distribute the power.[54]

From the outset, there was strong opposition to the project, particularly from wealthy property owners who had homes with views that would be obstructed by the forty-story-tall wind turbines.[55] Among the notable opponents of the project were Senator Ted Kennedy and William Koch, the billionaire brother of conservative activists Charles and David Koch.[56] As construction on the project stalled and project partners reneged on agreements (e.g., the Massachusetts utility companies canceled their power purchase agreements), the opposition took firmer root, and ultimately the project was halted altogether.[57] While the general public may not have much sympathy for wealthy landowners losing their unobstructed ocean views, the case provides another illustration of how the loss of place attachment can affect the viability of renewable-energy projects.

Of course, people's impressions of large-scale renewable-energy development may differ. For people like Kimberly Groth or William Koch, wind turbines may be an eyesore that irreparably changes the character of the rural landscapes or ocean views that they treasure. They may further believe that such projects reduce property values and the quality of their local environments. Yet others may view towering wind turbines as technological marvels that inspire awe and human progress. There is no "correct" view here. Beauty is in the eye of the beholder, as the saying goes. The crucial point is that people are not reacting in some sort of irrational way that can be explained by NIMBYism but are instead acting rationally to a sense of disruption and loss.

A second important factor contributing to opposition to renewable-energy development is that people often lack trust in the companies that propose and promote the projects. Project developers tout the benefits of projects, typically focusing on addressing climate change, the provision of clean energy, local job creation, and long-term financial benefits that often accompany projects through both landowner and community compensation. At the same time, they tend to minimize the impacts—real or perceived—noted by local citizens, such as impacts to human health, quality of life (e.g., noise, shadow flicker, and other nuisances), property values, economic composition, and the changes to the landscape as previously discussed.[58]

Trust is built over time through multiple interactions, and new relationships may not begin with a foundation of trust. In fact, the Gallup data presented earlier suggests that people may start with distrustful opinions about companies in the energy sector. Moreover, relationships are influenced by historical interactions, even if those interactions are not among the same two entities. For example, if someone had a previous bad interaction with an energy developer, they may start from a place of mistrust with a new developer who shows up in their town, even if it is years later and the proposed project differs in significant ways.

It is crucial to emphasize here that what matters the most about project impacts are perceptions, regardless of whether perceptions actually match reality. For example, although people often lament the loss of property value associated with proximity to wind projects, at least in the US context, most research has found little evidence that home prices actually decline with the installation of wind turbines.[59] And when studies do find decline, they appear to be short-lived and limited to homes within especially close proximity.[60] Regardless of what the research says, however, people's perceptions of lost home value are likely to drive their perceptions of projects.

Complicating matters is that there is growing misinformation around renewable-energy development, often fueled by organized campaigns designed to raise opposition to new projects. Misinformation includes claims that wind and solar projects *contribute* to climate change, poison local water, and result in health impacts such as cancer for nearby residents. In another example, offshore wind turbines are often alleged to be the reason for whale beachings. Such misinformation often festers on social media platforms like Facebook and X and can be backed by big money, including from fossil fuel companies trying to slow down the clean-energy transition.[61]

Moreover, local communities often bemoan site selection processes undertaken without sufficient community engagement, further sowing distrust. Quite often, communities are engaged late in the process, and only after an energy developer has identified a place as desirable and economically and technically feasible for a project. In other cases, opportunities for community engagement are lackluster or happen to fall during regular business hours when people are at work or during family evening hours, without any consideration for alternative child care.

In one study of landowners in Michigan, researchers found that people who believed that the wind siting planning process was fair—meaning, there were ample opportunities to provide input, community input was acted upon, and project developers acted openly and kept promises—were much more likely to view completed projects as beneficial to them and their communities. By contrast, those who viewed the planning process as unfair were more likely to see adverse impacts of the completed projects in the form of visual and noise problems, loss of property values, and impacts to human health.[62]

The implication of this study and others like it is that procedural justice matters for acceptance of energy infrastructure projects; that is, if community members are engaged in decision-making and planning processes that they deem to be fair and inclusive, they are more likely to support a project. Although the types of communities in question may differ, historical environmental-justice communities have insisted upon this same principle for decades. Given that people often begin with low levels of trust in the companies that are asking them to host projects, perceptions of fairness and inclusivity are essential.

For their part, companies are starting to understand the challenge they face. Dahvi Wilson, vice president for public affairs at Apex Energy, told NPR reporter Julia Simon that local engagement is becoming increasingly difficult and that just providing information is insufficient: "I think for a long time, and maybe still in some places, developers thought, 'Well, we just need to give better information. We just need to give more information.' And it's like, it's so not about that at all! It's about who you trust and if anybody's going to believe you if you're a company."[63]

A good process may not assuage local concerns or resolve issues of mistrust, but it is certainly a start. However, it is also important to recognize that there can exist a tension between procedural justice and making the transformations to the energy system that are required to meet the climate crisis. That is, more and better community engagement does not necessarily lead to more acceptance of projects. Community self-determination can be as much a barrier to climate action as something that facilitates it.

A third factor that contributes to growing opposition to the local siting of renewable-energy facilities is traditional partisan politics. Although public-opinion polls show that Americans of broad political stripes support renewable energy, Democrats and Republicans are deeply divided on the importance of climate change as a motivating factor and on how much priority should be given to moving the country away from fossil fuels. With respect to climate change, although large majorities of the American public express concern about climate change, when one analyzes these opinions in more detail, one quickly sees enormous differences between Democrats and Republicans.[64]

As one example of this general phenomenon, consider responses to a question that Gallup asked in its annual survey about the environment in March 2023. Overall, 39 percent of respondents told Gallup that they were concerned "a great deal" about climate change. However, the disparity between self-identified Democrats and Republicans was large—65 percent of Democrats noted a great deal of concern, compared to just 8 percent of Republicans. Independents were in the middle at 42 percent.[65]

As further evidence, consider how Americans classify different issues that they would like to see the president and Congress prioritize. In a survey conducted by the Pew Research Center in February 2023, 59 percent of Democratic respondents identified climate change as an issue that they wanted to see the federal government address that year, compared to just 13 percent of Republican respondents.[66] This 46 percentage-point gap—and the 47 percentage-point gap on protecting the environment—was the largest for any of the issues that Pew asked about in its survey.

The partisan divide on climate change runs deep, and it helps explain growing polarization on other environmental issues as well.[67] Moreover, the divide between Republicans and Democrats on climate change extends to the types of policies that might be employed to hasten the shift from fossil fuels to renewable technologies, such as carbon taxes, clean energy mandates, inclusion of the social cost of carbon in government impact assessments, and renewable-energy subsidies.[68]

Partisan division among citizens does not necessarily play out directly into support or opposition to energy infrastructure projects. For example, these public-opinion data do not simply translate to Democratic support and Republican opposition to renewable-energy projects, and it is important to emphasize that some of the "reddest" states produce the most electricity from renewable sources, including the top state, Texas. Empirical research on the factors that explain opposition to wind farm projects, both in the United States and Canada, suggests that partisan affiliations are uncorrelated with opposition, once demographic and other factors are accounted for.[69] Some studies of other types of energy infrastructure have reached similar conclusions, such as one analysis of attitudes toward the Keystone XL pipeline that found that partisan affiliations were less important in explaining support and opposition to the pipeline for people living in close proximity to the proposed project.[70]

Nonetheless, politics can still work to shape the discourse around renewable energy generally and with regard to the siting of specific projects. One pathway is through elections. Partisan polarization at the national level is well documented, and the sharp partisan division on energy and climate policy is as sharp as it is on other issues like immigration and abortion. Nearly without exception, Democratic officials in Washington, DC, support policies to move the United States away from fossil fuels. By contrast, Republican officials tend to express continued enthusiasm for fossil fuels, sometimes shrouded in support for "all of the above" approaches to energy policy. As just one example at the federal level, not a single Republican in Congress voted in favor of the 2022 Inflation Reduction Act, which was generally characterized as the most consequential climate change legislation ever enacted in the United States.

As policy debates at the state and local level increasingly mirror those at the national level,[71] partisan divisions on energy and climate change are also prevalent in state legislatures and the halls of county and city government. The implications, however, are somewhat different. While partisan division in Washington generally produces legislative gridlock, at the state and local level it tends to produce policy extremes. In the area of renewable energy, many "blue" states have enacted mandates that their utility companies produce a sizable portion of their electricity come from solar, wind, and other non-fossil fuel sources, while fewer "red" states have such requirements. Similar policy differences exist between Democratic and Republican controlled counties and cities.

Partisan differences in perceptions about the need to address climate change, and the urgency of doing so, also likely affects the salience of energy infrastructure projects. The fact that Republicans are more dismissive of or agnostic toward climate change—and generally opposed to policies or actions that would reduce greenhouse gas emissions—likely diminishes the salience of clean-energy projects for those who share the same political preferences. In this way, the climate change benefits of these projects are less important to these individuals and communities, and they are often insufficient as a motivating factor to offset the concentrated costs of the local siting of a project. These communities are being asked to take on the burden of solving a problem that may be of far less importance and salience to them. This dynamic is not that different from that of coal miners in Appalachia or autoworkers in the Midwest who are being asked to take on the burden of finding new employment opportunities while their industries shrink as the country moves away from fossil fuels and internal-combustion-engine cars, respectively.

The overlap of these three factors—the uneven geography of project costs and benefits, distrust in energy companies, and deep partisan differences on the problem of climate change and the need for clean sources of energy—combine to slow progress on the energy transition. These dynamics are not unique to wind and solar power but similarly play out with other types of energy infrastructure, such as electricity transmission and carbon capture and storage.

Transmission

A defining feature of the US clean-energy transition is the complicated geography of electricity generation. For example, the best locations for wind tend to be in the Midwest, the Rocky Mountain West, and offshore, while those for solar are in the Southwest and Southeast.[72] In some parts of the country, existing transmission networks are congested, which further complicates adding new capacity and prevents the moving of electricity across long distances and different sections of the grid. This is the case in the Northeast, where it is difficult to move electricity between New England states (and Canada) and the densely populated states of New Jersey and New York.[73] Yet US companies will need to build tens of thousands of miles of new transmission lines to move electricity, especially the new, clean electricity, to the places where people live. Not only will these developments require hundreds of billions (or even several trillions) of dollars of investment,[74] but they will also require countless governments—federal, state, and local—to permit the new projects and for communities to accept this new infrastructure.

Should one anticipate similar siting controversies and public opposition to transmission lines as has been the case for renewable-energy development? Some of our research on the topic might lead one to be optimistic that such controversies are not inevitable. As part of our surveys of the US public about the types of energy infrastructure that they will accept, we conducted an experiment. We asked a nationally representative sample of Americans if they would support a decision to build new transmission lines in their communities as a way to connect different types of new sources of electricity to meet growing demand. To measure differences in preferences, we randomly sorted people into one of four groups: a control group, to whom we told nothing about what the lines would connect to; and a coal group, a natural-gas group, and a renewable energy group, each of whom was told that the transmission lines would connect to a new energy facility using their group's fuel type.

In general, people favored the new transmission line in their community, with the average level of support being halfway between neutral and slightly favorable, on a five-point scale. However, preferences differed among those who learned about the specific energy source that would be connected. Specifically, as shown in figure 6.2, levels of support increased in the cases of renewables and declined for natural gas and coal.[75] Our study suggests that people might be willing to accept transmission lines if they know that the lines are intended to bring

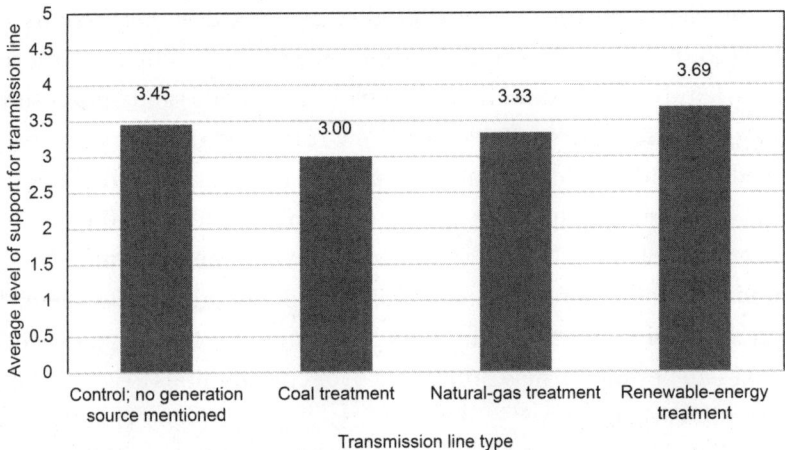

Figure 6.2 Average response to survey experiment questions about support or opposition to proposed local transmission lines, by line type, 2017 Notes: Support levels fall on a 5-point rating scale, where a 1 indicates strong opposition and a 5 indicates strong favor. Source: Adapted from Carley, Ansolabehere, and Konisky, "Are All Electrons the Same?"

more clean energy onto the grid. This study asked people to think abstractly, however, and the findings may not coincide with how people might respond to an actual project proposed in their community.

In other research, we took a more direct approach. Specifically, we interviewed county government officials; representatives of important stakeholder groups such as unions, farm bureaus, businesses, and environmental groups; and affected landowners about a proposed high-voltage transmission line in the rural Midwest.[76] The line would move electricity generated from wind turbines in one state, across two other states, for ultimate delivery to a fourth state. In other words, people affected by the project included those in communities in states that would not directly benefit from the clean energy being produced. These types of long-distance, interstate transmission lines will be necessary to enable a clean-energy transition.

People with whom we spoke who were opposed to the project tended to emphasize how the transmission line would disrupt their sense of place. Much like Kimberly Groth in Honey Creek, opponents often characterized their agricultural community as a rural respite and expressed concern that the transmission line would negatively affect their viewsheds and natural areas. In this way, they depicted the transmission line as an unwanted threat. Place sentiments were also important for project supporters, the key difference being that they perceived their communities and landscapes in a different way—more as rural-industrial landscapes than as untouched, undeveloped rural oases. The proposed transmission line for them did not threaten their sense of place. One supporter told us: "You can't see very far without seeing cell phone towers; you can't see very far without seeing other utility poles."

Supporters also often gave more weight to the national (i.e., non-local) benefits of the transmission line project, taking pride in providing a public good for others. One project supporter said: "I think it's almost patriotic for me to be able to provide to people that we're not only providing their food, we're not only able to provide them quality product for them to eat, but we're also able to help them to have electricity, to have energy, that that's something that we are able to provide here in rural [state] that they can't do in the cities because it's just not feasible."[77]

But this pride was not shared by all. In fact, some of those skeptical about the project commented on the uneven and, in their view, unfair distribution of costs and benefits of the project. Several people noted concerns about how the project would affect their property values. Others interviewed emphasized the personal sacrifice that their communities were being asked to accept to benefit people living elsewhere. One interviewee noted that farmers had already done enough and should not be asked to accept infrastructure on their land to primarily benefit others. This sentiment is quite similar to what we heard from coal miners in Appalachia, who also expressed notions of personal sacrifice for the rest of the public. In a similar vein, opponents tended to evaluate the potential benefits in mostly local terms, often characterizing it as unnecessary. One told us: "This RTO [Regional Transmission Operator] was not . . . Doesn't need the energy [from the transmission line]. We have plenty of cheap abundant electricity as it is and we're also utilizing wind energy right here . . . that we make here."[78]

The case here illustrates that place attachments affect people's willingness to accept transmission lines in similar ways as it affects their views of renewable-energy generation projects. Another similarity pertains to the decentralized nature of decision-making around transmission projects. Although many often talk about the "grid" as if it is a single system, in reality it is divided into three main sections that are then managed by twelve separate transmission planning regions. The three sections, known as the Eastern Interconnection, the Western Interconnection, and the Electric Reliability Council of Texas (ERCOT), are mostly isolated from each other, meaning electricity does not flow among them. To further complicate matters, the high-voltage lines that are central to these interconnection systems are typically planned more locally.[79] The twelve transmission planning regions also operate independently—under the regulatory oversight of the Federal Energy Regulatory Commission (FERC), except for ERCOT—with the main goal of achieving reliability within their regions. Importantly, they historically have focused very little on moving electricity across the three regions, which could help transport electricity generated far away to more homes and businesses. Exacerbating this disconnected system is the fact that, historically, the transmission system was mostly built by individual utility providers seeking to meet their own power distribution needs rather than thinking holistically about the needs of the region, let alone the nation as a whole.[80]

The fragmented nature of the US electric grid—and the sheer number of private actors and layers of government that are required to develop long-range transmission lines—creates many potential veto points. At the federal level alone, a project might require approvals from numerous agencies beyond FERC, especially if the transmission line crosses government-owned land, such as lands managed by the Bureau of Land Management, the Forest Service, the Army Corps of Engineers, and the Fish and Wildlife Service, which collectively manage more than one-fourth of the nation's land.[81] The list of state and local government agencies is even longer and includes relevant state environmental, natural-resource, transportation, and historic-preservation agencies, along with county, municipal, and tribal authorities.[82] With so many venues for decisions, there are countless opportunities for partisan or other politics to shape outcomes, and a high likelihood for legal contestation among aggrieved parties who object in some way to a project.

The TransWest Express Transmission project provides an illustrative example. This more-than-seven-hundred-mile transmission line will carry electricity from the 3,000-megawatt Chokecherry and Sierra Madre Wind Energy project in south central Wyoming, across Colorado, Utah, and Nevada, to provide electricity to people in Los Angeles and other parts of California. Perhaps ironically, the wind farm is located in Carbon County. The wind farm itself first sought federal and state permits in 2008[83] and once completed will be among the largest in the United States, comprising about a thousand turbines spread over 220,000 acres (about 340 square miles).[84] The TransWest Express Transmission line is a $3 billion project that has been twenty years in the making, first proposed by the Arizona Public Service Co. in 2005.

The project was long stalled by opposition from some environmental organizations and communities and landowners who were concerned about impacts to nature, wildlife, and private property. At the groundbreaking ceremony in 2023, the chief operating officer was quoted as saying that the project's long road "literally took the support and contribution of hundreds of people at all levels of federal, state, and local governments. [I]t took communities; it took labor unions; environmental organizations; it took tribes; it took contractors and consultants; it took individual landowners to get us to this milestone."[85] TransWest Express is one project. The US clean-energy transition requires hundreds, perhaps thousands, more like it.

Carbon Capture and Storage

If siting and distributing new *clean* energy is difficult, what about making existing *dirty* energy cleaner? Many projected pathways to net-zero carbon emissions assume that emissions reductions will come, in some measure, from continuing to burn fossil fuels.[86] The catch is that companies will first develop and deploy technologies to capture the carbon dioxide that is released during combustion and then store it underground rather than releasing it into the atmosphere. If successful, carbon capture and storage (CCS) technology would contribute to climate change mitigation and, perhaps in a significant way, reduce the need to build out other new infrastructure or replace reliance on fossil fuels with other energy sources like clean hydrogen.

Perhaps not surprisingly, fossil fuel companies and other industries often view CCS technology as viable and a critical part of the US energy future. If developed and deployed on a widespread basis, CCS may provide a lifeline for fossil fuel power plants and other large industrial sources of carbon dioxide emissions (e.g., concrete plants, oil refineries, steel manufacturers). Some climate experts also see CCS as a necessary technology for industries where there are not suitable substitutes, and recent federal policy has set aside billions of dollars in tax incentives and subsidies to help companies develop and deploy CCS technologies.[87]

Capturing and storing carbon is not a new idea, but it remains an early-stage, developing technology that has not been widely demonstrated at scale. As of 2023, only one industrial-scale facility was using carbon capture in the United States—the Petra Nova project at the NRG Energy–owned coal and natural-gas power plant near Houston, Texas. The project began in late 2016 and resumed in 2023 after a three-year hiatus. In this specific case, the carbon dioxide captured is used for enhanced oil recovery as part of the storage.[88]

In addition to the technical and economic uncertainty associated with CCS, this approach to addressing carbon dioxide emissions is controversial for other reasons. Many in the environmental community point to a moral hazard. The use of CCS would create incentives to continue the use of fossil fuels and potentially divert resources and investment away from alternative sources of energy.[89] With CCS in place, fossil fuel power plants and other industries that emit large amounts of emissions might continue to operate for decades, or longer. And, although their carbon dioxide emissions might come under control, there are other impacts to consider. For example, current forms of CCS are energy intensive, so overall energy use may increase, and some studies show it may also increase other types of air pollution.[90] In addition, the adoption of CCS will require the development of an extensive network of pipelines to move the captured carbon dioxide to permanent storage locations. Siting pipelines, however, creates its own set of controversies.

More broadly, the same types of challenges that make siting renewable-energy facilities and transmission lines difficult complicate efforts around CCS. First consider the geography of the costs and benefits. The benefits of CCS—a reduction in carbon dioxide emissions—are primarily global. Reducing these emissions is beneficial to everyone, regardless of whether one lives near or far away from a CCS project. However, the costs are largely concentrated on those who live nearby. CCS—when deployed at coal and natural-gas power plants or other large industrial emitters—means the continued operations of facilities that tend to have large environmental impacts, including the continued emissions of other pollutants that adversely affect the health and quality of life of nearby residents. In other words, an energy transition that includes the continued operation of these industrial facilities will adversely affect the same marginalized communities that have for decades experienced the environmental and health burdens of the current fossil fuel–based energy system.

For these reasons, efforts to encourage CCS in some parts of the United States have not been received well by many environmental-justice advocates. Beverly Wright, the executive director of the Deep South Center for Environmental Justice, described the problem for *The Washington Post*: "In the real world, this is an experiment. And this experiment is going to be conducted on the same communities that have suffered from the oil and gas industry."[91]

The historic burdens that facilities have imposed on frontline communities, moreover, has generated mistrust, which raises residents' skepticism about the possible use of CCS. Chad Ross, a resident of Donaldsonville, Louisiana, told a reporter for *The Washington Post* who was investigating early developments of CCS in Louisiana: "It is called Cancer Alley, and that's part of the reason we don't trust them. It's still not so good to have all these plants, so many of them, all around us. Anything could happen."[92] Donaldsonville is home to the world's largest ammonia and nitrogen facilities, operated by CF Industries, which is also Louisiana's largest single source of greenhouse gas emissions and a place where CCS could be employed. Another Donaldsonville resident, Ashley Gaignard, expressed a similar sentiment: "Don't do it my neighborhood. Do it where you live. Right about now, it's politics over people. And I don't think they give a damn about people."[93] These sentiments are not dissimilar to perceptions among many environmental-justice advocates that marginalized groups have long been forced to live in "sacrifice zones."

The politics around CCS may further intensify reactions to the technology. CCS tends to be more popular among fossil fuel advocates in Washington, DC, and in states that are heavily invested in either the development or use of fossil fuels. Many of these advocates are conservative Republicans—but not all of them, with notable counterexamples such as former Democratic Senator Joe Manchin—who view

CCS as a way to simultaneously address climate change emissions while maintaining reliance on fossil fuels. For states like Texas, Louisiana, West Virginia, Kentucky, and Indiana, all of which maintain large use of fossil fuel in their electricity sectors or host large industrial sources of carbon dioxide emissions, the use of CCS would enable a "fossil-fuel-friendly" approach to the energy transition. For many Democratic officials, especially those who think about climate change in existential terms, the investment and potential reliance on CCS is anathema. There is not much common ground, creating a fraught political environment for CCS that is bound to get more contentious in coming years.

● ● ●

To meet the climate change challenge, the United States needs enormous amounts of new energy infrastructure. In this chapter, we have highlighted several of the major challenges that this requirement generates, including the uneven geography of project costs and benefits, mistrust of the energy companies that will be largely responsible for developing new projects, and divisive politics that can stymie progress. Layered on to the difficult decisions facing communities, companies, and government agencies is the reality that the existing energy system, built on the back of fossil fuels for two centuries, already imposes uneven burdens. Failure to progress only means allowing historical inequities to persist. Moving forward requires confronting the trade-offs and recognizing that the infrastructure investments needed will not satisfy everyone. There are no easy solutions.

We next turn our attention to the transportation sector and electric vehicles, where the trade-offs are not any easier.

CHAPTER 7

The Life Cycle of an Injustice

The first modern electric vehicles in the United States, the Chevy Volt and the Nissan Leaf, hit showroom floors in early 2010. The Chevy Volt, priced at $41,000 pre-tax credits, was a plug-in hybrid electric vehicle (PHEV) that could run on both electricity and gas. Once the battery is drained, a PHEV switches over to operating on gasoline. The Nissan Leaf, priced at $32,780 pre-credit, was a fully electric vehicle, otherwise referred to as a battery electric vehicle. In the European Union, Mitsubishi's i-MiEV beat all electric vehicles to market in 2011, while in China, it was BYD's e6 electric-vehicle model, also in 2011.

Readers may recall auto companies' early marketing campaigns, which attempted to motivate would-be buyers with environmental messages. Among perhaps the most memorable was a Nissan commercial that featured a roaming polar bear who leaves the melting Arctic, treks over mountains and across valleys, to eventually greet the driver—an affluent-looking young man—with a massive bear hug when he steps out of his house to get into his Leaf. The tag line concludes: "Innovation for the Planet, Innovation for All."

Although it is notable that an automobile company embraced an eco-friendly message that featured a giant bear endangered by the impacts of climate change, we suspect that Nissan soon realized its mistake. To attract consumers to electric vehicles, the polar bear–

hugging, environment-saving approach would not work, and might even turn off consumers who prioritized brand loyalty, cargo space, or vehicles that do not require one to change their behavior. An environmental appeal would surely not be enough to keep modern electric vehicles from experiencing the same fate as 1990s versions of electric vehicles, as featured in the documentary *Who Killed the Electric Car?*—namely, a concern over vehicle range and sticker price.[1]

Electric vehicle marketing has come a long way. Companies soon pivoted to commercials that showcased the functionality of the electric vehicle, the long driving range, the appeal of a quiet vehicle, and even the mundane day-to-day activities that a driver can do in their vehicle as they commute to work. Some may recall, for example, a young and smitten cartoon mailman who wants to sneak past the vicious dog to connect with the dog's attractive owner. The mailman, after several unsuccessful attempts, eventually finds his way into a Renault Kangoo Z. E. electric vehicle, by which he silently bypasses the dog and is rewarded by the electrifying touch of the dog's owner as she reaches for the package. Here, the animal is the antagonist and the agenda is romance, in juxtaposition to the earlier Nissan Leaf's polar bear as a harbinger of climate resurrection.

The second wave of ads sought to demonstrate to consumers that the electric vehicle is no different from a standard conventional vehicle in its functionality but comes with extra perks such as a lack of noise and the ability to switch to autonomous vehicle mode—and, in some lucky cases, that it may also spark an energetic romance.

Vehicle performance has consistently improved as well. The early models of the Nissan Leaf and Chevy Volt had ranges of 100 and 40 miles, respectively, and charging the cars with a standard Level 1, 120-volt charger took all night. Just ten years later, a new Nissan Leaf had a range of 215 miles, and a now fully electric Chevy Bolt has a range of 259 miles, with a much wider range of charging options. The longest-range vehicles are typically the most expensive. Consider, for example, the Lucid Air, which has a range of 520 miles and a sticker price of $170,500, and the Tesla Model S, which has a range of 405 miles and a sticker price beginning at $76,630.[2] However, a growing number of manufacturers are designing vehicles with a range over 200 miles on a single charge and at a cost that is approaching parity with their internal-combustion-engine-equivalent vehicle, and even favorability when applying available tax credits.[3]

Tesla has been an EV market leader for the past twenty years and has been responsible for pushing the limits of the EV market in range, compatibility, and design.[4] Founded in Palo Alto, California, in 2003, Tesla introduced its first model, a two-seat sports car called the Roadster, in 2006.[5] Subsequent models included the Model X and the Model S, the latter of which won *Motor Trend*'s 2013 Car of the Year—as the first EV to do so—and was recognized as the highest scoring car by *Consumer Reports*.[6] Since that time, Tesla's offerings have grown to five vehicles. With battery ranges between approximately 250 and 500 miles depending on vehicle and battery options, sticker prices range between $40,000 for the Model 3 sedan[7] to over $100,000 for the Cybertruck,[8] a social-media sensation in its early days with proposed monikers that ranged from "Beep Boop Bop Truckitty Truck Truck Tron Tron" to the "Starship Pooper."[9]

Consumer interest in electric vehicles has taken off. It took eight years after the Leaf and Volt made their appearance for the United States to hit one million electric-vehicle sales in 2018, and then only another three-and-a-half years to reach a total of two million sales.[10] Globally, rates of interest in electric vehicles are even higher. According to a poll conducted by Ernst and Young in 2022, 73 percent of Italian residents are interested in purchasing an electric vehicle, 69 percent of Chinese, 63 percent of South Koreans, 61 percent of Norwegians, 38 percent of Australians, and 29 percent of Americans.[11]

Policymaker and auto company commitments confirm that the electric vehicle is here to stay. Commitments to a zero-gasoline future are all the rage among auto companies. Bentley and Volvo plan to do away with gasoline vehicles by 2030 and Ford by 2035; BMW plans to make electric vehicles half their sales by 2030; Ford to be carbon neutral by 2050; and Honda to phase out gasoline car sales by 2040.[12]

In many cases, especially in the United States, auto companies are moving faster than policymakers. But policy is starting to catch up. Forty countries, twenty-seven of which are in the European Union; eight US states; and the District of Columbia have banned gasoline vehicles in coming years.[13] In the case of California, this ban is set for 2035. Eleven US states have zero-emission vehicle mandates that require that a portion of all cars sold in those states must be produced by zero-emissions technologies, a category that includes battery electric vehicles and hydrogen fuel cell vehicles. Federal policy uncertainty following the 2024 elections created headwinds, but the auto industry will continue toward electric, even if on a slower timeline.

A zero-gasoline future is good news for our planet. The transportation sector accounts for 37 percent of all emissions across the world[14] and 28 percent within the United States.[15] A switch to a combination of electric, hydrogen, and biofuel vehicles could save 23 GT of carbon, which is in line with climate targets of 1.5 degrees Celsius from preindustrial levels.[16]

Yet a transition to electric vehicles is not without complications, especially with respect to equity and justice. The case of electric vehicles illustrates some of the key challenges and opportunities that the energy transition brings. A focus on this single technology reveals the multiple dimensions of energy justice along the full life cycle of the technology, or the "embodied injustices"[17] of a technology, ranging from the distribution of harms associated with making the EV and its components (e.g., through the extraction of critical minerals such as those containing lithium and cobalt used to make batteries and the eventual disposal of these commodities in dumps); to the economic impacts for autoworkers and communities (e.g., dislocation from changes to the structure and geographic footprint of the manufacturing supply chain, loss of employment in dealer-associated and third-party maintenance and repair shops, shifts in sales-related employment); to the affordability of and access to electric vehicles; and to participation in decisions that help communities navigate through the massive transition of transportation systems.

What Is a Technology Life Cycle?

In chapter 2, we discussed the burdens that different segments of the US population have experienced from the production, distribution, and consumption of fossil fuels. Here, we extend this life cycle analysis to the EV. Cataloging burdens through the life cycle of an energy system is a helpful way to demonstrate the far-reaching effects of our energy choices.

The mining and extraction phase for any vehicle includes activities to procure the materials that are used to construct the vehicle base (e.g., steel or aluminum), the tires (e.g., rubber), and the external and internal components such as the motor, electronics, battery, and transmission (e.g., plastic, copper, various metals).[18]

Every product that is sold in a marketplace has a life cycle. This life cycle starts with raw material extraction, moves to production, is often transported to a market where it is consumed or used, and then is sent to waste or recycling. In the case of energy technologies, a life cycle typically also includes refinement of raw materials before production and, with technologies such as automobiles, may have a particularly long, intricate supply chain that spans countries and continents. Figure 7.1 presents a life cycle for an electric vehicle from the extraction phase to the manufacturing centers, throughout the manufacturing supply chain, and from consumption to waste disposal. For simplicity, we omit the transportation of inputs and finished products but note that these transportation requirements can produce significant emissions and also have their own set of equity and justice challenges (e.g., one may consider who lives next to the roadways that long-haul

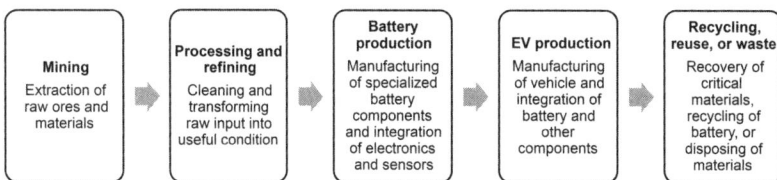

Mining	Processing and refining	Battery production	EV production	Recycling, reuse, or waste
Extraction of raw ores and materials	Cleaning and transforming raw input into useful condition	Manufacturing of specialized battery components and integration of electronics and sensors	Manufacturing of vehicle and integration of battery and other components	Recovery of critical materials, recycling of battery, or disposing of materials

Figure 7.1 Electric-vehicle supply chain

trucks drive along to transport materials). We also omit component production, noting that each component has its own life cycle as well.

For an electric vehicle, the mining and extraction phase is especially important because of the battery, which typically contains an assortment of metals, including several critical metals. The most common electric-vehicle battery types are lithium-ion batteries, although nickel-metal hydride batteries have been commonly used in hybrid electric vehicles, as have lead-acid batteries in limited applications.[19] A lithium-ion battery, the same type of battery found in most laptops and cell phones, contains thousands of individual battery cells.[20]

The mining and extraction phase of an electric vehicle is inclusive of all these earth metals, as well as the other components of the car, such as steel, aluminum, plastics, and rubber. These materials are processed and refined and sent to battery and other parts manufacturers. The supply chain eventually ends with the assembly manufacturers and then the distribution to dealerships, where consumers then usually go to purchase their electric vehicles.

Most cars have long lifespans. Following their initial sale or lease, most vehicles are resold, sometimes more than once. At the end of the vehicle's life, its parts are recycled or reused, including the battery, which is commonly sent to waste facilities for disposal or, increasingly, recycled.[21]

Along this life cycle is a complex web of challenges. We present examples of such challenges through four case studies: cobalt mining in the Democratic Republic of the Congo; automobile manufacturing transitions in the midwestern United States; electric-vehicle consumption patterns across the world; and disposal of electronic waste in Ghana. These case studies track the life cycle of an electric vehicle, from extraction of resources through disposal of waste, and all that happens in between. These cases also highlight the very real and observable realities of a growing clean-energy technology.

These case studies showcase how energy injustices can transcend geographic borders, including country borders, where one population can benefit from the use of a technology while others are burdened by both its production and disposal. Importantly, even if we were to move the full life cycle of the electric vehicle into the United States, it might perpetuate some of the same environmental-justice burdens of the past, particularly for Indigenous communities that inhabit lands that contain critical minerals and for disadvantaged communities that would likely serve as the storing grounds for e-waste.

The case studies also reveal just how complicated the set of trade-offs are in the clean-energy transition—while adopting lower-carbon technologies is necessary if we are to rapidly decarbonize our economy, it is difficult to do so without imposing risks to the health, well-being, and social development of certain communities and populations. The challenges outlined in these case studies will only become more severe as we move from early-stage adoption of electric vehicles to widespread use by most Americans.

Case 1: Cobalt Mining in the Democratic Republic of the Congo

Cobalt mining in the Democratic Republic of the Congo (DRC) is marked by exploitation, corruption, theft, and geopolitics, all set in an impoverished but resource-rich country. As the demand for batteries—for both electric vehicles as well as other common consumer devices such as cell phones—continues to grow, so too will the challenges of sourcing critical minerals. A battery for a modern Tesla vehicle requires over four hundred times the amount of cobalt to make than a cell phone,[22] so as electric vehicle sales grow, these challenges are certain to accelerate.

The DRC is a landlocked country in the heart of Africa that has experienced conflict for decades. These conditions have had profound impacts on child development, health, and poverty within the country. The DRC is the second most populous African country with a population of nearly a hundred million. Ranked within the bottom ten of all countries on the 2020 Human Capital Index, the DRC has a poverty rate of over 60 percent. The country also has low rates of resource access: about half the population have access to water for basic needs and only a fifth have access to reliable electricity.[23]Access to sanitation and healthcare is also low; less than a third of the population have access to sanitation, while the country has an average of 0.28 doctors and 1.19 nurses for every ten thousand people.[24]

The DRC is also one of the world's largest producers of cobalt. In 2021, the country produced about 70 percent of all the cobalt used across the world—more than the next producer, Russia, by a factor of sixteen.[25] It also contains half the world's cobalt reserves, far surpassing those known to exist in other countries, including the United States (see figure 7.2).[26]

Cobalt is an element found in the Earth's crust that, with processing, appears as a shiny metal. Cobalt is most commonly used in lithium-ion batteries but is also used in alloys, such as in gas turbine engines for aircraft.[27]

Cobalt is typically mined in two ways: through large-scale industrial mines owned by individual companies and through smaller-scale artisanal mines. The former involve more centralized and capital-intensive operations, with a single or small number of open pits. Artisanal mines, on the other hand, involve more decentralized operations and an assortment of mostly handheld tools such as shovels and

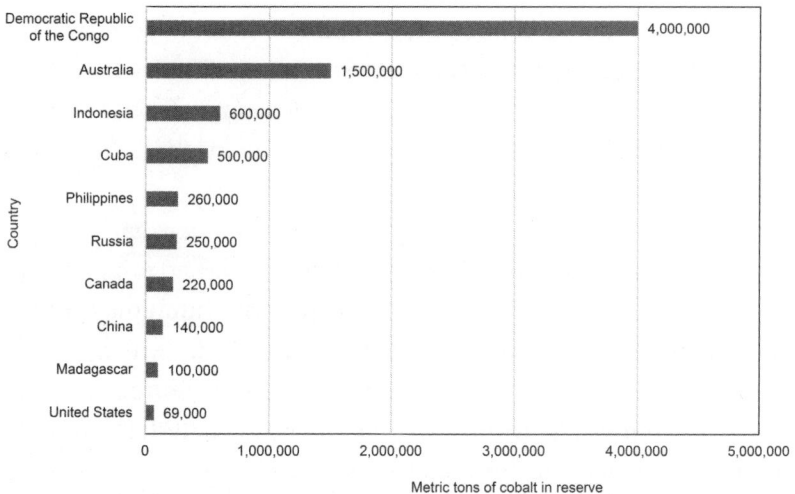

Figure 7.2 Top ten global cobalt reserves by country, 2022
Source: Adapted from Basov, "World's Largest Cobalt Reserves by Country in 2022."

chisels, and other "low-tech" mining and processing techniques. Artisanal mining also includes individuals who pick through mined rock and debris, or tailings, to extract any remaining cobalt by hand. About 20 percent of all cobalt comes from the artisanal sector.[28]

Miners in artisanal mines are referred to as "creuseurs," a French word for "diggers."[29] These mines typically run about thirty to forty meters underground and with many open pits or tunnels that branch off in different directions. Some artisanal mines connect to the larger-scale mines through underground tunnels.

Miners who gather cobalt at the various artisanal mines sell the ore to nearby buying houses, who pay the miners on a per-pound basis. These buying houses are typically owned by foreign nationals, often from China.[30] The buying houses then sell the ore to international trading companies, which have refinery facilities within the DRC. After refining, the ore is shipped overseas, where it undergoes further refinement and processing as it is made into manufacturing components. The resulting supply chain is long, involves many middlemen who profit off the miners' efforts,[31] and also involves companies that span across the world, ultimately feeding into the supply chain for electric vehicles.

The environmental conditions of cobalt mines vary but typically contain a highly concentrated amount of dust and debris in the air, especially during the dry season. This dust contains byproducts of the mining activities, including rock and heavy metals. Similar to dust from coal mining operations in Appalachia, the particles are inhaled by those who work or live near the cobalt mines. Long-term exposure to this dust can lead to a fatal lung disease, referred to as "hard metal lung disease," as well as other respiratory complications such as asthma and decreased pulmonary function.[32] Studies have found high concentrations of heavy metals in children's and others' blood and urine who live and work around mining facilities,[33] and high rates of babies born with birth defects.[34] Cobalt exposure can also cause skin irritation. Yet workers are often without gloves or protective respiratory equipment, and the DRC's 2002 mining code and 2003 regulations do not provide guidance for miners on safety procedures.[35]

The artisanal cobalt mines in the DRC are also marked by dangerous work conditions and a lack of safety precautions.[36] Unlike the industrial operations, artisanal mines do not always have set plans backed by engineering and safety considerations. One report from 2015 of a mining operation follows the video recording taken by a miner:

> In one house, researchers saw a mine shaft leading down through the floor of a room. The owner said that he did not want to start a mine, but said he felt he had no choice when his neighbours started tunnelling under his plot of land . . . His footage shows a narrow shaft that descends deep underground, which the miners said was more than 30 metres deep. The miners climb down with bare feet, holding on to small ledges that have been cut out of the side. At the bottom of the pit, the miners have dug a tunnel that follows the heterogenite seam as it curves, rises and falls. At several points the tunnel meets those of neighbouring teams—the miners told researchers that they worry that at night when their mine is unguarded, their neighbours sometimes break into their tunnel to steal their ore. In some places, the tunnel is wide and tall so that the men can stand upright; in others it is extremely narrow, and they have to crawl. The tunnel is not held up by supports, but the miners say they know how to dig through the rock so that it does not collapse.[37]

Yet, recent reports have documented tunnel collapses, including a November 2021 collapse that was kept hidden from the international media, as an investigative journalist explained: "I tried to investigate the accident, but soldiers had already blocked access to the mine. The truth of what had happened must never be revealed, for it directly contradicts the assertions by consumer-facing tech and electric vehicle companies that their cobalt supply chains are clean."[38] Others have also documented high levels of radioactivity in surrounding cobalt mining areas, since cobalt tends to be mined in the same regions as uranium.[39]

Beyond the environmental and public-health aspects of artisanal cobalt mines, a significant body of research and field work reveals that they are also often sites of child labor and human rights abuses.[40] Artisanal miners within the DRC sit at the bottom of a chain of hierarchy, and are regularly exploited, if not worse, by Congolese police, political leaders, and other members of society.[41] In multiple investigations from Amnesty International and Afrewatch, the organizations have reported the presence of significant child labor, a lack of safety precautions, and human exploitation, as well as a lack of corrective action from DRC government officials.[42]

The DRC government neither provides social services to those who work in and around the mines nor sets clear, enforceable health and safety regulations. The regulations are defined by one scholar as an "assemblage of heterogeneous government techniques"[43] that span more formal, corporate operations; authorized artisanal extraction; and unauthorized extraction. The cobalt industry within the DRC is still relatively new and thus immature in its rules and regulations, market functions, and pricing transparency.[44]

Approximately 110,000 to 150,000 artisanal miners work the cobalt mines in the DRC.[45] A significant fraction of these workers are children and women. Children often work to help support their families and to cover their education expenses. A 2014 UNICEF report estimated that about forty thousand children worked in mines—including but not limited to cobalt mines—across the southern DRC. Children pick through the rock and rubble for those chunks that may still contain cobalt, and they also help wash and prepare the ore and material before it is sent to the buying houses.

Children often work long days, typically around twelve hours when not in school, with some children not in school working up to twenty-four hours at a time.[46] Children are typically paid about $1–2 per day. Although the cobalt is usually paid by the pound, when children deliver large sacks of cobalt to the traders, they have to simply accept what they are handed.[47] In addition to working long hours for little pay in precarious conditions, reports from human rights organizations have uncovered other abuses. Field investigators from Amnesty International and Afrewatch found that children were beaten by security officers for trespassing in these industrial mining territories.[48]

By one estimate, approximately 25 to 50 percent of artisanal workers across Africa are women.[49] Studies of cobalt mining communities have found the presence of both sexual exploitation of women and incidence of prostitution.[50]

Women and children are not the only ones to experience exploitation at these mining sites. Several site visits by researchers and organizations through the years have confirmed that artisanal miners, including the male miners, experience significant exploitation by bosses and the police. During a visit to an artisanal mine by Amnesty International, the visitors found officials serving as mining police or security, who required payment from the visitors and also payment from the miners for each bag that they carried out of the site to deliver to traders.[51] Benjamin Sovacool described conditions in DRC cobalt mines as a form of modern-day slavery: "Miners have such few rights in the DRC they are treated little better than slaves. They toil and work their youth or adulthood away for the benefit of some government official, some mining firm, or another family member. It's a new form of African slavery, a subterranean slavery."[52]

WHAT'S NEXT FOR MINING AND AUTO COMPANIES?

Automobile companies in the United States as well as elsewhere across the world are not oblivious to the tragic human health conditions surrounding cobalt mining. Companies like Ford are spending billions to develop their own battery plants in America that rely on technologies other than cobalt, such as lithium iron phosphate. GM is partnering with LG Chem to build a battery factory in Ohio to manufacture their Ultium batteries, which require less cobalt; and Volkswagen, BMW, and Ford are investing in companies that make solid-state batteries.[53] In recent years, scientists have made advancements in a variety of cobalt-free alternatives, including those that replace cobalt with a combination of other elements, use organic materials, and use other techniques such as chemical doping. One leading technology is the lithium-ion phosphate battery, which is actually used in several modern EVs yet still has a lower energy density than lithium-ion batteries using cobalt as of the time of writing.[54] Automakers including

GM, Volkswagen, Ford, Volvo, and Toyota are also investing in companies that recycle batteries, such as Li-Cycle and Redwood.[55]

The US government is also in significant pursuit of alternatives to cobalt that can both be procured on domestic land and generate a boon to US industry. With grants from the Bipartisan Infrastructure Law, the Department of Energy awarded $2.8 billion to US manufacturers that are part of the electric-vehicle supply chain. That funding, along with the American Battery Materials Initiative and Defense Production Act (DPA) as well as electric-vehicle content requirements in the Inflation Reduction Act, aims to bolster US production of critical minerals and reduce reliance on Chinese production. The Inflation Reduction Act dedicated an additional $500 million to the DPA for domestic mining and $40 billion for the Innovative Technologies Loan Guarantee Program to support US production of critical minerals.[56] The DPA authorized the Department of Defense to process critical minerals, such as lithium, nickel, and cobalt, through domestic sustainable mining and to invest in other elements of production.[57] The CHIPS and Science Act more broadly dedicated $40.5 million to establish a nonprofit Foundation for Energy Security and Innovation, which will work with private-sector companies to support innovation in US energy production.[58]

Though the automotive manufacturing sector is seeking to move away from cobalt use in EVs, the extraction processes of many other critical minerals is fraught with similar labor, geopolitical, and legal concerns.[59] As a result, there is growing interest in expanding US domestic mineral extraction. Several of potential sites of mineral extraction provide immense opportunity for future energy developments. Idaho mining company Perpetua Resources has an Idaho mine that yields antimony and is said to have the potential to power a million homes with battery storage and solar farms. Reserves within the Santa Rita Mountains in Arizona, home of the Tohono O'odham, Pascua Yaqui, and Hopi people, also contain significant deposits of copper.[60]

Yet these pursuits are not without their own set of complications. Most critical minerals within the United States are located on native lands. One estimate puts 97 percent of all nickel, 89 percent of all copper, 79 percent of all lithium, and 68 percent of all cobalt reserves within a thirty-five mile radius of Native American reservations.[61] This significant potential for new sources of battery components comes with potential costs, especially to local environments, by threatening and yielding damage to Indigenous communities' water resources, land, wildlife and native species, and culturally sensitive places, such as sacred burial and ceremonial sites.

The move to produce domestic minerals for batteries thus has the potential to recreate the types of disproportionate burdens that have come with past and current energy systems, especially burdens within extractive economies. Moreover, they create similar trade-offs for communities that must decide between access to jobs, and well-paying ones at that, and possible sacrifices to their land and environment, in the same way that they have faced trade-offs between coal, gas, and uranium jobs and preservation of their native lands. These same trade-offs characterize the ongoing situation in the cobalt mining operations of the DRC: On the one hand, cobalt mining provides an important resource for a clean energy future; on the other hand, it comes at the expense of the environmental, social, and public health of those who live and work around the mines. As the demand for these commodities increases, these tensions—in Africa as well as elsewhere across the world—will become even more acute.

As a means to mitigate the potentially negative social and environmental consequences of mining, efforts are underway to develop strategies to maximize sustainable principles. Operating on an international basis but headquartered in Seattle, Washington, the Initiative for Responsible Mining Assurance was founded by a coalition of nongovernmental organizations, mining companies, affected communities, businesses, and labor unions to assess social and environmental performance at mines using an internationally recognized standard.[62]

The United States may also learn from its northern neighbor. In Canada, the national government released *The Canadian Critical Minerals Strategy* in 2022 as a guideline for how Canada will use its mineral resources in coming years. The strategy contains five objectives: 1) Support economic growth and competitiveness; 2) promote climate action and strong environmental management; 3) advance reconciliation with Indigenous peoples; 4) foster diverse and inclusive workforces and communities; and 5) enhance global security and partnership with allies.[63] These objectives advance holistically beneficial principles for encouraging investment and trade while creating high-paying, high-value jobs that ensure participation and benefit to Indigenous peoples while upholding human rights for all. The strategy promotes environmental safeguards from a local to a global scale, and it intends for actions to enhance international cooperation and advancement of shared ideals and best practices.[64]

Case 2: Automobile Supply Chains in the Industrial Heartland

Auto companies span the world, but they tend to concentrate their production in certain regions. In China, for example, most auto companies reside in a few counties, such as Jilin, Hubei, Shanghai, and the Yangtze River delta.[65] In the United States, a historical hub of the auto supply chain—from component parts production to assembly—are located in the three-state region of Michigan, Ohio, and Indiana. In the 1980s, this region expanded to include more southeastern states such as Kentucky, Alabama, Texas, and the Carolinas, due to both domestic and foreign companies (e.g., BMW, Volkswagen, Mercedes, Toyota, Honda, and Nissan) seeking lower labor costs associated with nonunion shops and right-to-work policies in these states.[66] Broader trends of globalization and offshoring of labor and capital costs have also affected the geography of the US auto supply chain, mainly by expanding the geography well beyond the United States. The North American Free Trade Agreement (NAFTA), as agreed to by the United States, Canada, and Mexico in 1994, also had a tremendous impact on the auto industry's labor, costs, and manufacturing processes, as it shifted production from the United States to Mexico.[67]

Many of the communities in these Midwestern and Southern regions have a complicated history with auto companies, not too dissimilar in some ways from the experiences of fossil fuel communities discussed in chapter 3. When the companies are doing well, and specifically when the cars and trucks that they assemble are in high demand, the companies can hire and pay good wages to a large number of employees from surrounding communities. These employees, in turn, support other local businesses such as restaurants and convenience stores. When demand for a specific vehicle declines—or even in anticipation of future changes in consumer preferences—auto companies must decide whether to switch production to another vehicle or to close the plant entirely. Switching may be as easy as running a new model down the assembly line, but it may also require retooling the factory, during which time assembly workers may lose hours or be furloughed.

When factories close, or when some of the workers are laid off, not only are the workers affected, but so too are the surrounding communities. Workers may choose to relocate to another facility within the same company, uprooting their families in the process, or they may look for new positions in the local region. Yet without a robust employer that serves as an economic anchor, surrounding businesses will suffer and more companies will inevitably close. With failed visions of job security and economic development come resentment and disappointment. The histories of both Detroit, Michigan, and Lordstown, Ohio, two of the country's most commonly recognized automobile hubs, exemplify these trends.

DETROIT, MICHIGAN: "WE HOPE FOR BETTER THINGS; IT WILL RISE FROM THE ASHES"

Detroit has had a long and checkered history with automotive manufacturing. The city's relations with the auto industry began in 1903, when Henry Ford started Ford Motor Company. Even though Ford eventually moved its facilities out of Detroit, the city still rapidly became a car manufacturing hub. General Motors, which was founded in 1908, moved their headquarters from Flint to Detroit in the early 1920s. Cadillac, a subdivision of General Motors, consolidated plants that were scattered throughout the state into one large plant in Detroit in 1921. Four years later, Chrysler Corporation purchased an automobile company named Chalmers Automobile Company. Two years after that, they built a large manufacturing plant in Detroit.[68] With this plant, the companies that soon came to be known as the "Big Three"— Ford, GM, and Chrysler—made Detroit the center of American automobile manufacturing.

Despite the Great Depression, the Big Three still grew over the next decade. By the end of the 1930s, Chrysler was producing over fifty thousand cars per year and employed over ten thousand people.[69] After World War II, the population in Detroit reached nearly two million, making it one of the largest cities in the country, with approximately 296,000 manufacturing jobs, many of those in the auto industry. Detroit also became a destination for many African Americans seeking to escape the racism and lack of economic opportunity of the Jim Crow South.[70]

Despite the closure of the massive 3.5-million-square-foot Packard Motor Co. plant in 1958, the 1960s was a strong decade for the Detroit auto industry. The Big Three market share continued to rise, reaching 85 percent of the American market by 1968. United Auto Workers (UAW), the union representing workers for the Big Three and other auto companies, also peaked in the same year at 1.4 million members, making them an extremely powerful union.[71] In the early 1970s, UAW flexed its muscle in labor negotiations with the Big Three. Over four hundred thousand workers went on a sixty-seven-day strike against GM and landed the biggest settlement to that point in the history of the industry.[72]

Both the 1973 oil crisis and the 1978–1979 energy crisis were disastrous for US automakers and also for Detroit. The 1973 crisis encouraged the development of smaller, more fuel-efficient vehicles mostly designed in East Asia and Europe, and the subsequent 1978–1979 crisis brought the industry to its knees.[73] In particular, consumers began to view smaller, fuel-efficient Japanese-built automobiles as superior to similarly sized US-made cars in terms of reliability, comfort, and economy. These perceptions were likely shaped by high-profile failures of several of the Big Three's small, economically focused cars of the 1970s—including the catastrophically rust-prone Chevy Vega and the incendiary Ford Pinto.[74] The advantage enjoyed by Japanese automakers became so strong that the Reagan administration and the Japanese Ministry of International Trade and Industry reached an agreement in 1981 to restrict the number of vehicles exported to the American market.[75] Japanese auto manufacturers responded by building plants in North America to avoid these voluntary export restrictions. The fierce competition led to near-bankruptcy for Chrysler, forcing it to accept a $1.5-billion government bailout in the form of loan guarantees[76] and to lay off over one-third of its workforce and close twelve of its factories.[77]

The recession in the early 1980s only exacerbated problems for the auto industry, leading to massive layoffs across the Big Three. By 1986, only half the displaced workers had been able to find work, and those who found work had on average 30 percent lower wages.[78] This state of affairs was compounded by racial tensions throughout the city that had risen since the 1960s, and by 1980 Detroit had 35 percent fewer people than it had had in 1950, mainly due to "white flight." Leaders had been unable to assuage racial tensions, despite numerous mayoral candidates making it a main campaign pledge.[79]

In 1990, Chrysler closed one of its first plants, the Jefferson Avenue Assembly line. Although the company replaced this plant with a new one, the new plant required half as many workers.[80] Even in a decade of relatively uninterrupted growth, Detroit still lost well-paying auto industry jobs to offshoring and automation. Further employment reductions occurred during the 1980s and 1990s as each of the Big Three spun off parts manufacturing and outsourced production and support services, with some capabilities combined into new stand-alone companies.[81] Simultaneously, Japanese, South Korean, and German automobile companies moved into the United States and primarily settled into the Great Lakes region and the South. This relocation was accompanied by extensive anti-union campaigns and employee plans and benefits that challenged US automobile companies. This economic pressure from all sides continued into the first decade of the 2000s. In

2007, following a brief strike, UAW negotiated a new contract for workers. UAW membership had steadily dropped since the 1970s, so they had a weakened position in negotiations. As a result of this weakened position, new hires became subject to a two-tier wage structure and no longer received retiree healthcare, and legacy retiree workers were now part of a Voluntary Employee Beneficiary Association as opposed to receiving pensions from the company.[82]

The next year, the Great Recession hit the Big Three hard, and thus it also hit Detroit.[83] President Obama signed a stimulus package that included $80 billion in federal aid for GM and Chrysler, but both companies declared bankruptcy. GM reconsolidated with over thirty-four thousand fewer employees; Chrysler eventually merged with Fiat to become FCA, before becoming Stellantis a few years later. UAW attempted to negotiate new labor contracts with the Big Three in 2009, which resulted in wage freezes, no cost-of-living adjustments, and the loss of job banks for workers.[84]

Autoworkers' desires to reconcile the concessions made during the Great Recession as well as rising concerns that auto companies would begin seeking nonunionized labor for electric-vehicle and battery manufacturing led most recently to the 2023 UAW strikes. The resolution of the strikes, consisting of a five-year agreement reached in November 2023, included greater employer 401(k) contributions and higher wages—estimated to be an approximately 33 percent increase—including cost-of-living adjustments for longer-tenured workers.[85]

The historical struggles of the auto industry coincided with the decline of Detroit. At present, over one-third of Detroit's population lives in poverty, while over one-fourth of residential properties remain vacant and about one-fifth of the population remain unemployed. The small tax base has deprived Detroit of essential tax revenue, and the city has difficulty providing basic services.

Electric vehicles bring both hope and challenges to Detroit, abiding by the city's motto: "We hope for better things; it will rise from the ashes." The production of electric vehicles is growing rapidly in both the United States and the world at large, and some of the acceleration is happening in Detroit. Jeep recently opened the Mack Detroit

Assembly Complex, the first new auto factory in Detroit in thirty years. The factory was expected to bring five thousand jobs to the area, hiring mostly Detroiters, to build the electrified Grand Cherokee. The American Rescue Plan included a $52 million grant for UAW and the Big Three to collaborate on training workers and upgrading factories to become better suited to produce electric vehicles. Battery manufacturers are also coming to Detroit; in 2022, Governor Gretchen Whitmer announced two new electric-vehicle battery projects, providing nearly $4 billion in capital investment and over four thousand high-paying jobs.[86] The Inflation Reduction Act of 2022 also identified Michigan as a worthy source of EV-related investments, with about $19 billion earmarked for the state as of early 2024.[87]

However, electric vehicles are not necessarily Detroit's silver bullet. Volkswagen expects that electric vehicles will require 30 percent fewer workers to manufacture, as they require simpler parts than internal-combustion-engine vehicles. Ford announced three thousand layoffs in August 2022, despite their push for electrification. Additionally, Detroit would need to drastically alter its infrastructure locally to provide sufficient EV charging stations for Detroit residents to be able to even purchase and operate their own electric vehicles, a herculean task for a city that is billions in debt.

As goes the auto industry, so goes Detroit. The two have been intertwined since the turn of the twentieth century. The recent growth in electric vehicles represents a potential boom for the city, albeit one with risks. The city remains massively indebted, but residents are working closely with the statehouse to reduce the debt. While some pointed to Detroit as the prime example of the 2008 Great Recession, many believe that Detroit can re-emerge. As *National Geographic* put it: "Tough, Cheap, and Real, Detroit is Cool Again."[88]

LORDSTOWN, OHIO: "RIDE WITH LORDSTOWN"

Lordstown, Ohio's history with the auto industry is shorter lived than Detroit's, but it is just as complicated and intimate. A suburb of Youngstown, Ohio, Lordstown is a community of just over three thousand people. Lordstown is about an hour away from Cleveland and Pittsburgh and, like those two cities, has been a home to many industries. In 1964, GM built their Lordstown Assembly complex in the city, making GM and Youngstown Sheet and Tube the city's two biggest employers. Just six years after the completion of the Lordstown Assembly factory, the 1972 UAW autoworker went on strike for seventy-two days, part of the national GM strike described earlier, ultimately leading to pay raises and increased benefits for the factory workers.[89]

The post-strike honeymoon for Lordstown, however, did not last long. On September 19, 1977—a day that would be dubbed "Black Monday"—the Youngstown Sheet and Tube factory closed. This closure led to five thousand workers losing their jobs, and over the next five years a total of fifty thousand workers were laid off from factories in surrounding areas. GM's Lordstown Assembly factory remained open and provided jobs for some of the laid-off workers, but many of the retraining programs for non-factory work were unsuccessful.[90]

Lordstown Assembly survived the wave of automation in the 1990s as well as the Great Recession and subsequent impact on the auto industry, which included GM declaring bankruptcy and an employment decline from 10,600 workers in the early 1990s to 4,500 workers in 2017.[91] There were some warning signs of economic downturn, so the union agreed to $118 million in annual concessions to keep the Lordstown plant open.[92] However, on November 26, 2018, just eighteen months later, disaster struck. Management called union representatives to their offices early in the morning and informed them that the plant was going to be "unallocated" (a word not even recognized in the *Oxford English Dictionary*). In other words, GM was closing the factory.[93] On the same day, GM closed four other factories throughout the United States despite GM posting an over $8 billion profit in 2018.[94] Union leaders were given fifteen minutes to absorb the news before they were sent to inform the workers of the plant's impending closure.

With the "unallocation" of the factory, the last pillar of support for Lordstown fell. The remaining workers suddenly lost their jobs, and a study from Cleveland State University predicted that the closure would cost the surrounding area eight thousand jobs in total and

$8 billion in economic activity. Lordstown lost $3 million in tax revenue, over one-third of its annual total, and surrounding areas suffered similar economic impacts.[95]

Workers described heartbreaking consequences of the factory closing in an interview with the *New York Times*. One woman described how the factory closing meant she had to sell her family home, breaking a promise to her recently deceased parents. Another worker lamented that they would have to move up to nine hours away from Lordstown while their partner had to stay in the city, thereby separating the family. Multiple workers who had worked at the factory for over twenty years described how their loyalty to GM was ignored and how they felt like GM viewed the situation as a "game." Children of the factory workers also felt the situation. A middle school student said, "I was kind of scared and sad when I found out about the plant. . . . We know everyone . . . I'm scared about moving, but I'm trying to be brave about it. . . . I wasn't o.k. at first. I don't want to go. All my friends don't want me to go, and I don't want to leave friends."[96] Interviewees had different reasons for their analyses, but they all agreed on the same thing: GM "unallocating" was an act of betrayal with disastrous consequences for the Lordstown community.

GM sold its Lordstown plant to a startup called Lordstown Motors for $20 million and loaned the company another $40 million.[97] Lordstown Motors invested an additional $12.2 million into the facility with the goal of building electric trucks.

Yet Lordstown Motors has been plagued with problems. Originally slated to begin sales by the end of 2020, the COVID-19 pandemic resulted in a delay of manufacturing. In June 2020, founder and CEO Steve Burns announced that there were twenty thousand pre-orders for the flagship truck, the Endurance.[98] But after further delays, a much-awaited public test run of the Endurance in January 2021 proved disastrous, as the car caught fire ten minutes into the journey.[99] The explosive test run was not disclosed until a March report titled *The Lordstown Motors Mirage: Fake Orders, Undisclosed Production Hurdles, and Prototype Inferno* revealed many concerns about the Endurance, including severe design flaws and falsely inflated sales numbers. This revelation led investors in the now-$4 billion company to file a class action lawsuit claiming that Lordstown Motors had no sellable product. Two months later, Burns resigned from the company.[100]

In November 2021, Foxconn signed a $230 million contract for the Lordstown Motors assembly plant, giving many hope that Lordstown Motors might still survive. In addition to purchasing the plant, Foxconn also agreed to a contract manufacturing agreement to aid in the production of the Endurance.[101] However, less than two years later, the deal fell through over disagreements of exactly how many shares of Lordstown Motors stock Foxconn would buy. The deal's failure led to Lordstown Motors declaring bankruptcy and going up for sale in June 2023.[102]

Hope is not lost for Lordstown. GM's battery manufacturing partner, LG Chem, announced a $2.3-billion battery factory in the city in December 2020, after GM promised to bring jobs back to the area. The factory, which will produce Ultium Cells, is aiming to produce 30 gigawatt-hours of batteries each year, making it one of the largest battery factories in the world.[103] The project was slated to begin battery production in early 2023, and provide 1,100 jobs, many of them reserved for UAW members.[104] Although Lordstown has taken serious blows over the past few decades, "nine-lives Lordstown" looks to return to its manufacturing prominence.[105]

THE AUTO TRANSITION WITHIN DETROIT AND LORDSTOWN

On the eve of the transition to electric vehicles in automobile factories across the world, we conducted focus groups and interviews with autoworkers and community members in Detroit and Lordstown, respectively.[106] Our conversations focused on their impressions of the future of the industry, their jobs, and their communities in light of the transition to electric vehicles, yet their impressions of and lived experiences during the regions' complicated histories with the auto industry shadowed all discussions. The stories that we heard were laden with feelings of mistrust, abandonment, and unreturned loyalty, and they reflected a long—and in some cases multigenerational—history with their employers.

Dan, an autoworker in Detroit, experienced enough plant changes through the years to be able to reflect on those changes' impacts on the surrounding communities, both in the glory days of plant openings and after the heartbreak of plant closures. He told us:

> As one community falls apart, where they locate the battery plant like in Tennessee, you may see a lot of people moving from where the plant closed to new opportunities, especially if they're younger, and they can do that. They'll move and try to get into those new plants and now you have new roads, new school. As some of those plants take five thousand, six thousand people and all the other stuff around that, which is easy another four or five times that. To support that, the assembly plant. Then a whole new communities could rise up or they definitely expand. Housing prices will go up, right, at those places, just like the housing prices go downward where everybody left. I mentioned last time but I also worked at Clark Street plant in Detroit. When they closed that plant down and the body shop or the plant that made the vehicle bodies. You see what happened to all those businesses and the communities around them. I've driven by there and they're ghost towns now. That's unfortunate.

Rochelle is a former GM employee who lives south of Lordstown. In our interview with Rochelle, similarly to Dan, she reflected on the area around the former GM plant and the impact that the closure of the "golden egg" plant has had on the community:

> Directly in the vicinity of Lordstown I haven't had reason to really go up there too much anymore but when I have I've driven through to get to Warren or whatever. It's obviously a lot of business closures, barbershops, and some of the little local stores that were there that I could stop out on the way to work to pick up a bag of chips or some lunch or something and they're just not there anymore. I've seen in my area in particular the unem-

ployment rate is really high and a lot of houses for sale that are still for sale, not too many. I mean, some of them have sold but I see some brothers and sisters that are still waiting for their houses to sell and they've been out of the state for two years now. And I think that area's real estate has been really affected by that and there's just not a lot of economic growth in spite of the fact that General Motors is now building a battery plant up there . . . I'd say that the Youngstown area and the Warren, Trumbull County area, it's rough. A lot of poverty never recovered from the steel industry shutting down and General Motors was our golden egg and now it's gone.

In some locations, these conditions are experienced more suddenly with the closing of a plant. In others, the loss of employment and economic livelihood is more gradual, a slow drip as the company nears time to fully shut down a plant. One Lordstown autoworker with whom we spoke discussed how this process played out in Lordstown when the GM plant closed:

> I think it was probably more sudden, I think that happened over to, you know, that, I mean the plant closed, it has a date on its closing date, but you know, they had been cutting shifts prior to that years. So for the last five years prior to that plant closed and they started playing the game of, you know, eliminate the plant. So they were trying to force people to move before the last year. So I'm sure all of that had an effect.

This thorny history, and workers' uncertainty about whether they can expect to retain their jobs or need to pick up and move, or whether they can even have enough confidence in their job security to buy homes, has bred long-standing mistrust among both autoworkers and community members. Lori, a former GM autoworker who lives in Detroit, responded to our questions with outrage:

> One of the things that just gets me is, they rip apart our community and they divide families . . . And then over the holidays, they run a goddamn

ad about how General Motors is all about families. Liar, lar, liar. I had to grab the remote out of Dennis' hand and change the channels. I could not even deal with that.

These sentiments have evolved through decades of automobile companies' boom-and-bust cycles. In the last several decades, new challenges associated with automation in the assembly of vehicles and their components have exacerbated these sentiments, since automation has contributed to an approximately 37 percent decline in the automobile workforce since 1990.[107] These trends are occurring simultaneously to changes in workforce composition, specifically less reliance on unionized workforce and a greater proportion of autoworkers carrying hourly, temporary, and lower-paid positions.[108] A community member in Flint, down the road from Detroit and also in a community with a long, difficult history with the auto industry, told us that these various changes through the years had inflicted post-traumatic stress disorder (PTSD):

> We've been devastated. We've been suffering from PTSD since the '70s. Different things, the shop shutting down, people going from seven-day work week twelve hours a day, to five days a week eight hours a day. Men losing homes and vehicles and et cetera, families. It went real deep and still there's only like ten thousand jobs here in Flint with General Motors or whatever, the three main manufacturers. So there are still some healing that needs to be done. And I don't know that some are ever going to receive it.

The newest challenge facing these communities is the transition to electric vehicles. Electric vehicles require many different components than traditional internal-combustion-engine cars and trucks. Internal combustion engines contain an engine, a gas tank, a crankshaft, smart plugs, a catalytic converter, and a generator, among other parts. The transmission system is highly complicated and contains its own set of parts. The electric vehicle has fewer moving parts and contains a motor, battery, charger, and power control unit. Estimates vary on the difference between them, but one source puts the internal combustion engine at 1,200 components, in contrast to the electric vehicle's two hundred components.[109]

Fewer parts mean fewer workers. While researchers disagree about the cumulative effects on employment for the auto industry, many studies find that expanding electric-vehicle production will require a contraction of the assembly workforce. Companies such as Ford and Bosch predict that manufacturing needs will contract by about 30 percent.[110] Others put this number much lower,[111] or about equal,[112] which demonstrates the considerable uncertainty about the impacts of electric vehicles on the workforce. Policymakers are clearly concerned about the possibility of lower job numbers, however, and have tried to combat this possibility with domestic sourcing and content requirements to offset job losses.

It is undeniable, however, that electric vehicles will fundamentally change the way in which cars are made and who makes them. The different components of electric vehicles—such as their semiconductors and batteries—means that they will require different supply chains as well as different skill sets to manufacture and assemble. Electric vehicles also have a simple motor, one that can more easily be made by a robot than an internal combustion engine can.

All these changes, and the unknowns about them, have understandably created concerns among autoworkers that their jobs will soon become obsolete, concerns that arguably fueled the fire of the 2022–2023 UAW strikes. Even those who are more familiar with upcoming plans for the factories express concern about the future for autoworkers. Tom, a Detroit autoworker for over three decades, gave us his take on how things would change with the impending electric vehicle transition:

> So I'm very familiar with all the plants where some of the transition going on and it's pretty dramatic. A lot of plants are going to get totally retooled to accommodate electric. There's safety, there's . . . The battery packs will be built outside and they'll be brought in but they're huge and they're very heavy. But so are engines and transmission, so there's a lot of swapping

from engine, gas stuff to the electric. As it's going to change how the vehicles are built dramatically and people too. They're not going to need, at least to assemble. An engine will be assembled, the hard part, and the block, and so on at an engine plant. But the vehicle assembly plants dress the engine, they put all the power steering, the alternators. They do the wiring, attach it to the transmission at the vehicle assembly plants. Well, that's all going away with electric vehicle, there'll be electric motor at the wheel, between the wheels. The battery pack will come in and it gets lifted up and connected with just a few connections. So I think assembly it's going to be really a major change for the workers, where the workers work. I think they're already looking at like 30 percent, 40 percent less labor in this vehicle assembly plant as a result of electric vehicle. So the work or load will be shifted to the motor plants and the battery plants.

Autoworkers and community members anticipate this shift in production, and the possibility that their jobs will disappear. Yet, they also believe that they should be included in the transition. At the heart of this conviction is a sentiment that they have more than proven their loyalty to the companies through years of hard work on the factory floors, weathering the ups and downs of the industry. As a Lordstown factory worker, Pete, explained to us:

> There's no more big gains in fuel economy. So we have to find a different fuel system. And I think electric, I would have liked, you know, I have personal feelings that, you know, I felt I earned instead of the people I worked with on, the right to build that next car. That's where I am on it. I mean, I earned it. I have knees that don't work now because of it because I gave my life to that company.

The majority of autoworkers with whom we spoke in Lordstown and Detroit expressed this same skepticism about whether they would be the ones to build the next generation of electricity powered cars. After a long history of feeling burned by their employers, many assume that they will be left behind in this transition. Further, many of them also commented that, even if they remained employed and were able to assemble electric vehicles, they would still not be able to own one due to its high cost and the lack of charging infrastructure in their communities.

Phil lives in Lordstown, where he has observed the GM plant for decades. With new investment in the old GM facility for electric-vehicle manufacturing, he expressed cautious optimism about the potential impact on the community, but that optimism was shrouded in distrust from decades of feeling neglected:

> There is still a bitterness with respect to what happened with the Chevy Cruze. That vehicle had three turns, they were producing those cars and selling those cars at breakneck speed. And there is an undercurrent, a sentiment that for whatever reason, the company made the decision at some executive level to transition that car off of their portfolio services. And there's a sentiment that occurred through advertising and promotions and incentivizing the sale of those cars because let's face it, that's what moves cars, incentivizing and advertising and in driving the market. So when the company made the decision to close that plant, it burned a lot of families, it ripped families apart. My brother-in-law works up in Detroit and my sister-in-law lives two miles away from me. She commutes up to see her husband. Her kids has just graduated from high school, so it's had a tremendous impact emotionally, financially and whatnot on those families. So when we look at Lordstown Motors and LTM, there's excitement to match credit. But I think there's also a skepticism that we've been burned so many times in this area.

These concerns expressed by autoworkers from Detroit and Lordstown are not too dissimilar to those expressed by coal miners in Appalachia. These concerns include feelings of abandonment by decision makers and the rest of the country, and unreciprocated loyalty to both their company and its products. This loyalty was built on the workers' history—both through their own history of employment as well as the generations of those before them—and on sacrifice, a sacrifice of their bodies (i.e., their knees, their backs) and their mental health. And this loyalty, as perceived by both autoworkers and coal miners alike, in their view should at the very least lead government and industry employers to help them be a part of the next thing, the next wave of the energy transition.

Case 3: Who Owns the Electric Car?

While the supply side of electric vehicles is fraught with histories of disenfranchisement and abandonment, the demand side also has its own set of complications. In short, electric vehicles are not accessible to everyone, and it is often those who could benefit the most who often have the least access.

The most common profile of an electric-vehicle consumer is a college-educated male, with higher-than-average income, and who already owns other vehicles. Picture the Gen-Xer in the Leaf commercial who received the giant polar bear hug. They tend to have enough disposable income to afford an expensive electric vehicle, as well as to keep their other vehicle that they use for longer distant trips beyond their electric vehicle's range.

Research reveals that Black, Latino, and low-income households are the least likely to own electric vehicles.[113] As of 2016, 90 percent of all electric-vehicle tax credits were claimed by households in the top income quartile, which comprises those who make over $141,000 per year.[114] In a California study, the mean income of electric-vehicle buyers was $190,000. In addition, 81 percent of them owned their homes and 81 percent were college graduates.[115] In California, researchers found that although 39 percent of the population is Latino, only 10 percent of all EV owners were Latino. In the United States, 14 percent of the population is Black, yet they only represent 7 percent of EV owners.[116]

For consumers, electric vehicles still come with important drawbacks, such as limited range, long charge times, and high up-front costs. While some models have significantly reduced these barriers over time, what is more important for electric-vehicle sales and deployment is the perception of these barriers—and perceptions have not changed dramatically since 2010.[117] These drawbacks are particularly pronounced for low-income users.[118] The price of an electric vehicle tends to be several thousand dollars more than those of comparable internal-combustion-engine vehicles. The range of an electric vehicle is often too short, or perceived as too short, to get drivers where they need to go without triggering "range anxiety," which is a fear of getting stranded on the side of the road with a dead battery.[119] Limited public charging stations across the country contribute to range anxiety. Finally, the time it takes to charge one's vehicle can prolong travel times during long-distance trips and potentially commutes to work.

While these factors may be a nuisance for some potential car buying consumers, these drawbacks can be particularly prohibitive for low-income and otherwise disadvantaged consumers. The up-front cost, for example, makes an electric vehicle completely unobtainable for most low-income households. And while tax credits can help bring down the cost of the car, low-income households are typically not able to take advantage of them because their tax liability is not high enough to recover the full credit. The tax credit offered in the United States between 2010 and 2022 was up to $7,500 for an electric vehicle.[120] In order to carry tax liability rendering them eligible for this much of a credit, a taxpaying household required an income of $66,000 per year for an individual and $91,000 for joint filers.[121] This tax credit—unlike an on-the-spot rebate[122]—also necessitated that a buyer have the money at point of sale and then be able to wait until they filed their annual taxes the next year. If a buyer were unable to pay that full up-front cost, then they would need to access financing, which might also be difficult since many low-income buyers are credit constrained and thus likely to be denied such financing.

Low-income households are more likely to purchase used cars.[123] Yet it is less common for tax credits to apply in the secondary automobile market. Policy in this regard changed in the United States with the passage of the Inflation Reduction Act in 2022, which added incentives of up to $4,000 for the purchase of a used electric vehicle or 30 percent of the original sales price, whichever is lower.

One can temper range anxiety by installing residential chargers or living or working near a public charging station. Yet not everyone has access to a garage or parking space where a charger can be installed. Low-income households are more likely to occupy rental units, meaning that potential electric-vehicle buyers must rely on property landlords to install the charging units. This challenge is especially pronounced with larger multifamily dwelling units, which may not have the physical space to add a charger. These households then must rely on public charging facilities, which are not yet plentiful in most parts of the country.[124]

For these and other reasons, Black and Hispanic households are significantly less likely to have access to public charging stations. One study found that, even after controlling for income, location relative to highways, and the presence of multifamily housing, census blocks in which the majority of residents are white are 1.5 times more likely to have public chargers than census blocks in which the majority are Black or Hispanic.[125]

In our research, we have found that designing new innovations like the electric vehicle for the populations with the most barriers to adoption will, over the long term, help the technology diffuse more widely. We heard these arguments in interviews we did in 2021 with experts working on making electric vehicles more accessible to traditionally disadvantaged populations. One expert explained that companies typically design products to appeal to customers in the "the middle of the bell curve." Yet, in this expert's opinion, when working on products that require mass adoption, like the electric vehicle for decarbonization objectives, "designing towards the use case of people, of residents, of users who might have different or higher needs . . . if you can figure out how to serve that community first, it'll benefit everyone."[126] Another expert explained: "If we're not considering all communities which include frontline and low-income communities into products or planning or anything related to electric transportation, then . . . as the market develops, it's going to miss them because they weren't involved up front."[127]

Case 4: The Burdens of E-Waste

When someone buys a new cell phone and throws their old one away, or when their hybrid or electric car eventually dies, these discarded products produce electronic waste, or "e-waste" for short. A 2020 United Nations report revealed that the average American creates about forty-four pounds of e-waste annually, compared to the global average of just over sixteen pounds. Most of this e-waste comes from computers, refrigerators, and air conditioners.[128] This number is expected to grow in coming years.[129]

E-waste is typically stored in landfills. Most electronics contain toxic chemicals and metals, which can seep into the groundwater when stored without some sort of impervious ground-protection layer. Of the forty-four pounds of e-waste that the average American produces each year, just 20 percent is disposed of in domestic recycling facilities, with the rest sent overseas to countries with more relaxed disposal laws, like Taiwan, Thailand, Pakistan, and nations in West Africa.[130] European countries also send much of their e-waste to West African countries, Nigeria and Ghana chief among them. These countries have neither the infrastructure nor the capacity to deal with the imported e-waste—leading to areas in these countries smothered with e-waste. Ghana—and more specifically, Agbogbloshie—exemplifies this trend.

A 2021 study estimated that Ghana imported about 165,350 tons of secondhand electronic goods per year, while recent estimates put this number closer to 250,000 tons.[131] Most of these goods are resold at a cheaper price, but between 30 and 40 percent of these goods (75,000–100,000 tons) are broken[132] and are then sent to Agbogbloshie.

Located just south of Ghana's capital, Accra, the wetlands of Agbogbloshie were not always the world's dumping ground for e-waste. In the early 1990s, Accra's yam market moved out of the central business district to Agbogbloshie's current location. The new yam market was quickly followed by the informal creation of a scrapyard around the market, providing spare parts to trucks that carried the yams to market.[133] Around 2000, the scrapyard started to become a popular site to dump imported e-waste, quickly overtaking Agbogbloshie's primary purpose as a market. E-waste tripled between 2003 and 2008, and currently there are between 13,000 and 17,000 tons of e-waste dumped there every year.[134]

These conditions led Agbogbloshie to be named one of the ten most polluted sites in the world. It inspired the 2018 film *Welcome to Sodom*, a movie whose title is based on Agbogbloshie's inhabitants' references to the area as "Sodom," a place where "the constant flames of burning plastic over the blackened ground of scorched computer parts accumulate into an apocalyptic sight that evokes visions of biblical infernos."[135] This movie has been acclaimed by critics for its dramatic cinematography of a dystopian world that reveals the "poisonous side of globalization."[136]

Around forty thousand people live in or around Agbogbloshie, many of whom are migrants from surrounding nations. There are between 4,500 and 6,000 individuals who work directly at the site and another 1,500 people indirectly involved in work at Agbogbloshie, such as in transporting scrap.[137] The workers rummage around the scrap heap, with the goal of stripping away precious metals like gold, cadmium, and lithium that they can then resell. The stripped metals are then sent to China, the United States, India, and Germany to be repurposed into "new" electronic goods.[138]

In order to get to the precious metals, workers must burn plastic off of the wires and burn other metals, which releases toxic fumes, specifically polycyclic aromatic hydrocarbons, although burning toxic waste is technically illegal in Ghana.[139] Just over 50 percent of workers spend their work time dismantling old electronics, about 24 percent sort the electronics into usable and unusable parts, 22 percent trade the parts, and just over 3 percent burn the parts, the last being the most dangerous of all.[140] Yet all workers are exposed to these harmful conditions and toxic deposits. A health survey of the workers at Agbogbloshie found that they had particularly high levels of cadmium and lead in their blood.[141] High cadmium in the blood can lead to cancer, organ failure, and cardiovascular failure, while high amounts of lead can lead to kidney and brain damage. These landfills are not only bad for human health due to the smoke and fine particulate matter that lingers in the air, but also due to the toxic chemicals that are saturated in the soil.

A story published in *Wired UK* about a young burner named Shaibu reveals the health implications of working in the waste fields of Agbogbloshie. Shaibu, who has worked as a burner for years, frequently coughs up blood. When his coughing fits get unmanageable, he buys a drink from a local medicine man that claims to "wash his heart." On his best days, Shaibu makes under $2.50, much of which he sends to his parents to support his two siblings. Shaibu's friend Ibrahim moved to Agbogbloshie from Northern Ghana after he dropped out of school and heard from his friend that he could earn a living working at the scrapyard. Ibrahim has similar coughing fits, and also experiences debilitating headaches and chest pains. Even when he can afford painkillers, they do nothing to help. Both Shaibu and Ibrahim acknowledge the serious health problems they have, but they also believe that they have no alternatives.[142]

Agbogbloshie also has a significant impact on the surrounding environment. Chicken eggs from around the site have been found to have extremely high levels of brominated dioxins; eating just one egg would surpass the European Food Safety Authority's daily tolerable intake by 2,200 percent.[143] Although the health effects of brominated dioxins are not well understood, studies have identified links to high risks of liver damage.[144] Even though there is still a yam market nearby in Agbogbloshie, the ground around the area has become polluted with chemicals preventing any vegetable growth in the area.[145] Additionally, the fumes from the site permeate throughout Accra, exposing millions of people who live there to toxic air.

Prior to the surprise demolition of Agbogbloshie in July 2021, conditions were improving around the site. The Blacksmith Institute from the United States was donating equipment to help workers strip metals so that workers did not need to burn the e-waste.[146] Additionally, the German Development Agency built health posts, technical-training facilities, and a soccer pitch to help alleviate some of the problems in the area.[147]

On July 1, 2021, the government of Ghana brought in bulldozers to demolish Agbogbloshie without warning. This action left thousands without work, forcing many from their homes. The demolition did not erase e-waste burning from Ghana but instead decentralized the endeavor, leading to some being forced to burn e-waste inside their homes or burn e-waste in more residential areas of the city.[148] It also put the thousands of workers at Agbogbloshie, like Shaibu and Ibrahim, out of a job. Even with the benefits of no longer inhaling the fumes from toxic waste, these workers now have to find an entirely new field of work, despite their now-long-standing health complications.

●●●

In this chapter, we presented a set of case studies of locations and conditions—and statistics in the case of the electric vehicle consumer—across the world that are connected along a single technology's supply chain. These locations share common conditions, such as a history of poor working conditions and public health in the cases of cobalt mining and electronic waste. They are also all examples of sacrifice zones, where a single community of people bears a significant share of the burdens associated with an energy technology or service. The cases also illustrate a tendency for communities to be cast aside, replaced, or forgotten, as the world moves from one technology or service to another.

These case studies are not isolated incidents. Such sacrificing of human lives, professional aspirations, and personal well-being as they relate to energy systems are prevalent across the world. In energy scholar Benjamin Sovacool's work on vulnerable communities in mining and extraction, pulling from 198 studies that include over 300 case studies on climate mitigation activities (e.g., solar installation, wind installation, electric-vehicle deployment), he found significant incidence of enclosure (e.g., land grabbing or forced resettlement), exclusion (lack of due process), encroachment (e.g., on land, water), and entrenchment (e.g., poverty or vulnerability). He also found that, time and again, it is the same set of vulnerable groups and populations who face these conditions: local and host communities, rural poor, Indigenous groups, racial and ethnic minorities, and women.[149]

The global energy transition and its associated efforts to rapidly decarbonize the economies of the United States and other countries will inevitably change the economic, financial, and social landscape of communities across the world. While the use of fossil fuels has pernicious short- and long-term effects on people and the environment, and from an emissions perspective is significantly more harmful than electric vehicles and other internal combustion engine substitutes, low-carbon and renewable-energy technologies also contribute to social and environmental challenges. And the distribution of these challenges, as in our past and present energy systems, is highly uneven across geography and society. In the case of many modern renewable-energy technologies, especially those such as electric vehicles or solar panels, which rely on minerals sourced from around the world,[150] the geographical gap between where benefits and costs are experienced may span continents. Dr. Sovacool refers to this as the "decarbonization divide." As he explains: "Patterns of injustice and domination are embedded in existing processes of decarbonisation, in spite of the assumption that low-carbon trajectories represent a more just way of producing energy. While decarbonisation may thus contribute to cleaner air and cleaner production in the Global North, much of the environmental and social harm is simply made invisible and displaced, or spatially externalized, to the Global South."[151] To put this explanation in the context of the American consumer, those who are able to enjoy clean-energy technologies rarely face the hidden burdens of those technologies. As this chapter has made clear, many of those burdens fall on vulnerable populations who live elsewhere, often in the Global South, and in places that do not always have human and labor rights, nor significant environmental regulations.

When discussing Dr. Sovacool's conception of the decarbonization divide in the classroom with our graduate students, one of our students, looking forlorn, said, "Man, it took me years to talk my dad into buying an electric lawn mower and now I feel as though I should call him and tell him to return it. Or at least I should fess up to him about the implications of his purchase . . ."

Clearly the answer is not that he should place that call. Nor is the answer that he or his father eschew electric vehicles because they are accompanied by a complicated and in some cases tragic set of conditions. The reality is that there is no easy answer, just as there is no easy solution to climate change. Decarbonization is essential to the long-term stability of our planet. Therefore, the challenge before us is to improve these conditions and address these disparities, starting first with a recognition of the challenges.

CHAPTER 8

The Uneasy, Uneven Future

At the heart of this book are two inevitabilities: Climate change is here, and its impacts are severe and likely to worsen for decades to come; and the clean-energy transition is underway, and it is likely to accelerate for the foreseeable future. In our view, these inevitabilities are incontrovertible, regardless of any specific policy choice or US presidential election.

This is not to suggest that the path forward on climate change and the clean-energy transition will be linear or predictable. Take climate change first. There is much already known from established science. According to the most recent, authoritative reports from the Intergovernmental Panel on Climate Change (IPCC), the world has already experienced at least 1.2 degrees Celsius of warming. Given current and still-increasing levels of greenhouse gas emissions, the amount of warming will continue to increase. Climate science indicates that we are currently on a path to blow by the Paris Agreement's aspirational goal of 1.5 degrees Celsius of warming, likely headed to well past 2 degrees Celsius unless countries around the world take drastic measures to reduce emissions. The present trajectory does not look promising. New temperature records seemingly are set each year. According to the International Energy Agency, in 2023, global carbon dioxide emissions from energy combustion and industrial processes hit an all-time high, resuming the long-term growth pattern that was temporarily interrupted by the COVID-19 pandemic.[1] Stabilizing climate change requires the world not only to reverse this growth but reach net-zero emissions by mid-century.

Climate change also brings many unknowns. While scientists can say with confidence that climate change has already resulted in more extreme weather and natural disasters, it remains difficult to forecast exactly what future storms will bring and just how destructive wildfires, hurricanes, flooding events, and other climate-charged natural disasters will be. Similarly, climate impacts like sea-level rise are challenging to forecast and depend on shifting projections of ice melt. Adding to the uncertainty are what climate scientists refer to as tipping points, which are defined as thresholds in the climate system that, when surpassed, may produce accelerated and potentially irreversible changes.[2] Science writer Robert McSweeney characterizes climate change as a game of Jenga: "The gradual increase in global temperature sees block after block removed from the tower and placed on top. As time goes on, the tower becomes more and more misshapen and unstable. At some point, the tower can no longer support itself and it tips over. In the game of Jenga, the tower collapses in a split second. For a component of the Earth system, the shift to one physical state to another may take many decades or centuries. But the feature they have in common is that once the collapse has started, it is virtually impossible to stop."[3] A helpful example is the Greenland ice sheet. Scientists who monitor the extent of ice melt in Greenland have already noted record levels of ice melt in recent years.[4] The Greenland ice sheet is the result of hundreds of thousands of years of climatic history. Yet, as McSweeney points out, "if the Greenland ice sheet were to pass a tipping point that led to its disintegration, simply reducing emissions and lowering global temperatures to pre-industrial levels would not bring it back again. It would probably require another ice age to achieve that."

While predicting the future impacts of climate change is difficult, the damages to this point are already enormous. A recent study from two climate risk scholars analyzed data from all available studies of extreme event attribution—the science that analyzes how much of an extreme weather event can be directly attributed to climate change—to estimate global economic damages. They found annual global damages of $143 billion per year, the majority of which come from the loss of human life.[5] $143 billion is more than the entire GDP of countries like Morocco and Ecuador, and similar to the overall economy of the state of Mississippi. While these damages are a vast undercount of the full economic costs from climate change, the point is clear—climate change is already wreaking havoc.

There are also many knowns and unknowns regarding the clean-energy transition in the United States. Among the knowns, America's historical reliance on coal in the electricity sector is coming to an end. Twenty-five years ago, one-quarter of total energy consumption was from coal. By 2022, this percentage dropped by more than half to just 10 percent,[6] and its proportion will continue to dwindle. The fate of other fossil fuels is less certain. Despite all the attention to decarbonization in recent decades, the United States has actually increased both its production and consumption of natural gas, thanks in large measure to fracking and its effect on natural-gas prices. And while the consumption of oil has remained relatively steady for the past two decades, domestic production of oil has quickly grown to the point where the United States now produces more oil than any other nation in the world and the same level as Russia, Canada, and China combined.[7] Although many countries have committed in principle to phase out their use of fossil fuels, history suggests that these commitments are likely cheap talk, or at least that they make no guarantees on a time scale that matches the urgency of the climate crisis. The energy transition in the United States may be inevitable, but it has a very long way to go, and the path to anything approximating net-zero greenhouse gas emissions by 2050—or any similar emissions reduction path—is not certain.

At the risk of gross understatement, successful decarbonization of the US economy requires a massive and expensive transformation of the energy system. At the present time, 80 percent of the energy consumed in the United States is from fossil fuels.[8] Globally, the numbers are the same.[9] These are staggering statistics and a stark reminder of just how much progress remains to be made.

The United States, to this point, has made the most progress in reducing emissions in the electricity sector, relative to other major sectors of the economy like transportation. Between 1997 to 2022, the percentage of non-carbon-based electricity generation grew from 30 to 40 percent, with renewable energy such as wind and solar accounting for much of the increase. Yet, we still have a long way to go to achieve Paris climate commitments or reach net-zero emissions. The growth in renewables has been contemporaneous with the growth of natural gas; about 60 percent of electricity generation still comes from fossil fuels, and the addition of natural-gas capacity is growing, not shrinking.[10] So, even in the sector that has achieved the most carbon abatement, much remains to be done.

Progress in other sectors of the economy lags behind. Consider the transportation sector, which contributes the most greenhouse gas emissions of any part of the economy. In the aggregate, the level of total emissions from the transportation sector has remained relatively flat over the past two decades.[11] There is some progress, however, to report. The fuel economy of the US auto fleet has doubled over the past twenty years, and—pushed by federal policy—most major auto manufacturers are making improvements to reduce the greenhouse gas emissions from the cars that they sell.[12] And Americans are purchasing more and more electric vehicles, which over time will further reduce total emissions from this part of the economy. However, as of 2023, ninety-nine of a hundred cars on the road in the United States and nine of every ten vehicles purchased were still gas-powered.[13] The literal and figurative road ahead here is long, too.

Greenhouse gas emissions from industry and buildings have declined over the past ten years, but emissions from agriculture have remained relatively level.[14] Across all these sectors, it will be necessary for millions of companies, homeowners and landlords, and farmers to invest in technologies, materials, and processes to reduce their carbon footprints. Perhaps part of the solutions will come from the development and widespread deployment of carbon capture and storage or negative-emissions technologies, such as direct air capture. At this point, however, these technologies have questionable feasibility and enormous price tags, and they are fraught with moral hazard and other ethical challenges.

Putting it all together, the challenge of decarbonization is both large and certain. But there are also many unknowns about this transition. How fast will the transition happen, and will it happen in time to stave off a planetary collapse? Is the polarized and institutionally fragmented US political system capable of bringing about the sustained policy changes needed to facilitate the transition? Will individuals and companies embrace the challenge and make the investments—both financially and behaviorally—to reduce their carbon footprints, and will they do it willingly or only if compelled to do so by government mandates?

These are only some of the unknowns necessary to confront. And, of course, the challenges extend far beyond the borders of the United States. The world's climate fate requires global commitment, though the United States arguably has the most to contribute given its historical contributions of emissions.

The central argument of this book is that there is another known, which is that the clean-energy transition in the United States will inevitably create inequitable and unjust outcomes. This is not to suggest that it is impossible to mitigate some of these inequities and injustices, nor are we in any way suggesting that we throw our hands up in the air and surrender. Although there are no easy solutions, we discuss some important principles that, if embraced, we believe will help make progress. First, however, we review the main arguments and themes of the book as they emerged from the stories and cases we shared in the preceding chapters to further underscore the challenges that await.

On the Front Lines

Energy systems have always created winners and losers, and the same will be true for the energy systems that will emerge from the clean-energy transition. The political discourse about these future energy systems can be pollyannaish, with advocates for specific public policies or technologies promoting them as "win-wins" or solutions that are "good for everyone." As we have argued in this book, this is generally not the case, or at least it is a gross oversimplification. For many individuals and communities on the front lines of the energy transition, severe challenges lie ahead.

One illustrative example is the steady move away from coal as an electricity generating source. The United States currently consumes about the same amount of coal as it did fifty years ago, when the economy was about twenty times smaller, and just one-third of what it did at its peak roughly twenty years ago. [15] For the country as a whole, the benefits of reduced coal use are large and demonstrable, particularly in the form of improved air and water quality. From a climate change perspective, the reduced use of coal to generate electricity is the single most important step the United States has to this point taken to reduce its carbon emissions. In these ways, most, if not all, of US society has benefited from the shift away from coal.

Yet, as we discussed in chapter 3, the shift away from coal has had devastating effects for many communities, especially those in Appalachia that have been among the most deeply affected by the shrinking demand for this energy source. Countless mines in Kentucky, West Virginia, and other Appalachian states, as well as states in other regions of the country, have closed, and coal mining employment has dwindled; the total number of coal mining jobs is now about half what it was ten years ago, and just a quarter of what it was forty years ago. [16] Most major American coal companies have experienced bankruptcy over the past ten years, and their long-term prospects for survival are dim. Bankruptcies often come with further mine closures, loss of long-term health care benefits, and underfunded employee pensions. These types of economic shocks would be difficult for any local economy to absorb—just as is the case when any major employer leaves a community—but they are especially difficult for many coal communities that do not have diversified economies and that have long experienced persistent poverty, exploitation, and health complications connected to their local extractive economies.

Beyond the mines are the communities that hosted the coal-fired power plants that once powered homes and businesses across the country. Over the past decade or so, the energy industry has retired more than three hundred coal-fired power plants,[17] and analysts predict that up to another 40 percent will retire by the decade's end.[18] In addition to the loss of direct employment from these power plant closures, there is also the loss of tax revenue that these facilities provided to host communities.

From the outside looking in, it is easy to diminish the decline of coal as a minor economic shock that affects only a few hundred thousand workers in concentrated parts of the country, and indeed the decline has often been portrayed this way historically in political and policy debates. However, our conversations with former coal miners reveal something much deeper. Coal miners, many of whom are from families where generations have made their living in the mines, express feelings of loss and abandonment, and they often place the blame for the change in circumstances on faraway politicians more interested in climate change than in their well-being, or on the rest of America, who have used them for cheap energy and now abandoned them in the quest for a clean-energy future. Feelings of unrecognized and uncompensated sacrifice run deep. And the effects of the reduced demand for coal not only threatens the economic vitality of these communities but also tears at their social and cultural fabric.

In no uncertain terms, the coal communities—places home to power plants and the mines that sourced them—are on the front lines of energy transition, illustrating one example of those who are adversely affected by the clean-energy transition. Yet coal communities are not alone in this regard, as we highlighted throughout the book. The accelerating shift in the transportation sector from internal-combustion-engine cars to electric vehicles, and its effect on autoworkers and auto communities, provides another illustration.

Compared to the move away from coal, the shift away from internal combustion engine cars is still in its nascency. As noted above, at this time, about 90 percent of all cars sold in the United States still run on gasoline. However, relative to coal, the impact of the transition to electric vehicles is likely to be more deeply felt because of the much larger size of the auto industry. It is too early to forecast precisely which workers and communities will be adversely affected by the industry's turn to electric vehicles, since this outcome will depend on future public policies and investment decisions of the major auto manufacturers. Nonetheless, the impacts are likely to be severe for many.

In chapter 7 of the book, we told the story of Lordstown, Ohio, a community with a long and often challenging history with the auto industry. In 2019, GM shut down its Lordstown Assembly plant, ending a more than fifty-year run as a major assembler of Chevrolet cars and vans. The plant, one of the largest industrial employers in the area, provided a middle-class income for more than ten thousand people at its peak.[19] GM sold the facility to Lordstown Motors, a new entrant into the auto industry, which had hopes to assemble its electric pickup truck at the plant. The new venture brought hope to the community and to many of GM's former workers who had lost their jobs when the GM plant closed. This hope, however, was short-lived. Three years later, Lordstown Motors was bankrupt.

GM's decision to shutter its Lordstown Assembly had more to do with the unused capacity at the facility and the slow selling of the Chevrolet Cruze assembled there than with the company's stated commitment to an all-electric future.[20] Nonetheless, the shutdown of the facility is illustrative of the types of impacts that will come when auto companies shutter other assembly plants and other companies in the internal-combustion-engine-car supply chain go out of business as part of the transition to electric vehicles. The closure of Lordstown Assembly put 1,600 employees out of work and risked as many as eight thousand jobs in the surrounding area and $8 billion in economic activity. The community of Lordstown also was expected to lose $3 million in tax revenue, equating to one-third of its annual total.[21]

The autoworkers and community members we spoke to in Lordstown highlighted their economic hardship. But they also articulated a sense of loss, abandonment, and distrust that was similarly palpable in our conversations with out-of-work coal miners. And, like the coal miners, they expressed frustration about their lack of voice and agency in the decisions of the companies to which in their view they had given their lives. As one former Lordstown employee told us, "I felt I earned the right to build that next American car."

The transition to electric vehicles is poised to produce significant benefits as well. Every internal-combustion-engine car that is replaced with an electric vehicle means fewer emissions, both of planet-warming pollutants and conventional pollutants that adversely affect local air quality. With respect to the latter, people living in closest proximity to roads and highways will benefit the most, which means disproportionate benefits for many low-income Americans, as well as for racial and ethnic groups who tend to live near transportation infrastructure. But, even more directly, there will be tens of thousands of new jobs working to make electric vehicles, both from the major automakers of the past, such as Ford and GM, but also for the newer, electric-only entrants into the market like Tesla and Rivian. In addition, as companies in the United States, spurred on by federal tax credits, further develop the domestic supply chain for electric vehicles, there will be new jobs in different communities to produce batteries and mine the critical minerals they need.

As a final illustration of the likely burdens that the energy transition will bring for some people, consider the case of energy insecurity in America. To both meet decarbonization goals and upgrade the electric grid to meet modern needs, respectively, utility companies and other energy providers will need to make enormous capital investments in power generation, transmission and distribution, and other types of infrastructure. Even if the government subsidizes these investments, they are still likely to result in higher prices for consumers, especially in regulated markets where utility providers will seek to recoup their costs and reasonably profit through rate increases. Ten or twenty dollars more per month will go unnoticed for many people, some of whom might be willing to pay even more if it means their energy providers are moving toward cleaner sources. But for those who face energy insecurity, such as Darrin, whose story we told in chapter 4, ten or twenty more dollars per month may be the difference between his being able to pay his bill and his being disconnected.

As these examples highlight, the clean-energy transition will generate adverse outcomes for some individuals and communities. These outcomes, moreover, can be further understood by several additional themes, which we noted throughout the book. First, there is often an uneven geography of costs and benefits associated with the transition. Second, many of the individuals who experience the burdens and lack opportunities associated with our energy systems are the same individuals who often both literally and figuratively are without power. And third, individuals and communities on the front lines of our changing energy systems face difficult and complex trade-offs. We discuss each in turn.

The Uneven Geography of Costs and Benefits

The costs and benefits that will result from the energy transition will often not be experienced in the same geographic space. That is, the individuals and communities who will bear the costs of the transition will often differ from the individuals and communities who enjoy the benefits. In many—but not all—instances, this disjunction will arise because the costs of changes—for example, the siting of new energy infrastructure or the closing of a coal mine or fossil fuel power plant—will be geographically concentrated, while the benefits, such as reduced carbon emissions and improved air and water quality, will be more widespread.

The uneven geography of energy-related costs and benefits is also ubiquitous in the current energy system. As we discussed at length in chapter 2, many of the health and ecological impacts of long-term reliance on fossil fuels have been concentrated in certain places and imposed on specific communities. These externalities have often fallen disproportionately on historically marginalized populations, as has been documented by decades of research on environmental injustice. So, while Americans across the country have benefited from inexpensive fossil fuels, many communities have endured the pollution from the use of these sources of energy. The clean-energy transition is likely to produce similar outcomes.

Take, for example, the siting of new renewable-energy infrastructure. The energy transition requires the tremendous buildout of new, utility-scale wind and solar power, not to mention the expansion of the electric grid to bring this new energy to its users. The benefits of renewable energy are enjoyed widely, but the costs of their deployment are largely concentrated on the communities that host the facilities. In other words, the people who will benefit from a cleaner energy system will not have to endure any of the direct, adverse impacts.

The uneven geographic distribution of costs and benefits of wind projects has, in part, motivated a strong backlash against these projects in some parts of the country. In chapter 6, we discussed the actions of many local jurisdictions to put in local land-use ordinances to limit, if not ban altogether, the development of new renewable-energy infrastructure as a protection against the unwanted costs of the facilities. Much of this opposition has occurred in rural, Midwestern parts of the country, where local residents resist the perceived imposition of wind farms, despite potential new tax revenue, landowner lease agreements, and community benefits packages that provide compensation for the projects. Costs in these cases are more about damages to a sense of place and autonomy. For many people, large wind farms irreparably damage landscapes, converting serene, rural places into perceived industrial zones. In doing so, they disrupt the strong attachments that people have with the places where they live. Although opposition to new wind farms is often portrayed by project proponents and the media as selfish NIMBYism (and certainly, these cases do exist), we have found that these costs represent actual concern that the clean-energy transition imposes real harms on local communities. Yet, the benefits will mostly be enjoyed by people far away, people who will not have to look at towering wind farms dotting the landscape, listen to the constant hum of the rotating blades, or see the flickering lights in the night sky.

The uneven distribution of costs and benefits of the energy transition emerges in several other contexts, such as the case of carbon capture and storage (CCS). The wide-scale deployment of CCS as a climate mitigation strategy, assuming the technologies prove to be technologically and socially viable and financially feasible, would help utility companies and other major industrial sources reduce their carbon emissions. From a climate change standpoint, everyone would benefit. Yet the adoption of CCS technologies would mean keeping fossil fuel facilities online, where they are likely to continue to adversely affect those individuals who live nearby. Capturing carbon at the point of combustion addresses the emissions causing climate change, but it reduces neither the harms created from the mining and refining of the energy resources at the places of extraction and processing, nor the emissions of other pollutants where the fossil fuels are burned. Moreover, because so many fossil fuel facilities are located in low-income communities of color, the environmental injustice problems that epitomize the current energy system will persist.

Electrification of transportation and homes provide additional illustrations. Despite the climate and air-pollution abatement benefits that come with more electric vehicles, there will be concentrated costs for those communities where the extraction of the critical minerals occurs where and the electronic waste is eventually disposed, in terms of health and the environment. Electrification of people's homes also imposes costs, in this case the material type of costs. As people electrify their homes, for example, by replacing natural-gas furnaces with electric heat pumps or natural-gas stoves with electric ones, natural-gas utility providers will need to spread the same fixed costs on to fewer of their customers. In this way, the costs of legacy infrastructure shifts to an ever-shrinking number of people. Given the types of barriers to new technology adoption that we enumerated in chapter 5, these additional costs are most likely to fall to people with lower incomes and those from historically marginalized groups, such as racial and ethnic minorities. These higher costs will exacerbate energy insecurity, which already occurs at higher rates in the groups, and further reduce their ability to meet their basic energy needs. At the most extreme, higher household energy costs of this or any type creates the possibility of residents having to go without energy altogether.

Those Without Power

Flipping on a light switch, turning on the television, or charging a mobile phone are not just modern conveniences for most Americans, they are afterthoughts, done with such regularity that they barely register in most people's psyches. Most Americans are only reminded of their reliance on energy when they lose it. We opened the book with one such example of a time when many people were acutely aware of their lack of electricity, during the winter vortex that hit Texas in February 2021, which resulted in millions of Texans losing electricity and heat during an extreme cold-weather event, with tragic consequences.

Power outages from storm damaged transmission lines are relatively common but, for most people, they are short-lived. However, for millions of Americans, the risk of having to live without power due to nonpayment shutoffs is a near-constant state of affairs, and one that becomes a reality with some frequency. In chapter 4, we shared stories of individuals who have experienced energy insecurity and utility disconnections along with statistics of the problem. These personal accounts and aggregate data reveal that energy insecurity is an underacknowledged yet pervasive form of material hardship. Our analysis of available utility disclosures indicates that annually about three million electricity and natural-gas customers have their service shut off by their utility because of nonpayment of their bills. Because of sparse disclosure of such data nationally, the true number of disconnections is likely much higher. Some of America's biggest corporations, Fortune 500 giants like Duke, Southern, Pacific Gas and Electric, and American Electric Power disconnect tens of thousands of customers each year.

Undoubtedly, some of these shutoffs are for customers with the means to pay, who either deliberately or mistakenly missed payments. For most of these customers, the situation can quickly be rectified with a payment, leading to a rapid restoration of service. However, in the vast majority of cases, the evidence suggests that utility companies are also shutting off service to people who simply cannot afford to pay. In all likelihood, these individuals were already limiting their energy use, perhaps by living with dangerously cold or hot temperatures to avoid turning their heat or air conditioning up in an effort to reduce their bills. The electricity grid may create the possibility of near-universal service, but the United States does not treat connectivity as a basic right. Instead, whether or not people have access to energy in their homes is a decision that rests with utility providers.

The energy insecurity problem is partially about poverty and material hardship—that is, people with low incomes are more likely to face challenges paying their energy bills, just as they may struggle to make ends meet in other ways, such as paying for food or medicine. However, energy insecurity is not just about income. A growing body of social science, including our own studies of low-income US households, documents enormous disparities in who experiences energy insecurity, with respect to both being able to afford residential-energy services as well as being disconnected from service. What we have learned is that, even when accounting for income, people of color, families with young children, those who are medically vulnerable, and people living with poor housing conditions are all more likely to experience energy insecurity.

The same people who risk living without reliable energy are those who lack the opportunities associated with the evolving energy landscape. Evidence abounds that the pollution and associated health burdens generated by historic fossil-fuel production disproportionately falls on the same individuals who struggle to afford energy. The communities that have become synonymous with environmental injustice share many of the same sociodemographic characteristics as those that face higher risks of utility shutoffs. Similarly, as we argued in chapter 5, the same types of individuals often lack access to the new technologies that might alleviate the high energy burdens that contribute to their energy insecurity, and they are also not well-represented in the workforces of these new technologies. The burdens from the existing energy system—along with exclusion from the newly emerging energy system—illustrate the cascading effects of and intertwining ways that energy marginalization affects these individuals.

Being powerless means more than not having access to affordable and reliable energy in one's home. The notion also extends to the degree that someone has autonomy, control, leadership, or a voice in the energy decisions that affect them. The experience of coal miners and autoworkers provides an example of groups that express concern about a lack of participation in energy decisions. Early evidence suggests that workers in other fossil fuel–related industries such as the oil and gas industry share similar sentiments. In addition to perceptions that their economic well-being is being sacrificed as part of the transition to cleaner sources of energy, workers in these sectors feel powerless to stop the transition and often near-powerless to adapt to the changes. As we highlighted in the stories woven throughout the book, we heard these sentiments both from people who have already lost their jobs and from those who expect to experience a similar fate in the coming years. These individuals and workers on the front lines see the changes as being imposed on them, often by people far away—both literally and ideologically—with divergent interests and priorities. Recent work by political scientists Alexander Gazmararian and Dustin Tingley highlights that people often distrust those offering solutions, finding them to lack credibility.[22]

The lack of agency and opportunity to participate in decision-making is an important factor in explaining opposition and skepticism toward changing energy systems. The successful cases of renewable-energy siting in rural areas provides a good reminder here. As we discussed in chapter 6, research reveals that communities, including the landowners most directly affected by new infrastructure siting, are much more likely to embrace projects when energy developers engage with them early and often and take their needs into account. The deficit of trust and confidence that many Americans have toward both government agencies and energy companies—in many cases, for very real and deeply historical reasons—creates conditions where feelings of lack of power and agency turn into hostility and opposition to the projects needed for the clean-energy transition. These conditions are a formidable challenge that must be addressed as part of decarbonization efforts.

Impossible Trade-Offs

Individuals and communities on the front lines of the energy transition face difficult and complex trade-offs. Many of the same examples highlighted above apply here as well, such as the conditions under which energy insecure households make decisions. The most likely choice a struggling household faces is not whether to pay their electric or natural-gas bill, with a looming threat of disconnection but, rather, the choice is often whether to pay their bill or pay for other basic necessities like food, medicine, clothing, or school supplies. As discussed in chapter 4, in a typical year about one in five US households indicates that they make this type of "heat-or-eat" choice.[23]

Households in these dire circumstances face other types of trade-offs, too. We have learned from our household surveys that people who cannot afford their utility bills pursue a wide array of efforts to cope. Some of these efforts are seemingly benign, such as turning down the thermostat a few degrees to save money while putting on an extra layer of clothing or using blankets to stay warm. But other coping strategies are more dangerous. The survey data reveal that over one-fourth of low-income households use potentially dangerous coping strategies, such as keeping warm by using their ovens, space heaters, or clothes dryers as a heating source as an alternative or complement to running their utility-based heat. They also report financial measures that they find it necessary to take, such as the accumulation of debt, bill balancing, and use of high-interest payday loans. As the United States remakes its energy system, costs will go up, which means the types of trade-offs that energy insecure households face today will persist, and more people may find themselves confronting similar decisions.

Others on the front lines of the energy transition face different types of trade-offs. The current employment shocks associated with the move away from fossil fuels to alternative energy sources create difficult choices for individuals and communities. For example, many of the Appalachian coal miners with whom we spoke were clear-eyed about their long-term career prospects. Many had already lost their jobs and had begun to participate in retraining programs to transition to different occupations. Others who still had their jobs recognized that they too would soon need to find alternative employment. One specific challenge they face is that the new jobs are unlikely to be located where they live now, a situation exacerbated by the historical lack of economic diversification of their communities. Workers in other types of extraction communities, such as oil workers in West Texas and North Dakota and natural-gas workers in Pennsylvania and Colorado, will face similar challenges in the future. We heard from some former coal miners that they now travel for weeks at a time as truck drivers or commute long distances to work jobs as electrical linesmen. Left behind are their families and broader social networks. Although many people expressed being willing to make such trade-offs to care for their loved ones, the losses are nonetheless difficult to absorb.

Trade-offs exist at the community level as well. The case of siting new energy infrastructure is illustrative. Neighborhoods, towns, cities, counties, and other local jurisdictions need to balance competing interests and priorities, such as economic development and environmental protection, while simultaneously managing divergent opinions about what type of infrastructure is acceptable for their community. As we discussed in chapter 6, opposition is already mounting in some parts of the country, creating new barriers to what is often already a challenging process of siting and building industrial-scale operations in the United States.

Yet, despite these challenges, the infrastructure is necessary. The clean-energy transition is going to require hundreds of billions, if not more, of new investments in renewable-energy installations, transmissions and distribution lines, battery systems, and, perhaps, carbon capture and storage. How this development unfolds will depend on how, and which, communities manage the trade-offs. Will more rural, Midwestern communities decide to accept the financial benefits of large-scale wind farms in exchange for what they may perceive as sacrifices to their agricultural landscapes? Will communities, urban and rural alike, be willing to host the pipelines and store the carbon dioxide captured from faraway fossil-fuel facilities to help reduce emissions into the atmosphere? Will communities be willing to permit the use of their land for transmission lines, disrupting viewsheds and fragmenting land so that clean energy can be distributed from one part of the country to another, even though they may not themselves benefit from that same energy? These trade-offs are not easy for communities to manage, yet the pace and ultimate success of the energy transition depends on the decisions they make.

Working Toward an Equitable and Just Energy Transition

The challenges of the clean-energy transition are immense; and perhaps the picture we have painted in this book seems grim and hopeless, leaving an impression that the price of completely overhauling the energy system is not worth pursuing. Stated unequivocally, we do not think this is the right conclusion. The benefits of shifting away from fossil fuels are enormous. Rebuilding the economy with cleaner, more efficient technologies will prevent millions of premature deaths from improved environmental quality, generate trillions of dollars of wealth, spur vast technological innovations that will spill over into other aspects of society, and provide countless individuals and communities with opportunities for an improved quality of life. Moreover, we do not have a choice. The severity and urgency of climate change requires us to make this transformation. The alternative by century's end is, in the words of David Wallace-Wells, an "uninhabitable Earth."[24]

Although we face no real option but to press ahead, that does not mean there are easy answers or clear choices about how to create an equitable and just energy transition. This situation is not simply a matter of getting the policy "right."

We wrote this book to provide a narrative of the many challenges that emerge when we center people in the energy transition. And, while the book is largely void of policy prescriptions to address these challenges, there are many potential policy approaches and programs. For example, for fossil fuel workers and communities who face current or future economic dislocation and social disruption, public, private, and nonprofit organizations can offer retraining programs and economic development initiatives. For households struggling to afford their essential household energy needs, regulators can offer rate reduction programs, legislators can expand their disconnection protections to better account for a changing climate, and government agencies can enhance the reach of energy assistance and weatherization programs. For people who are unable to access clean energy and energy-efficient technologies, whether they are rooftop solar panels, new appliances for their house or apartment, or a zero-emissions vehicle like an electric car, governments can subsidize their purchases—and can do so in ways that overcome some of the more challenging barriers to adoption. And, for communities who are being asked to accept industrial-scale energy infrastructure, government agencies and energy companies can redesign their decision-making processes to better engage and build trust with citizens, so their ideas and concerns are included in the decisions that deeply affect them and their communities.

These suggestions, of course, only scratch the surface of the types of policies and programs that might be pursued, and they do not convey the enormity of the public policy challenges, nor any sophistication of policy design, to appropriately and efficiently address these challenges. And, as we highlighted periodically throughout the book, many of these types of efforts are already underway. The 2022 Inflation Reduction, for instance, includes many initiatives that aim to assure that efforts to address and adapt to climate change are broadly accessible. The Biden administration's Justice40 Initiative—a commitment to direct 40 percent of the benefits of certain government programs to disadvantaged communities—is another example. These types of policies and programs certainly have a critical role to play in working toward an equitable and just energy transition. The retrenchment in these areas reflected in the first year of President Trump's second term reflects a change in ideological preferences, not an objective understanding of how those efforts can bring about positive and meaningful change.

In addition to being a policy priority that yields concrete policy actions, we also believe that building an equitable and just energy transition should be a universal aspiration that pervades all dimensions and aspects of the energy transition. A useful starting point for establishing the foundation of this aspiration is to come back to the three core principles of energy justice: distributive justice, procedural justice, and recognition justice. We discuss these principles in reverse order.

Recognition justice is the notion that there is a need both to account for past harms and to fully consider the structural and societal dynamics that keep certain populations persistently disadvantaged. We think it is essential that all energy transition stakeholders—energy companies, government agencies, advocacy groups, and elected leaders— consider not just the ways that their specific actions contributed to disproportionate burdens on some Americans, but also how the fossil fuel-based economy as a whole generated these burdens. The point here is not to assign culpability or to cast aspersions on specific actors or institutions, but more to begin to change the tenor of the conversation and to embrace a collective mentality about the importance of addressing these historic and ongoing inequities. In our view, debates about energy too often devolve into debates about one technology or another or one policy approach versus an alternative, which is not conducive to finding ways through decisions that require difficult trade-offs and acceptance of uneven costs and benefits.

We acknowledge that our emphasis here on recognition justice might come across as foolishly optimistic or simply cheap talk from the comforts of our professorial perches. While acknowledging our positionality, we sincerely believe in the importance of reckoning with the past as a way to better navigate the future. This is the reason we started this book with a look at the historical burdens of our energy system. Understanding the past, we think, is the right place to start before mapping out a future.

Procedural justice is another important tenet of energy justice, and it refers to ensuring access and opportunity for individuals and communities to participate in the energy decision-making processes that affect them. Procedural justice is a broad notion that everyone should have a voice that is heard and respected, as well as the opportunity to be involved and lead. Procedural justice is also a central pillar of broader environmental-justice frameworks.

The current energy system is replete with adversarial interactions, and this state of affairs will not change as the United States moves from a fossil fuel–based economy to one powered by alternative sources. Government agencies will still regulate energy companies; utility companies will still decide who loses access to energy when they cannot afford to pay their bills; auto companies and autoworkers will continue to renegotiate wages and labor conditions; and communities will continue to work with developers to determine which types of infrastructure they want and do not want.

These types of interactions are not always adversarial, however, and they are often filtered through a labyrinth of political and legal processes and institutions. Nonetheless, a real consequence is that decisions by energy stakeholders often occur in a context of mistrust, where one side is suspicious of the motives of the other, and vice versa. Recent controversies over pipelines, renewable-energy installations, electric vehicles, and policies to reduce greenhouse gas emissions illustrate the depth of tensions.

No process, no matter how transparent or inclusive, will overcome all disagreements, especially when disagreements reflect fundamental clashes in values. Yet, a commitment to honest exchange of information, consultative and deliberative processes where divergent perspectives are allowed and respected, and efforts to reach consensus and to create partnerships would go a long way to constructing a shared purpose and common understanding of how to bring about a clean-energy transition that is equitable and just.

Decisions about the future of energy climate, and society are no less contentious than decisions about the future of healthcare, education, or anything else. There is no easy path, and the deep political, economic, and cultural cleavages that make so many parts of modern life in America divisive will not be overcome by more public meetings or people just being kinder to one another. But there is also good reason, and lots of evidence, that people are more likely to accept outcomes when they believe the processes that led to them were fair, transparent, and inclusive. In the context of the energy transition, this human tendency pertains to decisions by government agencies about permitting and zoning, by energy companies about infrastructure siting choices, and by utility providers about prices. Organizational responsiveness—and treating people respectfully and with dignity—is also an important dimension. While not sufficient, building and sustaining trusting relationships across people and institutions is necessary for creating common purpose in meeting the decarbonization challenge in front of us.

Aspiring toward recognition and procedural justice, however, will only go so far in generating an equitable and just clean-energy transition. The third key principle of energy justice is distributive justice, which refers to how the benefits and burdens are spread across different populations, in which the objective is for no single population to bear a disproportionate share of the burdens and for all populations to have access to the benefits. Of these three types of energy justice, distributive justice is perhaps the most difficult. The stories and cases emphasized in this book have made it clear that there are uneven distributions of benefits and harms associated with energy, across both geographic space and population groups, and associated with both current and historic energy markets.

Yet it is not inevitable that the disparities of the past become the disparities of the future.

The energy transition is creating new opportunities as well: new opportunities to avoid concentrating the costs of the transition on just some individuals and communities; new opportunities to make energy technologies, and the jobs that produce them, more accessible to everyone; and new opportunities to shift priorities to ensure that everyone, and especially historically disadvantaged populations, reap the benefits of the affordable and reliable energy we all need to thrive and prosper in the twenty-first century.

How can the United States harness these types of opportunities? As a starting point, it is critical to recognize that any decision to promote a new technology—or design a public policy to shift the energy economy or individual energy-consumption behavior—is also implicitly a decision about who gets the benefits and who experiences the costs. Although it may be impossible to eliminate the uneven distributions of costs and benefits, a deliberate approach from the start will go a long way toward reducing them. For the government, such a deliberate approach means being mindful of and attentive to the distributional impacts of programs and policies, and building in redistributive, compensatory, or other measures at inception. For energy companies, it means the same, which requires that they adjust their business models and investment decisions, and if they are unwilling to do so, it will require government intervention to either create incentives or mandate changes.

These three principles—recognition justice, procedural justice, and distributive justice—are aspirational. Embracing them requires thinking in new ways, setting different priorities, and working collaboratively across energy producers, users, and policymakers. Building a future energy system, especially at the scale and on a time frame that effectively addresses climate change, requires this type of reimagination.

What approaches do we have in mind? First, solutions must be tailored to communities. The clean-energy transition will not lend itself to one-size-fits-all solutions that impose a single formula of decarbonization on everyone. Communities face different challenges and opportunities given their specific natural and geographic characteristics, unique histories, diverse economies, and varied social and cultural identities. Offshore wind farms, for example, may be a welcome development for some coastal communities, while for others the idea may be a nonstarter that jeopardizes their economic well-being and cultural identity. Similarly, economic transition for legacy fossil-fuel communities, such as coal mining towns or communities hosting retiring coal-fired power plants, will not be the same everywhere. Some places may welcome new industries, while others may desire a shift to a service-based economy. Workforce (re)training and economic development assistance should be structured in a way to match community needs and preferences.

Second, solutions must put power and agency in the hands of

people and communities. As emphasized throughout the book, most of the decisions that shape the US energy system are made by profit-oriented corporations. While there is oversight and regulation from federal, state, tribal, and local government agencies, decisions about sources of electricity generation, the location of energy infrastructure, and the accessibility and availability of technologies, among many other energy decisions, are largely in the hands of the private sector. Achieving more equitable and just energy decisions will require empowering individuals and communities with more authority and agency over these decisions, freeing them to use the resulting technologies and infrastructure to create wealth and opportunity for themselves.

Third, one way to better integrate communities into the decision-making that shapes the direction of the clean-energy transition is to seek cross-sectoral solutions that involve public, private, nonprofit, and community organizations. Multi- and cross-sectoral partnerships can help leverage financing and social acceptance of new energy solutions, spur innovation, and provide multiple objectives that can guide successful projects.

Fourth, solutions should include efforts to build resilience and adaptive capacity so people and communities are better able to manage changing climate and energy conditions. The ongoing clean-energy transition is disruptive, and it will continue to unsettle individuals, households, and communities in numerous ways—changes to employment opportunities, local economies, and, in some cases, social bonds and cultural identities. Climate change is also disruptive, and those disruptions are likely to get worse given current emissions trajectories. Adjusting and thriving through changes of this scale requires building localized resilience and adaptive capacity.

Fifth, solutions should seek dual opportunities of decarbonization and equity, and they should prioritize those decarbonization efforts that maximize equity outcomes. The decarbonization challenge encompasses significant changes in what energy we use and how we use it. Decarbonization is both a process and an end goal. The United States, and the rest of the world, remains at the very early stages of transforming their energy systems away from fossil fuels. Energy transitions are not fast, and despite the urgency of climate change, the current transition will take a long time to achieve. As this transition unfolds, policymakers, corporations, and communities have the opportunity to insist that it is done equitably, such that everyone has a chance to participate in its benefits. Doing so requires a deliberate focus on centering equity in efforts to both mitigate and adapt to climate change.

Finally, and most of all, a commitment to energy justice principles must refocus the discussion about the energy transition from one mostly about technology and public policy to one centered on people. People are at the core of energy systems, and an energy justice approach must recognize the dignity, vulnerability, and agency of people and craft solutions accordingly. Prioritizing people in the energy system will enable the United States to better achieve its energy and climate goals while also addressing more general social challenges, like racial injustice and income inequality. The future energy system will not solve those problems, but striving toward a more equitable and just transition will get us part of the way there. We believe it is imperative to try.

ACKNOWLEDGMENTS

We have many people to thank. Some of the research discussed in this book builds on prior studies we have completed at our Energy Justice Lab, and we are grateful to the current faculty, staff, and student colleagues at the Lab, as well as the more than three dozen Lab alumni. Alison Knasin, the Energy Justice Lab manager, works alongside us every day to keep us organized. We are especially grateful to two of our former PhD students, Michelle Graff and Trevor Memmott, who have been important collaborators in this work.

Among the many other colleagues with whom we have collaborated on portions of this research are Jeff Adams, Jacob Alder, Stephen Ansolabehere, Zoya Atiq, Shalanda Baker, Galen Barbose, Dominic Bednar, Parrish Bergquist, Gabe Chan, Keith Cooley, Kristin Dziczek, Caroline Engle, Tom Evans, Matt Flaherty, David Foster, Naomi Freel, Jason Gumaer, Christina Hajj, Laura Helmke-Long, Michael Jefferson, Shaun Khurana, Manuel Kwakye, Nick Land, Cristian Crespo Montañés, Destenie Nock, Eric O'Shaughnessy, Linus Platzer, Lucy Qiu, Daniel Raimi, Jen Silva, Benjamin Sovacool, Savannah Sullivan, Jalonne White-Newsome, Brett Wiley, Amanda Woodrum, Maddy Yozwiak, Nikos Zirogiannis, and the numerous collaborators for the Roosevelt Project. Several individuals also provided exceptional research assistance at all stages of this project, including Shreya Bansal, Forrest Levy, and Emma Schuster.

Several friends and colleagues provided critical feedback on the book. We are especially grateful to Michael Ash, Barry Rabe, Jennie Stephens, Michael Webber, and two anonymous reviewers who generously spent the time to read and comment on a full draft of the manuscript. We also benefited from helpful insights from numerous people who read draft chapters of the book, including Mijin Cha, David Foster, John Graham, Catie Hausman, Mike Hogan, Sarah Mills, Greg Nemet, Destenie Nock, David Pellow, Lucy Qiu, and Daniel Raimi.

Finally, we are grateful to Chad Zimmerman, our editor at the University of Chicago Press, who was an enthusiastic champion of this book from the beginning. Chad's clear vision and steadfast support were instrumental throughout the project. We also want to thank the production and promotional staff with the press, including Rosemary Frehe, Caterina MacLean, Michaela Luckey, and Jessica Wilson, who each contributed their time and expertise to bring the book to the finish line.

For us, this book is both a culmination and a beginning.

NOTES

CHAPTER 1

1. Federal Energy Regulatory Commission (FERC), *Final Report on February 2021 Freeze Underscores Winterization Recommendations* (FERC, November 16, 2021), https://www.ferc.gov/news-events/news/final-report-february-2021-freeze-underscores-winterization-recommendations.

2. Zack Budryk, "Almost 5 Million Without Power as Winter Storm Stresses Grid in Texas, 13 Other States," *Hill*, February 16, 2021, https://thehill.com/regulation/energy-environment/538953-almost-5-million-without-power-as-winter-storm-stresses-grid/.

3. Chris Stipes, "New Report Details Impact of Winter Storm Uri on Texans," University of Houston, March 29, 2021, https://www.uh.edu/news-events/stories/2021/march-2021/03292021-hobby-winter-storm.

4. Giulia McDonnell Nieto del Rio et al., "His Lights Stayed On During Texas' Storm. Now He Owes $16,752," *New York Times*, February 21, 2021, https://www.nytimes.com/2021/02/20/us/texas-storm-electric-bills.html.

5. Emily Foxhall and Erin Douglas, "Appeals Court Says State Agency Set Electricity Prices Too High During 2021 Winter Storm," *Texas Tribune*, March 17, 2023, https://www.texastribune.org/2023/03/17/puc-appeals-court-uri-prices/.

6. Patrick Svitek, "Texas Puts Final Estimate of Winter Storm Death Toll at 246," *Texas Tribune*, January 2, 2022, https://www.texastribune.org/2022/01/02/texas-winter-storm-final-death-toll-246/.

7. Peter Aldhous et al., "The Texas Winter Storm and Power Outages Killed Hundreds More People Than the State Says," *BuzzFeed News*, May 26, 2021, https://www.buzzfeednews.com/article/peteraldhous/texas-winter-storm-power-outage-death-toll.

8. Svitek, "Texas Puts Final Estimate of Winter Storm Death Toll at 246."

9. Andrea Salcedo, "An 11-Year-Old Boy Died in an Unheated Texas Mobile Home: Authorities Suspect Hypothermia.," *Washington Post*, February 19, 2021, https://www.washingtonpost.com/nation/2021/02/19/texas-boy-death-winterstorm-pavon/.

10. Katie Watkins, "As Winter Storm Death Toll Exceeds Hurricane Harvey, Scope of Loss Becomes Clearer," *Houston Public Media*, March 30, 2021, https://www.houstonpublicmedia

.org/articles/news/energy-environment/2021/03/30/394592/remembering-the-lives-lost-in-the-winter-freeze-as-death-toll-surpasses-that-of-hurricane-harvey/.

11. Giulia McDonnell Nieto del Rio et al., "Extreme Cold Killed Texans in Their Bedrooms, Vehicles and Backyards," *New York Times*, February 19, 2021, https://www.nytimes.com/2021/02/19/us/texas-deaths-winter-storm.html.

12. Zeal Shah et al., "The Inequitable Distribution of Power Interruptions During the 2021 Texas Winter Storm Uri," *Environmental Research: Infrastructure and Sustainability* 3, no. 2 (2023): 025011.

13. Cheng-Chun Lee et al., "Community-Scale Big Data Reveals Disparate Impacts of the Texas Winter Storm of 2021 and Its Managed Power Outage," *Humanities and Social Sciences Communications* 9, no. 1 (September 24, 2022): 1–12.

14. NOAA National Centers for Environmental Information, "U.S. Billion-Dollar Weather and Climate Disasters," 2023. https://www.ncei.noaa.gov/access/billions/.

15. Nat Bullard, "Decarbonization: Stocks and Flows, Abundance and Scarcity, Net Zero," *Nathanielbullard.com*, January 31, 2024, https://www.nathanielbullard.com/presentations.

16. Trevor Memmott et al., "Sociodemographic Disparities in Energy Insecurity Among Low-Income Households Before and During the COVID-19 Pandemic," *Nature Energy* 6, no. 2 (2021): 186–93, https://doi.org/10.1038/s41560-020-00763-9.

17. David M. Konisky et al., "The Persistence of Household Energy Insecurity During the COVID-19 Pandemic," *Environmental Research Letters* 17, no. 10 (2022): 104017, https://doi.org/10.1088/1748-9326/ac90d7; Sanya Carley et al., "Behavioral and Financial Coping Strategies Among Energy-Insecure Households," *Proceedings of the National Academy of Sciences* 119, no. 36 (September 6, 2022): e2205356119, https://doi.org/10.1073/pnas.2205356119.

18. US Energy Information Administration, "Nonfossil Fuel Energy Sources Accounted for 21% of U.S. Energy Consumption in 2022," *EIA.gov*, June 29, 2023, https://www.eia.gov/todayinenergy/detail.php?id=56980.

19. US Bureau of Labor Statistics, "Databases, Tables and Calculators by Subject: Employment Hours, and Earnings from the Current Employment Statistics Survey (National)," series ID CES1021210001, *Data.BLS.gov*, accessed February 23, 2023, https://data.bls.gov/timeseries/CES1021210001.

20. For a rich discussion of the origin and evolution of the term "sacrifice zone," see Ryan Juskus, "Sacrifice Zones: A Genealogy and Analysis of an Environmental Justice Concept," *Environmental Humanities* 15, no. 1 (March 1, 2023): 3–24.

21. Intergovernmental Panel on Climate Change, *Climate Change 2021: The Physical Science Basis: Contribution of Working Group 1 to the Sixth Assessment Report of the Intergovernmental Panel on Climate Change* (Cambridge University Press, 2021), https://www.ipcc.ch/report/ar6/wg1/chapter/summary-for-policymakers/.

22. K. Marvel et al., "Chapter 2: Climate Trends," in *Fifth National Climate Assessment*, eds. A. R. Crimmins et al. (US Global Change Research Program, 2023), https://doi.org/10.7930/NCA5.2023.CH2.

23. Crimmins et al., eds., *Fifth National Climate Assessment*.

24. US Geological Survey, "Fifty Years of Glacier Change Research in Alaska," USGS.gov, September 28, 2016, https://www.usgs.gov/news/national-news-release/fifty-years-glacier-change-research-alaska.

25. Dana Habeeb et al., "Rising Heat Wave Trends in Large US Cities," *Natural Hazards* 76, no. 3 (April 1, 2015): 1651–65, https://doi.org/10.1007/s11069-014-1563-z.

26. Jessica Boehm, "Phoenix Experienced Its Hottest and Driest Summer on Record," *Axios*, October 10, 2023, https://www.axios.com/local/phoenix/2023/10/10/2023-hottest -driest-summer-monsoon.

27. Consider solar as an example, which went down from 41.7 cents per kilowatt-hour in 2010 to 4.9 cents per kilowatt-hour in 2022. See International Renewable Energy Agency (IRENA), *Renewable Power Generation Costs in 2022*, IRENA, August 29, 2023, https://www .irena.org/Publications/2023/Aug/Renewable-Power-Generation-Costs-in-2022.

28. Zack Hale, "World Needs $14 Trillion in Grid Spending by 2050 to Support Renewables—Report," *S&P Global Market Intelligence*, February 23, 2021, https://www .spglobal.com/marketintelligence/en/news-insights/latest-news-headlines/world-needs -14-trillion-in-grid-spending-by-2050-to-support-renewables-8211-report-62816721.

29. Jennie C. Stephens, "Beyond Climate Isolationism: A Necessary Shift for Climate Justice," *Current Climate Change Reports* 8, no. 4 (December 1, 2022): 83–90, https://doi.org/10 .1007/s40641-022-00186-6.

30. David M. Konisky, ed., *Failed Promises: Evaluating the Federal Government's Response to Environmental Justice*. MIT Press, 2015.

31. Noel Healy et al., "Embodied Energy Injustices: Unveiling and Politicizing the Transboundary Harms of Fossil Fuel Extractivism and Fossil Fuel Supply Chains," *Energy Research and Social Science* 48 (February 1, 2019): 219–34, https://doi.org/10.1016/j.erss.2018.09.016; Gavin Bridge et al., "Geographies of Energy Transition: Space, Place and the Low-Carbon Economy," *Energy Policy* 53 (2013): 331–40.

CHAPTER 2

1. "Fifth Ward Neighborhood in Houston, Texas (TX), 77020, 77026 Subdivision Profile— Real Estate, Apartments, Condos, Homes, Community, Population, Jobs, Income, Streets," *City-Data.com*, accessed April 17, 2023, http://www.city-data.com/neighborhood/Fifth -Ward-Houston-TX.html.

2. Kelsey Brugger, "'It Should Not Have Taken This Long': Regan Confronts EJ Vows," *E&E News*, November 23, 2021, https://www.eenews.net/articles/it-should-not-have-taken-this -long-regan-confronts-ej-vows/.

3. US Environmental Protection Agency, "EPA Administrator Michael S. Regan to Embark on 'Journey to Justice' Tour Through Mississippi, Louisiana, and Texas," news release, November 6, 2021, https://www.epa.gov/newsreleases/epa-administrator-michael-s-regan -embark-journey-justice-tour-through-mississippi.

4. Darryl Fears, "'Cancer Has Decimated Our Community': EPA's Regan Vows to Help Hard-Hit Areas, but Residents Have Doubts," *Washington Post*, January 4, 2022, https://www .washingtonpost.com/climate-environment/2021/11/28/epa-regan-environmental-justice -tour/.

5. Robert D. Bullard, *Dumping in Dixie: Race, Class, and Environmental Quality*, 3rd ed. (Westview Press, 2008).

6. US Environmental Protection Agency, "EPA Administrator," *EPA.gov*, March 9, 2021, https://www.epa.gov/aboutepa/epa-administrator.

7. J. Mijin Cha et al., "Climate and Environmental Justice Policies in the First Year of the Biden Administration," *Publius: The Journal of Federalism* 52, no. 3 (July 1, 2022): 408–27, https://doi.org/10.1093/publius/pjac017.

8. US Environmental Protection Agency, "Where You Live," *EPA.gov*, March 2023, https://www.epa.gov/trinationalanalysis/where-you-live.

9. Richard Rothstein, *The Color of Law: A Forgotten History of How Our Government Segregated America* (Liveright Publishing, 2017).

10. Steve Neavling, "Struggling to Breathe in 48217, Michigan's Most Toxic ZIP Code," *Detroit Metro Times*, January 2020, https://www.metrotimes.com/news/struggling-to-breathe-in-48217-michigans-most-toxic-zip-code-23542211.

11. Bullard, *Dumping in Dixie*.

12. Dorceta Taylor, *Toxic Communities: Environmental Racism, Industrial Pollution, and Residential Mobility* (New York University Press, 2014), https://doi.org/10.18574/nyu/9781479805150.001.0001.

13. Steve Lerner, *Sacrifice Zones: The Front Lines of Toxic Chemical Exposure in the United States* (MIT Press, 2012).

14. Jonathan Colmer et al., "Disparities in PM2.5 Air Pollution in the United States," *Science* 369, no. 6503 (July 31, 2020): 575–78, https://doi.org/10.1126/science.aaz9353.

15. Ryan Wishart, "Class Capacities and Climate Politics: Coal and Conflict in the United States Energy Policy–Planning Network," *Energy Research and Social Science* 48 (February 2019): 151–65, https://doi.org/10.1016/j.erss.2018.09.005; Victor McFarland and Jeff D. Colgan, "Oil and Power: The Effectiveness of State Threats on Markets," *Review of International Political Economy* 30, no. 2 (2023): 487–510, https://doi.org/10.1080/09692290.2021.2014931; Kate Pride Brown, "In the Pocket: Energy Regulation, Industry Capture, and Campaign Spending," *Sustainability: Science, Practice and Policy* 12, no. 2 (2016): 1–15, https://doi.org/10.1080/2052546.2016.11949232.

16. Trevor Culhane et al., "Who Delays Climate Action? Interest Groups and Coalitions in State Legislative Struggles in the United States," *Energy Research and Social Science* 79 (September 2021): 102114, https://doi.org/10.1016/j.erss.2021.102114.

17. US Energy Information Administration, *Monthly Energy Review* (US Energy Information Administration, November 2024), https://www.eia.gov/totalenergy/data/monthly/pdf/mer.pdf.

18. US Energy Information Administration, "U.S. Energy Facts Explained—Imports and Exports," *EIA.gov*, August 9, 2023, https://www.eia.gov/energyexplained/us-energy-facts/imports-and-exports.php.

19. Daniel Raimi et al., "Mapping County-Level Vulnerability to the Energy Transition in US Fossil Fuel Communities," *Scientific Reports* 12 (September 21, 2022): 15748, https://doi.org/10.1038/s41598-022-19927-6; US Energy Information Administration, "Natural Gas Explained: Where Our Natural Gas Comes From," *EIA.gov*, updated December 21, 2023, https://www.eia.gov/energyexplained/natural-gas/where-our-natural-gas-comes-from.php; US Energy Information Administration, "Oil and Petroleum Products Explained—Refining Crude Oil—Refinery Rankings," *EIA.gov*, June 2022, https://www.eia.gov/energyexplained/oil-and-petroleum-products/where-our-oil-comes-from.php.

20. National Academies of Sciences, Engineering, and Medicine, *Monitoring and

Sampling Approaches to Assess Underground Coal Mine Dust Exposures (National Academies Press, 2018), https://doi.org/10.17226/25111.

21. US Department of Commerce, *1950 Census of Population*, advance report no. PC-14 (Bureau of the Census, July 1953), https://www2.census.gov/library/publications/decennial /1950/pc-14/pc-14-18.pdf.

22. US Bureau of Labor Statistics, "Databases, Tables and Calculators by Subject: Employment, Hours, and Earnings from the Current Employment Statistics Survey (National)," series ID CES1021210001, *Data. BLS.gov*, accessed February 23, 2023, https://data.bls.gov /timeseries/CES1021210001.

23. The shift from underground mining in places like Appalachia to strip mining in places like Wyoming also resulted from regulation of sulfur dioxide as part of the US Clean Air Act; the coal mined in Wyoming has much lower sulfur content.

24. US Mine Safety and Health Administration, "Coal Fatalities for 1900 Through 2023," US Department of Labor, 2024, https://arlweb.msha.gov/stats/centurystats/coalstats.asp.

25. WDTV, "Ex-Coal CEO Don Blankenship at End of Prison Term," *WDTV.com*, May 10, 2017, https://www.wdtv.com/content/news/Ex-coal-CEO-Don-Blankenship-at-end -of-prison-term-421849674.html.

26. Calculation is based on the average number of fatalities per year between 2000 and 2022, using data from US Mine Safety and Health Administration, "Coal Fatalities for 1900 Through 2023."

27. A. Scott Laney and David N. Weissman, "Respiratory Diseases Caused by Coal Mine Dust," *Journal of Occupational and Environmental Medicine / American College of Occupational and Environmental Medicine* 56, supplement 10 (October 2014): S18–22, https://doi .org/10.1097/JOM.0000000000000260.

28. Jacek M. Mazurek et al., "Coal Workers' Pneumoconiosis: Attributable Years of Potential Life Lost to Life Expectancy and Potential Life Lost Before Age 65 Years—United States, 1999–2016," *Morbidity and Mortality Weekly Report* 67 (August 3, 2018), https://doi.org/10 .15585/mmwr.mm6730a3.

29. David J. Blackley et al., "Continued Increase in Prevalence of Coal Workers' Pneumoconiosis in the United States, 1970–2017," *American Journal of Public Health* 108, no. 9 (September 2018): 1220–22, https://doi.org/10.2105/AJPH.2018.304517.

30. Nadja Popovich, "Black Lung Disease Comes Storming Back in Coal Country," *New York Times*, February 2018, https://www.nytimes.com/interactive/2018/02/22/climate /black-lung-resurgence.html.

31. John Sherlock, "The Price of Coal: The Coal Mine Health and Safety Act of 1969," *Catholic University Law Review* 20, no. 3 (January 1, 1971): 496–510.

32. US National Research Council, Ford Foundation, and National Academy of Engineering Study Committee on the Potential for Rehabilitating Lands Surface Mined for Coal in the Western United States, eds., *Rehabilitation Potential of Western Coal Lands: A Report to the Energy Policy Project of the Ford Foundation* (Ballinger Pub. Co., 1974).

33. US Department of Agriculture Economic Research Service, "Poverty Area Measures," *USDA.gov*, November 2022, https://www.ers.usda.gov/data-products/poverty-area -measures.

34. US Energy Information Administration, "U.S. Field Production of Crude Oil (Thou-

sand Barrels per Day)," *EIA.gov*, accessed April 17, 2023, https://www.eia.gov/dnav/pet/hist /LeafHandler.ashx?n=pet&s=mcrfpus2&f=a.

35. US Energy Information Administration, "Natural Gas Gross Withdrawals and Production," *EIA.gov*, accessed October 16, 2023, https://www.eia.gov/dnav/ng/ng_prod_sum _dc_NUS_mmcf_a.htm.

36. However, during times of high gas-price spikes, these facilities either drastically increase the cost of energy or, if the power plant is able to accommodate multiple sources of fuel, the operator will switch over to coal combustion until gas prices return to lower levels.

37. It is important to emphasize that the shift from coal to natural gas for electric-power generation may not have resulted in large aggregate declines in greenhouse gas emissions, because of the increase in methane emissions from the massive expansion of oil and gas production. See, for example, Daniel Raimi, "The Greenhouse Gas Effects of Increased US Oil and Gas Production," *Energy Transitions* 4, no. 1 (2020): 45–56.

38. Other examples include the Permian Basin.

39. Barack Obama, "Remarks by the President in the State of the Union Address," White House Office of the Press Secretary, February 12, 2013, https://obamawhitehouse.archives .gov/the-press-office/2013/02/12/remarks-president-state-union-address.

40. Bryan Walsh, "Exclusive: How the Sierra Club Took Millions from the Natural Gas Industry—and Why They Stopped," *Time*, February 2, 2012, https://science.time.com/2012 /02/02/exclusive-how-the-sierra-club-took-millions-from-the-natural-gas-industry-and -why-they-stopped/.

41. Katie Jo Black et al., "Economic, Environmental, and Health Impacts of the Fracking Boom," *Annual Review of Resource Economics* 13, no. 1 (2021): 311–34, https://doi.org/10.1146 /annurev-resource-110320-092648.

42. Nicole C. Deziel et al., "Unconventional Oil and Gas Development and Health Outcomes: A Scoping Review of the Epidemiological Research," *Environmental Research* 182 (March 1, 2020): 109124, https://doi.org/10.1016/j.envres.2020.109124.

43. Elaine L. Hill and Lala Ma, "Drinking Water, Fracking, and Infant Health," *Journal of Health Economics* 82 (March 1, 2022): 102595, https://doi.org/10.1016/j.jhealeco.2022.102595.

44. Janet Currie et al., "Hydraulic Fracturing and Infant Health: New Evidence from Pennsylvania," *Science Advances* 3, no. 12 (December 13, 2017): e1603021, https://doi.org/10 .1126/sciadv.1603021.

45. Black et al., "Economic, Environmental, and Health Impacts of the Fracking Boom."

46. US Energy Information Administration, "Nuclear Explained: Where Our Uranium Comes From," *EIA.gov*, accessed April 17, 2023, https://www.eia.gov/energyexplained /nuclear/where-our-uranium-comes-from.php.

47. Lance N. Larson, *The Front End of the Nuclear Fuel Cycle: Current Issues* (Congressional Research Service, July 29, 2019), https://sgp.fas.org/crs/nuke/R45753.pdf.

48. Ronald Reagan, *United States Uranium Mining and Milling Industry: A Comprehensive Review: A Report to the Congress* (US Department of Energy, 1984).

49. Doug Brugge and Rob Goble, "The History of Uranium Mining and the Navajo People," *American Journal of Public Health* 92, no. 9 (September 2002): 1410–19, https://doi .org/10.2105/AJPH.92.9.1410.

50. Judy Pasternak, *Yellow Dirt: An American Story of a Poisoned Land and a People Betrayed* (Simon and Schuster, 2010).

51. Brugge and Goble, "History of Uranium Mining and the Navajo People."

52. Suzanne Skadowski and Margot Perez-Sullivan, "$2 Billion in Funds Headed for Cleanups in Nevada and on the Navajo Nation from Historic Anadarko Settlement with U.S. EPA, States," EPA news release, April 15, 2016, https://www.epa.gov/archive /epa/newsreleases/2-billion-funds-headed-cleanups-nevada-and-navajo-nation-historic -anadarko-settlement.html.

53. Francie Diep, "Abandoned Uranium Mines: An 'Overwhelming Problem' in the Navajo Nation," *Scientific American*, December 30, 2010, https://www.scientificamerican.com /article/abandoned-uranium-mines-a/.

54. Brugge and Goble, "History of Uranium Mining and the Navajo People."

55. Brugge and Goble, "History of Uranium Mining and the Navajo People."

56. Traci Brynne Voyles, *Wastelanding: Legacies of Uranium Mining in Navajo Country* (University of Minnesota Press, 2015).

57. Dina Gilio-Whitaker, *As Long as Grass Grows: The Indigenous Fight for Environmental Justice, from Colonization to Standing Rock* (Beacon Press, 2019).

58. Gilio-Whitaker, *As Long as Grass Grows.*

59. Rose M. Mueller, "Surface Coal Mining and Public Health Disparities: Evidence from Appalachia," *Resources Policy* 76 (2022): 102567, https://doi.org/10.1016/j.resourpol.2022 .102567; Michael Hendryx and Benjamin Holland, "Unintended Consequences of the Clean Air Act: Mortality Rates in Appalachian Coal Mining Communities," *Environmental Science and Policy* 63 (2016): 1–6, https://doi.org/10.1016/j.envsci.2016.04.021; Sarah J. Surber and D. Scott Simonton, "Disparate Impacts of Coal Mining and Reclamation Concerns for West Virginia and Central Appalachia," *Resources Policy* 54 (2017): 1–8, https://doi.org/10.1016/j .resourpol.2017.08.004.

60. Aysha Bodenhamer, "King Coal: A Study of Mountaintop Removal, Public Discourse, and Power in Appalachia," *Society and Natural Resources* 29, no. 10 (2016): 1139–53, https:// doi.org/10.1080/08941920.2016.1138561.

61. US Energy Information Administration, "Natural Gas Explained: Natural Gas Pipelines," *EIA.gov*, updated March 19, 2024, https://www.eia.gov/energyexplained/natural-gas /natural-gas-pipelines.php.

62. US Department of Transportation, Federal Highway Administration, "Interstate Frequently Asked Questions," *Highways.dot.gov*, updated June 30, 2023, https://highways .dot.gov/highway-history/interstate-system/50th-anniversary/interstate-frequently-asked -questions#question4.

63. These statistics calculated from data available from the US Department of Transportation, "Pipeline Incident 20 Year Trends," Pipeline and Hazardous Materials Safety and Administration, August 21, 2023, https://www.phmsa.dot.gov/data-and-statistics/pipeline /pipeline-incident-20-year-trends.

64. US Department of Transportation, "Pipeline Incident 20 Year Trend."

65. Trevor Hunnicutt, "Schwarzenegger Tours Calif. Gas Line Blast Site," *Seattle Times*, September 16, 2010, https://www.seattletimes.com/business/schwarzenegger-tours-calif -gas-line-blast-site/.

66. Ryan E. Emanuel et al., "Natural Gas Gathering and Transmission Pipelines and Social Vulnerability in the United States," *GeoHealth* 5, no. 6 (2021): e2021GH000442, https:// doi.org/10.1029/2021GH000442.

67. Bill McKibben, "With the Keystone Pipeline, Drawing a Line in the Tar Sands," *YaleEnvironment360*, October 2011, https://e360.yale.edu/features/with_the_keystone_pipeline_drawing_a_line_in_the_tar_sands.

68. Kendra Pierre-Louis, "This Land Is (Still) Their Land: Meet the Nebraskan Farmers Fighting Keystone XL," *Popular Science*, September 15, 2017, https://www.popsci.com/keystone-xl-pipeline-nebraska-farmers/.

69. Timothy B. Gravelle and Erick Lachapelle, "Politics, Proximity and the Pipeline: Mapping Public Attitudes Toward Keystone XL," *Energy Policy* 83 (August 1, 2015): 99–108, https://doi.org/10.1016/j.enpol.2015.04.004.

70. John Henderson, quoted in Rebecca Hersher, "Key Moments in the Dakota Access Pipeline Fight," *NPR*, February 22, 2017, https://www.npr.org/sections/thetwo-way/2017/02/22/514988040/key-moments-in-the-dakota-access-pipeline-fight.

71. Michael Bennett Cohn, "Dakota Access Pipeline Standing Rock Standoff: Behind the Front Lines," *Newsweek*, November 28, 2016, https://www.newsweek.com/2016/12/09/dakota-access-pipeline-protest-standing-rock-sioux-525894.html.

72. Cohn, "Dakota Access Pipeline Standing Rock Standoff."

73. Robert Brulle and David Pellow, "Environmental Justice: Human Health and Environmental Inequalities," *Annual Review of Public Health* 27 (April 21, 2006): 103–24; Paul Mohai and Bunyan Bryant, "Environmental Injustice: Weighing Race and Class as Factors in the Distribution of Environmental Hazards," *University of Colorado Law Review* 63, no. 4 (1992): 921–32; Evan J. Ringquist, "Assessing Evidence of Environmental Inequities: A Meta-Analysis," *Journal of Policy Analysis and Management* 24, no. 2 (2005): 223–47, https://doi.org/10.1002/pam.20088.

74. US Energy Information Administration, "Frequently Asked Questions (FAQs)—When Was the Last Refinery Built in the United States?," *EIA.gov*, July 8, 2022, https://www.eia.gov/tools/faqs/faq.php?id=29&t=6.

75. US Energy Information Administration, "Frequently Asked Questions (FAQs)—How Many Gallons of Gasoline and Diesel Fuel Are Made from One Barrel of Oil?," *EIA.gov*, May 2023, https://www.eia.gov/tools/faqs/faq.php?id=327&t=9#:~:text=Petroleum%20refineries%20in%20the%20United,gallon%20barrel%20of%20crude%20oil.

76. See, for example, the lists compiled by the Political Economy Research Institute at the University of Massachusetts Amherst, which can be found at "Top 100 Polluter Indexes," *Peri.UMASS.edu*, 2024, https://peri.umass.edu/top-100-polluter-indexes.

77. Computed from EPA statistics found at "National Enforcement and Compliance History Online Data Downloads," Enforcement and Compliance History Online, last updated October 2, 2024, https://echo.epa.gov/tools/data-downloads#downloads.

78. Hazardous Substance Research Centers/South and Southwest Outreach Program, *Environmental Update #12: Environmental Impact of the Petroleum Industry*, June 2023, https://cfpub.epa.gov/ncer_abstracts/index.cfm/fuseaction/display.files/fileID/14522.

79. US Environmental Protection Agency, "Petroleum Refinery National Case Results," *EPA.gov*, December 2022, https://www.epa.gov/enforcement/petroleum-refinery-national-case-results; also cited in Cynthia Giles, *Next Generation Compliance: Environmental Regulation for the Modern Era* (Oxford University Press, 2022).

80. Mary Greene and Keene Kelderman, "Port Arthur, Texas: The End of the Line for an Economic Myth," *Environmental Integrity Project* (blog), August 2017, https://

environmentalintegrity.org/what-we-do/oil-and-gas/the-human-cost-of-energy-production/port-arthur-texas/.

81. US Census Bureau, "QuickFacts: Port Arthur City, Texas," *Census.gov*, July 2022, https://www.census.gov/quickfacts/portarthurcitytexas.

82. Saudi Aramco completed its purchase of the refinery from Motiva in 2022.

83. US Energy Information Administration, "Oil and Petroleum Products Explained."

84. Robert Nelson et al., "Mapping Inequality: Redlining in New Deal America," University of Richmond, accessed April 22, 2023, https://dsl.richmond.edu/panorama/redlining/.

85. Hilton Kelley, quoted in Ted Genoways, "Port Arthur, Texas: American Sacrifice Zone," Natural Resources Defense Council, November 13, 2014, https://www.nrdc.org/stories/port-arthur-texas-american-sacrifice-zone.

86. Nikolaos Zirogiannis et al., "Understanding Excess Emissions from Industrial Facilities: Evidence from Texas," *Environmental Science and Technology* 52, no. 5 (March 6, 2018): 2482–90, https://doi.org/10.1021/acs.est.7b04887.

87. Texas Commission on Environmental Quality, "TCEQ Air Emission Event Reports Database," web application, Texas Commission on Environmental Quality, 2023, https://www2.tceq.texas.gov/oce/eer/index.cfm?fuseaction=main.searchForm&newsearch=yes.

88. Centers for Disease Control and Prevention, "2020 Adult Asthma Data: Prevalence Tables and Maps," *CDC.gov*, March 22, 2022, https://www.cdc.gov/asthma/brfss/2020/tableL1.html; Centers for Disease Control and Prevention, "PLACES: Local Data for Better Health," *CDC.gov*, July 13, 2023, https://www.cdc.gov/places/index.html.

89. Katie Watkins, "Port Arthur Residents Call for Civil Rights Probe into How Texas Has Handled Air Pollution in Their Neighborhood," *Houston Public Media*, August 18, 2021, https://www.houstonpublicmedia.org/articles/news/energy-environment/2021/08/18/406282/port-arthur-residents-ask-epa-to-open-a-civil-rights-investigation-into-how-texas-has-handled-air-pollution-in-their-neighborhood/; Colin Cox, "EPA Agrees to Investigate Texas for Alleged Civil Rights Violations Caused by Air Pollution from Port Arthur Plant," *Environmental Integrity Project* (blog), October 15, 2021, https://environmentalintegrity.org/news/epa-agrees-to-investigate-texas-for-alleged-civil-rights-violations-caused-by-air-pollution-from-port-arthur-plant/.

90. American Lung Association, "Key Findings," American Lung Association *State of the Air* report, 2024, https://www.lung.org/research/sota/key-findings.

91. World Health Organization, "Ambient (Outdoor) Air Pollution," *WHO. int*, December 19, 2022, https://www.who.int/news-room/fact-sheets/detail/ambient-(outdoor)-air-quality-and-health.

92. Richard Burnett et al., "Global Estimates of Mortality Associated with Long-Term Exposure to Outdoor Fine Particulate Matter," *Proceedings of the National Academy of Sciences* 115, no. 38 (September 18, 2018): 9592–97, https://doi.org/10.1073/pnas.1803222115; Sumil K. Thakrar et al., "Reducing Mortality from Air Pollution in the United States by Targeting Specific Emission Sources," *Environmental Science and Technology Letters* 7, no. 9 (September 8, 2020): 639–45, https://doi.org/10.1021/acs.estlett.0c00424.

93. Centers for Disease Control and Prevention, "Injury Data Visualization Tools," WISQARS (Web-Based Injury Statistics Query and Reporting System), 2020, https://wisqars.cdc.gov/data/non-fatal/home.

94. Centers for Disease Control and Prevention, "Suicide Prevention: Facts About Suicide," *CDC.gov*, April 6, 2023, https://www.cdc.gov/suicide/facts/index.html.

95. National Highway Traffic Safety Administration, "Newly Released Estimates Show Traffic Fatalities Reached a 16-Year High in 2021," United States Department of Transportation press release, May 17, 2022, https://www.nhtsa.gov/press-releases/early-estimate-2021 -traffic-fatalities.

96. Rob Nixon, *Slow Violence and the Environmentalism of the Poor* (Harvard University Press, 2011); Erik Kojola and David N. Pellow, "New Directions in Environmental Justice Studies: Examining the State and Violence," *Environmental Politics* 30, no. 1–2 (2021): 100–18; Thom Davies, "Slow Violence and Toxic Geographies: 'Out of Sight' to Whom?," *Environment and Planning C: Politics and Space* 40, no. 2 (2022): 409–27.

97. Kevin P. Josey et al., "Air Pollution and Mortality at the Intersection of Race and Social Class," *New England Journal of Medicine* 388, no. 15 (April 13, 2023): 1396–404, https://doi .org/10.1056/NEJMsa2300523; Christopher W. Tessum et al., "Inequity in Consumption of Goods and Services Adds to Racial–Ethnic Disparities in Air Pollution Exposure," *Proceedings of the National Academy of Sciences* 116, no. 13 (March 26, 2019): 6001–6, https://doi.org /10.1073/pnas.1818859116.

98. American Lung Association, "Key Findings"; Colmer et al., "Disparities in PM2.5 Air Pollution in the United States."

99. US Environmental Protection Agency, "Clean Air Power Sector Programs: Progress Report—Affected Communities," *EPA.gov*, March 15, 2023, https://www.epa.gov/power -sector/progress-report-affected-communities.

100. Redlining was a routine practice of the federal Home Owners' Loan Corporation (HOLC) beginning in the 1930s to demarcate certain areas as too "hazardous" for investment. In general, hazardous areas were the areas in cities where Black residents lived, often forcibly so due to legal or de facto segregation. As a result, Black Americans were unable to access home loans backed by government insurance programs, making home ownership impossible for many. These practices continued virtually unabated until the enactment of the 1968 Fair Housing Act. For more, see Rothstein, *Color of Law*.

101. Lara J. Cushing et al., "Historical Red-Lining Is Associated with Fossil Fuel Power Plant Siting and Present-Day Inequalities in Air Pollutant Emissions," *Nature Energy* 8, no. 1 (January 2023): 52–61, https://doi.org/10.1038/s41560-022-01162-y.

102. David M. Konisky, "Inequities in Enforcement? Environmental Justice and Government Performance," *Journal of Policy Analysis and Management* 28, no. 1 (2009): 102–21; David M. Konisky and Christopher Reenock, "Regulatory Enforcement, Riskscapes, and Environmental Justice," *Policy Studies Journal* 46, no. 1 (2018): 7–36.

103. US Environmental Protection Agency, "Sources of Greenhouse Gas Emissions," *EPA.gov*, last updated October 22, 2024, https://www.epa.gov/ghgemissions/sources -greenhouse-gas-emissions.

104. Calculations based on National Tier 1 CAPS Trends, found at US Environmental Protection Agency, "Air Emission Inventory: Air Pollutant Emissions Trends Data," *EPA.gov*, March 31, 2023, https://www.epa.gov/air-emissions-inventories/air-pollutant-emissions -trends-data.

105. Gabe Samuels and Yonah Freemark, *The Polluted Life Near the Highway* (Urban Institute, November 2022); Sofie Van Roosbroeck et al., "Long-Term Personal Exposure to

PM2.5, Soot and NOx in Children Attending Schools Located Near Busy Roads, a Validation Study," *Atmospheric Environment* 41, no. 16 (May 1, 2007): 3381–94, https://doi.org/10.1016 /j.atmosenv.2006.12.023; Doug Brugge et al., "Near-Highway Pollutants in Motor Vehicle Exhaust: A Review of Epidemiologic Evidence of Cardiac and Pulmonary Health Risks," *Environmental Health* 6, no. 1 (August 9, 2007): 23, https://doi.org/10.1186/1476-069X-6-23; R. Baldauf et al., "Impacts of Noise Barriers on Near-Road Air Quality," *Atmospheric Environment* 42, no. 32 (October 1, 2008): 7502–7, https://doi.org/10.1016/j.atmosenv.2008.05.051.

106. Rachael Dottle et al., "What It Looks Like to Reconnect Black Communities Torn Apart by Highways," *Bloomberg.com*, July 28, 2021, https://www.bloomberg.com/graphics /2021-urban-highways-infrastructure-racism/; Tyler Fenwick, "How I-65/70 Shattered Black Neighborhoods," *Indianapolis Recorder*, December 27, 2018, https://indianapolisrecorder .com/c57d7e9c-09e4-11e9-b276-83369364ae4e/; Liam Dillon and Ben Poston, "The Racist History of America's Interstate Highway Boom," *Los Angeles Times*, November 11, 2021, https://www.latimes.com/homeless-housing/story/2021-11-11/the-racist-history-of -americas-interstate-highway-boom; Farrell Evans, "How Interstate Highways Gutted Communities—and Reinforced Segregation," *History*, September 21, 2023, https://www .history.com/news/interstate-highway-system-infrastructure-construction-segregation.

107. Robert Doyle Bullard et al., eds., *Highway Robbery: Transportation Racism and New Routes to Equity* (South End Press, 2004); Eric Avila, *The Folklore of the Freeway: Race and Revolt in the Modernist City* (University of Minnesota Press, 2014).

108. US Environmental Protection Agency, "National Priorities List (NPL) Sites—by State," *EPA.gov*, August 14, 2015, https://www.epa.gov/superfund/national-priorities-list -npl-sites-state.

109. US Environmental Protection Agency, "Coal Ash Basics," *EPA.gov*, February 27, 2023, https://www.epa.gov/coalash/coal-ash-basics; Stephen Ritter, "A New Life for Coal Ash," *Chemical and Engineering News*, February 15, 2016, https://cen.acs.org/articles/94/i7/New -Life-Coal-Ash.html.

110. Ritter, "New Life for Coal Ash."

111. Kathleen Masterson, "Q&A: Examining the Tennessee Coal Ash Spill," *NPR*, January 8, 2009, https://www.npr.org/2009/01/08/99134153/q-a-examining-the-tennessee-coal -ash-spill.

112. Ritter, "New Life for Coal Ash."

113. Ritter, "New Life for Coal Ash."

114. John Russell, "Getting Ready to Kick the Coal Habit at IPL's Harding Street Station," *Indy Star*, May 29, 2015, https://www.indystar.com/story/money/2015/05/29/getting-ready -kick-coal-habit-ipls-harding-street-station/28085729/; AES Indiana, "IPL Burns Coal for the Last Time at Harding Street Station," press release, March 25, 2016, https://www.aesindiana .com/press-release/ipl-burns-coal-last-time-harding-street-station.

115. AES Indiana, "Harding Street Station," *AESIndiana.com*, March 11, 2020, https://www .aesindiana.com/harding-street-station.

116. Abel Russ et al., *Poisonous Coverup: The Widespread Failure of the Power Industry to Clean Up Coal Ash Dumps* (Environmental Integrity Project and EarthJustice, November 3, 2022), https://environmentalintegrity.org/wp-content/uploads/2022/11/Poisonous -Coverup-Final.pdf.

117. US Census Bureau, "American Community Survey 5-Year Estimates," *Census Reporter*

profile page for Uniontown, AL, 2021, http://censusreporter.org/profiles/16000US0177904 -uniontown-al/.

118. "Arrowhead Landfill: Proven, Efficient and Environmentally Superior," Arrowhead Environmental Partners, 2024, https://arrowheadenvironmentalpartners.com/.

119. Debra Mayfield, "Five Years Later and the Story of the TVA Spill Continues," *Earthjustice.org*, December 23, 2013, https://earthjustice.org/article/five-years-later-and -the-story-of-the-tva-spill-continues.

120. Earthjustice, "Ashes: A Community's Toxic Inheritance," *Earthjustice.org*, September 1, 2014, https://earthjustice.org/feature/photos-a-toxic-inheritance.

CHAPTER 3

1. The full quote continues: "More than half of the renewable capacity added in 2019 achieved lower power costs than the cheapest new coal plants. Fossil fuels-related air pollution causes 1 in 5 of all deaths globally each year. And coal's economic viability is declining." See António Guterres, "Secretary-General's Video Message to Powering Past Coal Alliance Summit," *United Nations, March* 2, 2021, https://www.un.org/sg/en/content/sg/statement /2021-03-02/secretary-generals-video-message-powering-past-coal-alliance-summit.

2. Daniel A. Hawkins, "Two Miles to Hell: A Miner's Story," *Appalachian Voice* (blog), April 3, 2010, https://appvoices.org/2010/04/03/two-miles-to-hell-a-miners-story/.

3. Susan F. Tierney, *The U.S. Coal Industry: Challenging Transitions in the 21st Century* (Analysis Group, Inc., September 26, 2016): ES-2. https://www.analysisgroup.com /globalassets/insights/publishing/2016-tierney-coal-industry-21st-century-challenges.pdf.

4. Edward P. Louie and Joshua M. Pearce, "Retraining Investment for U.S. Transition from Coal to Solar Photovoltaic Employment," *Energy Economics* 57 (June 2016): 295–302, https:// doi.org/10.1016/j.eneco.2016.05.016.

5. Bureau of Labor Statistics data, available at US Bureau of Labor Statistics, "Databases, Tables and Calculators by Subject: Employment, Hours, and Earnings from the Current Employment Statistics Survey (National)," series ID CES1021210001, *Data. BLS.gov*, accessed February 23, 2023, https://data.bls.gov/timeseries/CES1021210001.

6. Linda Lobao et al., "Poverty, Place, and Coal Employment Across Appalachia and the United States in a New Economic Era," *Rural Sociology* 81, no. 3 (September 2016): 343–86, https://doi.org/10.1111/ruso.12098; Robert N. Stavins, "What Can We Learn from the Grand Policy Experiment? Lessons from SO2 Allowance Trading," *Journal of Economic Perspectives* 12, no. 3 (August 1, 1998): 69–88, https://doi.org/10.1257/jep.12.3.69.

7. Global Energy Monitor, "Global Coal Plant Tracker," *GlobalEnergyMonitor.org*, April 25, 2022, https://globalenergymonitor.org/projects/global-coal-plant-tracker/.

8. Reuters, "Factbox: U.S. Coal-Fired Power Plants Scheduled to Shut," *Reuters.com*, October 28, 2021, https://www.reuters.com/business/energy/us-coal-fired-power-plants -scheduled-shut-2021-10-28/.

9. US Energy Information Administration, "Electricity: Form EIA-860 Detailed Data with Previous Form Data (EIA-860A/860B)," *EIA.gov*, September 22, 2023, https://www.eia.gov /electricity/data/eia860/.

10. International Energy Agency, "Global Coal Demand Expected to Decline in Coming Years," *IEA.org*, December 15, 2023, https://www.iea.org/news/global-coal-demand

-expected-to-decline-in-coming-years; Govind Bhutada, "Charted: Energy Consumption by Source and Country (1969–2018)," *Visual Capitalist*, September 9, 2020, https://www.visualcapitalist.com/energy-consumption-by-source-and-country-1969-2018/.

11. International Energy Agency, "World Total Coal Production, 1971–2020," *IEA.org*, updated July 29, 2021, https://www.iea.org/data-and-statistics/charts/world-total-coal-production-1971-2020.

12. President Biden later rejoined the international agreement and, as of the 2023 Conference of the Parties (COP) in Dubai, there is now at least a stated commitment across most countries to draw down fossil fuels.

13. Maura Grogan et al., *Native American Lands and Natural Resource Development* (Revenue Watch Institute, 2011). Several additional tribal communities for oil and gas include Blackfeet (Montana), Assiniboine and Sioux (Montana), Cheyenne River Sioux (South Dakota), and various Pueblo tribes (New Mexico), among others. See US Department of the Interior, Indian Affairs, "Atlas of Oil and Gas Plays on American Indian Lands," *Bia.gov*, accessed October 29, 2024, https://www.bia.gov/bia/ots/demd/oil-gas-plays.

14. Daniel Raimi et al., "Mapping County-Level Vulnerability to the Energy Transition in US Fossil Fuel Communities," *Scientific Reports* 12, no. 1 (September 21, 2022): 15748, https://doi.org/10.1038/s41598-022-19927-6.

15. US Energy Information Administration, "U.S. Production of Petroleum and Other Liquids to Be Driven by International Demand," *EIA.gov*, April 4, 2023, https://www.eia.gov/todayinenergy/detail.php?id=56041.

16. US Energy Information Administration, "U.S. Production of Petroleum and Other Liquids to Be Driven by International Demand."

17. International Energy Agency, *World Energy Outlook 2021* (International Energy Agency, October 2021), https://www.iea.org/reports/world-energy-outlook-2021.

18. International Energy Agency, *Global EV Outlook 2020* (International Energy Agency, June 2020), https://www.iea.org/reports/global-ev-outlook-2020.

19. Nat Bullard, "Decarbonization: Stocks and Flows, Abundance and Scarcity, Net Zero," *Nathanielbullard.com*, January 31, 2024, https://www.nathanielbullard.com/presentations.

20. Robert Pollin and Brian Callaci, "The Economics of Just Transition: A Framework for Supporting Fossil Fuel–Dependent Workers and Communities in the United States," *Labor Studies Journal* 44, no. 2 (June 2019): 93–138, https://doi.org/10.1177/0160449X18787051.

21. Tom Guevara et al., *Economic, Fiscal, and Social Impacts of the Transition of Electricity Generation Resources in Indiana* (Indiana University Public Policy Institute, August 2020).

22. This trend can be seen in the rapid growth of miners without a pension fund: in 2016, about forty thousand miners had no pension, while that number doubled to eighty thousand by 2018. Many coal miners rely on the United Mine Workers of America (UMWA), a multi-employer pension fund. However, as of 2017, the plan had only $3 billion in assets and around $9 billion in liabilities, mostly due to coal companies going bankrupt. UMWA was scheduled to run out of funds by 2023, but in 2019 Congress authorized the treasury to use funds from the Abandoned Land Mines Reclamation Fund to spend up to $750 million annually on underfunded pensions. As of 2019, the average coal miner pension was $7,150 annually. For comparison, that is 52 percent of the federal poverty level, or the equivalent to annual earnings if you work for $3.42 per hour, forty days per week, fifty-two weeks per year. See Mary Williams Walsh, "Congress Saves Coal Miner Pensions, but What About Oth-

ers?," *New York Times*, December 24, 2019, https://www.nytimes.com/2019/12/24/business /coal-miner-pensions-bailout.html.

23. Pollin and Callaci, "Economics of Just Transition."

24. Lobao et al., "Poverty, Place, and Coal Employment across Appalachia and the United States in a New Economic Era."

25. Michael R. Betz et al., "Coal Mining, Economic Development, and the Natural Resources Curse," *Energy Economics* 50 (July 2015): 105-16, https://doi.org/10.1016/j.eneco.2015 .04.005.

26. Shannon Bell, *Fighting King Coal: The Challenges to Micromobilization in Central Appalachia* (MIT Press, 2016), https://mitpress.mit.edu/9780262528801/fighting-king-coal/.

27. Lobao et al., "Poverty, Place, and Coal Employment across Appalachia and the United States in a New Economic Era."; William R. Freudenburg, "Addictive Economies: Extractive Industries and Vulnerable Localities in a Changing World Economy," *Rural Sociology* 57, no. 3 (September 1992): 305-32, https://doi.org/10.1111/j.1549-0831.1992.tb00467.x.

28. Sanya Carley et al., "Adaptation, Culture, and the Energy Transition in American Coal Country," *Energy Research and Social Science* 37 (March 2018): 133-39, https://doi.org /10.1016/j.erss.2017.10.007.

29. Kelli F. Roemer and Julia H. Haggerty, "The Energy Transition as Fiscal Rupture: Public Services and Resilience Pathways in a Coal Company Town," *Energy Research and Social Science* 91 (September 2022).

30. Lobao et al., "Poverty, Place, and Coal Employment across Appalachia and the United States in a New Economic Era."

31. Katherine Locke, "Navajo Generating Station Shuts Down Permanently," *Navajo-Hopi Observer News*, November 18, 2019, https://www.nhonews.com/news/2019/nov/18/navajo -generating-station-shuts-down-permanently/.

32. Sean O'Leary, *The Natural Gas Fracking Boom and Appalachia's Lost Economic Decade* (Ohio River Valley Institute, February 12, 2021).

33. Stratford Douglas and Anne Walker, "Coal Mining and the Resource Curse in the Eastern United States," *Journal of Regional Science* 57, no. 4 (September 2017): 568-90. https:// doi.org/10.1111/jors.12310.

34. Mark D. Partridge et al., "Natural Resource Curse and Poverty in Appalachian America," *American Journal of Agricultural Economics* 95, no. 2 (2013): 449-56.

35. Adam Mayer et al., "Fracking Fortunes: Economic Well-Being and Oil and Gas Development Along the Urban-Rural Continuum," *Rural Sociology* 83, no. 3 (September 2018): 532-67, https://doi.org/10.1111/ruso.12198.

36. Jason Beckfield et al., *How the Gulf Coast Can Lead the Energy Transition* (Roosevelt Project, MIT Center for Energy and Environmental Policy Research, Harvard University, April 2022).

37. Mayer et al., "Fracking Fortunes."

38. Sari Kovats et al., "The Health Implications of Fracking," *Lancet* 383, no. 9919 (March 2014): 757-58, https://doi.org/10.1016/S0140-6736(13)62700-2.

39. For an overview of these studies, see Adam Mayer, "Quality of Life and Unconventional Oil and Gas Development: Towards a Comprehensive Impact Model for Host Communities," *Extractive Industries and Society* 4 (2017): 923-30.

40. Timothy M. Komarek, "Labor Market Dynamics and the Unconventional Natural Gas

Boom: Evidence from the Marcellus Region," *Resource and Energy Economics* 45 (August 2016): 1–17, https://doi.org/10.1016/j.reseneeco.2016.03.004.

41. Jameson K. Hirsch et al., "Psychosocial Impact of Fracking: A Review of the Literature on the Mental Health Consequences of Hydraulic Fracturing," *International Journal of Mental Health and Addiction* 16, no. 1 (February 2018): 1–15, https://doi.org/10.1007/s11469 -017-9792-5.

42. Darrick Evensen and Rich Stedman, "'Fracking': Promoter and Destroyer of 'the Good Life,'" *Journal of Rural Studies* 59 (April 2018): 142–52, https://doi.org/10.1016/j.jrurstud .2017.02.020.

43. Complicating and magnifying these trade-offs is the phenomena of the "social multiplier effect," which sociologists Jason Beckfield and Daniel Evrard describe as "a situation wherein, although a relatively small proportion of a community may be employed by it, the industry receives a disproportionate share of support because many residents possess strong family and social ties to energy workers." See Jason Beckfield and Daniel Alain Evrard, "The Social Impacts of Supply-Side Decarbonization," *Annual Review of Sociology* 49, no. 1 (July 31, 2023): 155–75, https://doi.org/10.1146/annurev-soc-031021-012201.

44. Jennifer Bowman and Kelly Johnson, *2015 Stream Health Report: An Evaluation of Water Quality, Biology, and Acid Mine Drainage Reclamation in Five Watersheds: Raccoon Creek, Monday Creek, Sunday Creek, Huff Run, and Leading Creek* (Voinovich School of Leadership and Public Affairs at Ohio University, June 30, 2016); Alex Meyer, "Appalachia's Orange Stain," *Post*, accessed October 27, 2023, https://projects.thepostathens.com /SpecialProjects/AcidMines-AMeyer/. In this specific case, a team of Ohio University professors devised a way to extract the iron oxide from the water and make it into paint pigment.

45. Gareth Evans, "A Toxic Crisis in America's Coal Country," *BBC News*, February 11, 2019, https://www.bbc.com/news/world-us-canada-47165522.

46. Evans, "Toxic Crisis in America's Coal Country."

47. Beckfield and Evrard, "Social Impacts of Supply-Side Decarbonization."

48. J. Mijin Cha, "A Just Transition for Whom? Politics, Contestation, and Social Identity in the Disruption of Coal in the Powder River Basin," *Energy Research and Social Science* 69 (November 1, 2020): 101657, https://doi.org/10.1016/j.erss.2020.101657.

49. She also posits a third theory, though less relevant to our discussion: that the coal industry's campaigns both introduce and reinforce a "local hegemonic masculinity" that serves to quell discontent and results in activism that falls along specific gendered lines, mainly female activism. See Bell, *Fighting King Coal*.

50. Bethel W. Tarekegne et al., *Coal-Dependent Communities in Transition: Identifying Best Practices to Ensure Equitable Outcomes* (Pacific Northwest National Laboratory, September 2021).

51. Tarekegne et al., *Coal-Dependent Communities in Transition*.

52. Sarah Bowman, "Indiana Supreme Court: Duke Can't Make Customers Pay $212 Million to Clean up Coal Ash," *Indianapolis Star*, March 11, 2022, https://www.indystar.com /story/news/environment/2022/03/11/court-duke-energy-cant-make-customers-pay-212 -m-clean-up-coal-ash/7001908001/.

53. Isabelle Chapman, "Gambling 'America's Amazon,'" *CNN*, December 5, 2021, https:// www.cnn.com/interactive/2021/12/us/coal-ash-ponds-plant-barry/.

54. Beckfield et al., "How the Gulf Coast Can Lead the Energy Transition."

55. US Global Change Research Program, *Climate Science Special Report: Fourth National Climate Assessment, Volume 2: Impacts, Risks, and Adaptation in the United States* (US Global Change Research Program, 2018), https://nca2018.globalchange.gov; Intergovernmental Panel on Climate Change, *Climate Change 2022—Impacts, Adaptation and Vulnerability: Working Group 2 Contribution to the Sixth Assessment Report of the Intergovernmental Panel on Climate Change*, 1st ed. (Cambridge University Press, June 2023), https://doi.org/10.1017/9781009325844.

56. Beckfield et al., "How the Gulf Coast Can Lead the Energy Transition."

57. For our peer-reviewed articles related to this fieldwork, see Carley et al., "Adaptation, Culture, and the Energy Transition in American Coal Country"; Michelle Graff et al., "Stakeholder Perceptions of the United States Energy Transition: Local-Level Dynamics and Community Responses to National Politics and Policy," *Energy Research and Social Science* 43 (September 2018): 144–57, https://doi.org/10.1016/j.erss.2018.05.017.

58. One study finds that every $1 million in investments in clean energy yields about seventeen jobs across the economy, while $1 million in investments in the fossil fuel industry yields about five jobs (see Robert Pollin et al., *Green Growth: A U.S. Program for Controlling Climate Change and Expanding Job Opportunities* (Center for American Progress and Policy Economy Research Institute, September 2014), https://cdn.americanprogress.org/wp-content/uploads/2014/09/PERI.pdf.

59. E. Mark Curtis et al., "Workers and the Green-Energy Transition: Evidence from 300 Million Job Transitions," working paper no. 31539 (National Bureau of Economic Research, August 2023), https://doi.org/10.3386/w31539.

60. Max Vanatta et al., "The Costs of Replacing Coal Plant Jobs with Local Instead of Distant Wind and Solar Jobs Across the United States," *iScience* 25, no. 8 (August 2022): 104817, https://doi.org/10.1016/j.isci.2022.104817.

61. Beckfield and Evrard, "Social Impacts of Supply-Side Decarbonization"; Beckfield et al., "How the Gulf Coast Can Lead the Energy Transition."

62. Shannon Elizabeth Bell and Richard York, "Community Economic Identity: The Coal Industry and Ideology Construction in West Virginia: Community Economic Identity," *Rural Sociology* 75, no. 1 (January 19, 2010): 111–43, https://doi.org/10.1111/j.1549-0831.2009.0000.x.

63. For an extensive history of the Just Transition movement, refer to Dimitris Stevis et al., "Introduction: The Genealogy and Contemporary Politics of Just Transitions," in *Just Transitions: Social Justice in the Shift Towards a Low-Carbon World*, eds. Dimitris Stevis et al. (Pluto Press, 2020), 1–31, https://doi.org/10.2307/j.ctvs09qrx.6.

64. Stevis et al., "Introduction: The Genealogy and Contemporary Politics of Just Transitions."

65. J. Mijin Cha and Manuel Pastor, "Just Transition: Framing, Organizing, and Power-Building for Decarbonization," *Energy Research and Social Science* 90 (August 1, 2022): 102588, https://doi.org/10.1016/j.erss.2022.102588.

66. Michael Aklin and Johannes Urpelainen, "Enable a Just Transition for American Fossil Fuel Workers Through Federal Action," *Brookings*, August 2022, https://www.brookings.edu/articles/enable-a-just-transition-for-american-fossil-fuel-workers-through-federal-action/.

67. Aklin and Urpelainen, "Enable a Just Transition for American Fossil Fuel Workers Through Federal Action."

68. Evergreen Action, "How States Are Centering Workers in the Clean Energy Transition," *Evergreen Action* (blog), January 24, 2024, https://www.evergreenaction.com/blog/how-states-are-centering-workers-in-the-clean-energy-transition.

69. US Department of Energy, "Communities LEAP (Local Energy Action Program)," *Energy.gov*, 2024, https://www.energy.gov/communitiesLEAP/communities-leap.

CHAPTER 4

1. We interviewed Darrin in 2021 about his experience with utility disconnection. Like all individuals with whom we conducted interviews or focus groups, his name has been changed in this book to protect his identity and anonymity.

2. Sam Gorski, "What's the Average Cost of Utilities in West Virginia?," *12WBOY*, April 3, 2023, https://www.wboy.com/news/west-virginia/whats-the-average-cost-of-utilities-in-west-virginia/.

3. US Energy Information Administration, "Residential Energy Consumption Survey (RECS)," *EIA.gov*, 2023, https://www.eia.gov/consumption/residential/.

4. Sanya Carley and David Konisky, "Utility Disconnections Dashboard," Energy Justice Lab, 2023, https://utilitydisconnections.org/. During COVID, an additional ten states had reporting requirements. The Utility Disconnection Dashboard contains data for these states as well as for nine other states from which they were able to source data. For more detail on COVID reporting and disconnection data, see Jean Su and Christopher Kuveke, *Powerless in the Pandemic 2.0: After Bailouts, Electric Utilities Chose Profits over People* (Center for Biological Diversity, April 2022), https://bailout.cdn.prismic.io/bailout/ddebd6e2-b136-4dc8-a1da-f6d4583b4c24_Powerless_Report2022_final.pdf.

5. Our work focuses on households that are within 200 percent of the federal poverty line.

6. Sanya Carley and David Konisky, *Survey of Household Energy Insecurity in Time of COVID: Preliminary Results of Wave-4, and Waves 1 Through 4 Combined* (Indiana University Energy Justice Lab, July 5, 2021), https://energyjustice.indiana.edu/doc/wave_4_report.pdf.

7. Peter A. Kahn et al., "Characterization of Prescription Patterns and Estimated Costs for Use of Oxygen Concentrators for Home Oxygen Therapy in the US," *JAMA Network Open* 4, no. 10 (October 19, 2021): e2129967–e2129967, https://doi.org/10.1001/jamanetworkopen.2021.29967.

8. Trevor Memmott et al., "Sociodemographic Disparities in Energy Insecurity Among Low-Income Households Before and During the COVID-19 Pandemic," *Nature Energy* 6, no. 2 (February 2021): 186–93, https://doi.org/10.1038/s41560-020-00763-9; Carley and Konisky, *Survey of Household Energy Insecurity in Time of COVID*.

9. Ariel Drehobl et al., *How High Are Household Energy Burdens? An Assessment of National and Metropolitan Energy Burden Across the United States* (American Council for an Energy-Efficient Economy, September 2020), https://www.aceee.org/sites/default/files/pdfs/u2006.pdf.

10. US Department of Energy and Environmental Protection Agency, "Benefits of ENERGY STAR Qualified Windows, Doors, and Skylights," *EnergyStar.gov*, accessed January 17, 2023, archived at https://web.archive.org/web/20240125115518/https://www.energystar.gov/products/building_products/residential_windows_doors_and_skylights/benefits.

11. Save on Energy Team, "Electricity Bill Report: March 2024," *SaveOnEnergy.com*,

March 2024, https://www.saveonenergy.com/resources/electricity-bills-by-state/; Gorski, "What's the Average Cost of Utilities in West Virginia?"

12. Ariel Drehobl and Lauren Ross, *Lifting the High Energy Burden in America's Largest Cities: How Energy Efficiency Can Improve Low-Income and Underserved Communities* (American Council for an Energy-Efficient Economy, April 2016), https://www.aceee.org/research-report/u1602.

13. Michelle Graff et al., "Opportunities to Advance Research on Energy Insecurity," *Nature Energy* 8, no. 6 (June 2023): 550–53, https://doi.org/10.1038/s41560-023-01265-0; US Energy Information Administration, "Residential Energy Consumption Survey (RECS)."

14. Court Appointed Special Advocates (CASA) employee, in phone conversation with Sanya Carley, January 2021.

15. David M. Konisky et al., "The Persistence of Household Energy Insecurity During the COVID-19 Pandemic," *Environmental Research Letters* 17, no. 10 (September 2022): 104017, https://doi.org/10.1088/1748-9326/ac90d7.

16. Shuchen Cong et al., "Unveiling Hidden Energy Poverty Using the Energy Equity Gap," *Nature Communications* 13, no. 1 (May 4, 2022): 2456, https://doi.org/10.1038/s41467-022-30146-5.

17. Luling Huang et al., "Inequalities Across Cooling and Heating in Households: Energy Equity Gaps," *Energy Policy* 182 (November 1, 2023): 113748, https://doi.org/10.1016/j.enpol.2023.113748.

18. Cong et al., "Unveiling Hidden Energy Poverty Using the Energy Equity Gap."

19. Bryan Ke, "Devastating House Fire Claims Lives of 3 Children, Grandmother During Texas Winter Storm," *Yahoo News*, February 22, 2021, https://www.yahoo.com/now/devastating-house-fire-claims-lives-170707185.html.

20. US Consumer Product Safety Commission, "Seasons Change, but Fire and Carbon Monoxide Safety Is Year-Round; Warm Up to CPSC's Tips for Staying Safe During Colder Weather," news release, December 14, 2021, https://www.cpsc.gov/Newsroom/News-Releases/2022/Seasons-Change-but-Fire-and-Carbon-Monoxide-Safety-Is-Year-Round-Warm-Up-to-CPSCs-Tips-for-Staying-Safe-During-Colder-Weather.

21. Kelly McCleary et al., "All 17 Victims of Bronx Apartment Fire, Including 2-Year-Old, Died of Smoke Inhalation, NYC Medical Examiner Rules," *CNN*, January 13, 2022, https://www.cnn.com/2022/01/11/us/new-york-bronx-apartment-fire-tuesday/index.html.

22. We include the following in the category of risky temperature strategies: use of space heaters, the fireplace, the oven, the dryer vent, or burning trash to generate heat.

23. Mark Wolfe, "Press Release: Gasoline and Home Heating Will Cost More Than Christmas Gifts This Winter," *National Energy Assistance Directors' Association* (blog), November 22, 2021, https://neada.org/pr-gasolineandheat/.

24. US Energy Information Administration, "Residential Energy Consumption Survey (RECS)."

25. Jayanta Bhattacharya et al., "Heat or Eat? Cold-Weather Shocks and Nutrition in Poor American Families," *American Journal of Public Health* 93, no. 7 (July 2003): 1149–54, https://doi.org/10.2105/ajph.93.7.1149.

26. John T. Cook et al., "A Brief Indicator of Household Energy Security: Associations with Food Security, Child Health, and Child Development in US Infants and Toddlers," *Pediatrics* 122, no. 4 (October 2008): e867–75, https://doi.org/10.1542/peds.2008-0286.

27. Congressional Research Service, *LIHEAP: Program and Funding* (Congressional Research Service, updated June 22, 2018), https://www.everycrsreport.com/files/20180622 _RL31865_85805bac2287a504f2a4eb05e4637a3cd21eaa2e.pdf.

28. LIHEAP Clearinghouse, "LIHEAP and WAP Funding," *Liheapch.acf.hhs.gov*, 2023, https://liheapch.acf.hhs.gov/Funding/funding.htm.

29. Libby Perl, *LIHEAP: Program and Funding* (Congressional Research Service, May 22, 2015), https://digital.library.unt.edu/ark:/67531/metadc807073/.

30. Michelle Graff et al., "Climate Change and Energy Insecurity: A Growing Need for Policy Intervention," *Environmental Justice* 15, no. 2 (April 2022): 76–82, https://doi.org/10 .1089/env.2021.0032.

31. Sanya Carley et al., "Behavioral and Financial Coping Strategies Among Energy-Insecure Households," *Proceedings of the National Academy of Sciences* 119, no. 36 (September 6, 2022): e2205356119, https://doi.org/10.1073/pnas.2205356119.

32. This estimate is based on a high-flow concentrator, a 700-watt model, at twenty-four-hour use. A lower-flow concentrator and less frequent use of the machine would yield smaller annual costs. For example, a 350-watt, medium-flow concentrator, at twelve hours of use would yield $161 more in energy costs each year. See Kahn, et al., "Characterization of Prescription Patterns and Estimated Costs for Use of Oxygen Concentrators for Home Oxygen Therapy in the US."

33. Edmond D. Shenassa et al., "Dampness and Mold in the Home and Depression: An Examination of Mold-Related Illness and Perceived Control of One's Home as Possible Depression Pathways," *American Journal of Public Health* 97, no. 10 (October 2007): 1893–99, https://doi.org/10.2105/AJPH.2006.093773; Mark J. Mendell et al., "Respiratory and Allergic Health Effects of Dampness, Mold, and Dampness-Related Agents: A Review of the Epidemiologic Evidence," *Environmental Health Perspectives* 119, no. 6 (June 2011): 748–56, https:// doi.org/10.1289/ehp.1002410; Tony G. Reames et al., "Exploring the Nexus of Energy Burden, Social Capital, and Environmental Quality in Shaping Health in US Counties," *International Journal of Environmental Research and Public Health* 18, no. 2 (January 13, 2021): 620, https:// doi.org/10.3390/ijerph18020620.

34. Diana Hernández, "Understanding 'Energy Insecurity' and Why It Matters to Health," *Social Science and Medicine* 167 (October 1, 2016): 1–10, https://doi.org/10.1016/j.socscimed .2016.08.029.

35. Jonathan Teller-Elsberg et al., "Fuel Poverty, Excess Winter Deaths, and Energy Costs in Vermont: Burdensome for Whom?," *Energy Policy* 90 (March 1, 2016): 81–91, https://doi .org/10.1016/j.enpol.2015.12.009.

36. US Environmental Protection Agency and Centers for Disease Control and Prevention, *Climate Change and Extreme Heat: What You Can Do to Prepare* (US Environmental Protection Agency, October 2016), archived at https://web.archive.org/web/20161216180653 /https://www.cdc.gov/climateandhealth/pubs/extreme-heat-guidebook.pdf.

37. Kate R. Weinberger et al., "Estimating the Number of Excess Deaths Attributable to Heat in 297 United States Counties," *Environmental Epidemiology* 4, no. 3 (April 23, 2020): e096, https://doi.org/10.1097/EE9.0000000000000096.

38. Shahzeen Z. Attari et al., "Public Perceptions of Energy Consumption and Savings," *Proceedings of the National Academy of Sciences* 107, no. 37 (September 14, 2010): 16054–59, https://doi.org/10.1073/pnas.1001509107.

39. US Department of Energy and Environmental Protection Agency, *Consumer Messaging Guide for ENERGY STAR Certified Appliances* (US Environmental Protection Agency, December 2019), https://www.energystar.gov/sites/default/files/asset/document/ES_Consumer_Messaging_Guide_19-20-508.pdf.

40. US Department of Energy and Environmental Protection Agency, "Super-Efficient Water Heater," *EnergyStar.gov*, accessed September 6, 2023, https://www.energystar.gov/products/energy_star_home_upgrade/super_efficient_water_heater.

41. Attari et al., "Public Perceptions of Energy Consumption and Savings."

42. Laura Kier, "How Much Is 1° Worth?," *EnergyHub* (blog), May 16, 2012, https://www.energyhub.com/blog/how-much-is-one-degree-worth/.

43. Consortium for Energy Efficiency, *CEE Annual Industry Report: 2020 State of the Efficiency Program Industry: Budgets, Expenditures, and Impacts* (Consortium for Energy Efficiency, September 2021), https://cee1.org/images/uploads/2020_AIR_Final.pdf.

44. Mike Specian et al., *2023 Utility Energy Efficiency Scorecard* (American Council for an Energy-Efficient Economy, August 24, 2023), https://www.aceee.org/research-report/u2304.

45. US Department of Energy, "Weatherization Assistance Program," *Energy. gov*, December 2019, https://www.energy.gov/scep/wap/weatherization-assistance-program.

46. Memmott et al., "Sociodemographic Disparities in Energy Insecurity Among Low-Income Households."

47. Drehobl et al., "How High Are Household Energy Burdens?"

48. Drehobl et al., "How High Are Household Energy Burdens?"

49. Drehobl et al., "How High Are Household Energy Burdens?"

50. Andrew Jones et al., "Climate Change Impacts on Future Residential Electricity Consumption and Energy Burden: A Case Study in Phoenix, Arizona," *Energy Policy* 183 (December 1, 2023): 113811, https://doi.org/10.1016/j.enpol.2023.113811.

51. Indiana Housing and Community Development Authority, "Weatherization/Energy Conservation," *Indiana. gov*, March 30, 2021, https://www.in.gov/ihcda/homeowners-and-renters/weatherizationenergy-conservation/.

52. US Department of Energy and Energy Efficiency and Renewable Energy, *Weatherization Energy Auditor Single Family: WAP Health and Safety Guidance* (Weatherization Assistance Program standardized curriculum, December 2012), https://www.energy.gov/sites/default/files/2016/07/f33/0_9_wap_health_safety_guidance_v2.0.pptx.

53. West Virginia Office of Economic Development, *West Virginia Weatherization BIL State Plan 7/1/2022–6/30/2027* (State of West Virginia Development Office, 2021), https://wvcad.org/assets/files/wap/Draft-Weatherization-BIL-State-Plan.pdf.

54. Michelle Graff, "Reducing Administrative Burdens in an Energy Bill Assistance Program," *Public Management Review (2024): 1–26*.

55. Pamela Herd and Donald P. Moynihan, *Administrative Burden: Policymaking by Other Means* (Russell Sage Foundation, 2018), https://doi.org/10.7758/9781610448789; Michelle Graff, "Unpacking the Determinants and Burdens of Energy Assistance in the United States" (PhD diss., Indiana University, 2021), https://www.proquest.com/docview/2566108736/abstract/D68EC01092434CE8PQ/1.

56. National Council on Aging, "Energy and Utility Assistance: How Do I Apply for LIHEAP?," *NCOA.org*, November 11, 2022, https://www.ncoa.org/article/how-do-i-apply-for-liheap; Community Action Partnership, "Weatherization—Apply for Services," Com-

munity Council of South Central Texas, 2023, https://www.ccsct.org/weatherization-apply-for-services/.

57. Tony Gerard Reames, "A Community-Based Approach to Low-Income Residential Energy Efficiency Participation Barriers," *Local Environment* 21, no. 12 (December 1, 2016): 1449–66, https://doi.org/10.1080/13549839.2015.1136995.

58. Reames, "Community-Based Approach to Low-Income Residential Energy Efficiency Participation Barriers."

59. Sanya Carley et al., *Electric Utility Disconnections: Legal Protections and Policy Recommendations* (Indiana University and University of Pennsylvania Energy Justice Lab, June 2023), https://utilitydisconnections.org/doc/electric-utility-disconnections-legal-protections-and-policy-recommendations.pdf.

60. Trevor Memmott, Sanya Carley, Michelle Graff, and David M. Konisky, "Utility Disconnection Protections and the Incidence of Energy Insecurity in the United States," *Iscience* 26, no. 3 (2023): 106244.

CHAPTER 5

1. Intergovernmental Panel on Climate Change, *Global Warming of 1.5°C: IPCC Special Report on Impacts of Global Warming of 1.5°C Above Pre-Industrial Levels in Context of Strengthening Response to Climate Change, Sustainable Development, and Efforts to Eradicate Poverty*, 1st ed. (Cambridge University Press, 2022).

2. Jonas Meckling et al., "Policy Sequencing Toward Decarbonization," *Nature Energy* 2, no. 12 (November 13, 2017): 918–22, https://doi.org/10.1038/s41560-017-0025-8.

3. An alternative to technology adoption is, of course, energy conserving behavior.

4. Edgar G. Hertwich and Glen P. Peters, "Carbon Footprint of Nations: A Global, Trade-Linked Analysis," *Environmental Science and Technology* 43, no. 16 (August 15, 2009): 6414–20, https://doi.org/10.1021/es803496a.

5. Ghislain Dubois et al., "It Starts at Home? Climate Policies Targeting Household Consumption and Behavioral Decisions Are Key to Low-Carbon Futures," *Energy Research and Social Science* 52 (June 2019): 144–58, https://doi.org/10.1016/j.erss.2019.02.001; Hertwich and Peters, "Carbon Footprint of Nations."

6. Benjamin K. Sovacool et al., "Equity, Technological Innovation and Sustainable Behaviour in a Low-Carbon Future," *Nature Human Behaviour* 6, no. 3 (January 31, 2022): 326–37, https://doi.org/10.1038/s41562-021-01257-8.

7. US Energy Information Administration, "How Much of U.S. Carbon Dioxide Emissions Are Associated with Electricity Generation?," *EIA.gov*, updated May 1, 2023, https://www.eia.gov/tools/faqs/faq.php?id=77&t=11.

8. Joint Research Centre (European Commission) et al., *GHG Emissions of All World Countries: 2023* (Publications Office of the European Union, 2023), https://data.europa.eu/doi/10.2760/953322; Crippa et al., *CO2 Emissions of All World Countries: JRC/IEA/PBL 2022 Report* (Publications Office of the European Union, 2022), https://op.europa.eu/en/publication-detail/-/publication/6c10e2bd-3892-11ed-9c68-01aa75ed71a1/language-en.

9. Climate Action Tracker, "Net Zero Targets," *ClimateActionTracker.org*, 2021, https://climateactiontracker.org/countries/usa/net-zero-targets/.

10. Dubois et al., "It Starts at Home?"

11. US Census Bureau, "Highlights of 2023 Characteristics of New Housing," *Census.gov*, June 1, 2023, https://www.census.gov/construction/chars/highlights.html.

12. The set of *Breaking Bad* is an actual three-bedroom home in Albuquerque, New Mexico, which the producers rented from the homeowners when filming the series. See Ioana Neamt, "We Are the Ones Who Knock—on Walter White's Fictional Door in Breaking Bad," *FancyPantsHomes.com* (blog), February 2, 2024, https://www.fancypantshomes.com/movie-homes/walter-white-house-in-breaking-bad/.

13. US Census Bureau, "2019: American Community Survey, ACS 1-Year Estimates Subject Tables, Table S2502: Demographic Characteristics for Occupied Housing Units," *Data.census.gov*, accessed October 23, 2023, https://data.census.gov/table/ACSST1Y2019.S2502?q=Owner/Renter+(Householder)+Characteristics&t=Housing.

14. Sydney Forrester et al., *Residential Solar-Adopter Income and Demographic Trends: November 2022 Update*, (Lawrence Berkeley National Laboratory, November 2022), https://emp.lbl.gov/publications/residential-solar-adopter-income-1.

15. Sydney Forrester et al., *Residential Solar-Adopter Income and Demographic Trends: 2023 Update* (Lawrence Berkeley National Laboratory, December 2023), https://emp.lbl.gov/publications/residential-solar-adopter-income-2.

16. Deborah A. Sunter et al., "Disparities in Rooftop Photovoltaics Deployment in the United States by Race and Ethnicity," *Nature Sustainability* 2, no. 1 (2019): 71–76.

17. Severin Borenstein, "Private Net Benefits of Residential Solar PV: The Role of Electricity Tariffs, Tax Incentives, and Rebates," *Journal of the Association of Environmental and Resource Economists* 4, no. S1 (September 2017): S85–122, https://doi.org/10.1086/691978.

18. Forrester et al., *Residential Solar-Adopter Income and Demographic Trends: November 2022 Update*.

19. Dan Power, "Here's What We Know About Energy Efficiency Access in Low-Income Communities," Alliance to Save Energy, June 15, 2021, https://www.ase.org/blog/heres-what-we-know-about-energy-efficiency-access-low-income-communities.

20. Consortium for Energy Efficiency, *CEE Annual Industry Report: 2019 State of the Efficiency Program Industry: Budgets, Expenditures, and Impacts* (Consortium for Energy Efficiency, August 2020), https://cee1.org/images/pdf/2019_AIR_Report_1.pdf.

21. KFF, "State Health Facts: Distribution of the Total Population by Federal Poverty Level (Above and Below 200% FPL)," *KFF.org*, 2022, https://www.kff.org/other/state-indicator/population-up-to-200-fpl/.

22. Power, "Here's What We Know About Energy Efficiency Access in Low-Income Communities."

23. Consortium for Energy Efficiency, *CEE Annual Industry Report: 2020 State of the Efficiency Program Industry: Budgets, Expenditures, and Impacts* (Consortium for Energy Efficiency, September 2021), https://cee1.org/images/uploads/2020_AIR_Final.pdf.

24. David P. Brown, "Socioeconomic and Demographic Disparities in Residential Battery Storage Adoption: Evidence from California," *Energy Policy* 164 (May 2022): 112877, https://doi.org/10.1016/j.enpol.2022.112877.

25. Sanya Carley, "Normative Dimensions of Sustainable Energy Policy," *Ethics, Policy and Environment* 14, no. 2 (June 1, 2011): 211–29.

26. Everett M. Rogers, *Diffusion of Innovations*, 3rd ed. (Free Press, 1983).

27. Ian Prasad Philbrick, "Why Isn't Biden's Expanded Child Tax Credit More Popular?," *New York Times*, January 5, 2022, https://www.nytimes.com/2022/01/05/upshot/biden-child-tax-credit.html.

28. Sarah Lozanova, "Solar Payback Period," *GreenLancer*, May 18, 2022, https://www.greenlancer.com/post/solar-payback-period.

29. Eartheasy, "LED Light Bulbs: Comparison Charts," *Eartheasy.com*, accessed October 22, 2023, https://learn.eartheasy.com/guides/led-light-bulbs-comparison-charts/.

30. Note that most studies on smart thermostats find that households do not use them fully as intended.

31. Robert Lou et al., "Smart Wifi Thermostat-Enabled Thermal Comfort Control in Residences," *Sustainability* 12, no. 5 (March 3, 2020): 1919, https://doi.org/10.3390/su12051919.

32. Samantha Lile, "How Can Efficient HVAC Save You Money?," *Motili*, June 17, 2023, https://www.motili.com/blog/how-can-efficient-hvac-save-you-money/.

33. Michael Thomas, "How Much Money Do Heat Pumps Save?," *Carbon Switch*, December 2022, https://carbonswitch.com/heat-pump-savings/.

34. Jeremy Hsu, "Solar Power's Benefits Don't Shine Equally on Everyone," *Scientific American*, April 4, 2019, 2023, https://www.scientificamerican.com/article/solar-powers-benefits-dont-shine-equally-on-everyone/.

35. Hanzelle Kleeman et al., "Effects of Redesigning the Communication of Low-Income Residential Energy Efficiency Programs in the U.S.," *Energy Policy* 178 (July 1, 2023): 113568, https://doi.org/10.1016/j.enpol.2023.113568.

36. Brad Penney and Phil Kloer, *Shelter Report 2015: Less Is More: Transforming Low-Income Communities Through Energy Efficiency* (Habitat for Humanity, 2015), https://www.habitat.org/sites/default/files/2015-habitat-for-humanity-shelter-report.pdf.

37. Although the government has set energy-efficiency standards for over sixty consumer appliances, these standards have not been updated in over a decade, and proposed standard revisions through the Biden administration introduce a delay in implementation. Anisha Kohli, "The U.S. Government Is Requiring Washers, Refrigerators and Freezers to Be More Efficient: Here's What It Means for You," *TIME*, February 11, 2023, https://time.com/6254877/energy-department-efficiency-home-appliances/.

38. US Census Bureau, "2019: American Community Survey, ACS 1-Year Estimates Subject Tables, Table S2502."

39. Brett Holzhauer, "Here's the Average Net Worth of Homeowners and Renters," *CNBC*, February 27, 2023, https://www.cnbc.com/select/average-net-worth-homeowners-renters/.

40. Naïm R Darghouth et al., "Characterizing Local Rooftop Solar Adoption Inequity in the US," *Environmental Research Letters* 17, no. 3 (March 1, 2022): 034028, https://doi.org/10.1088/1748-9326/ac4fdc.

41. Jacob Channel et al., "More Than 14 Million Households Across U.S. Don't Have Internet," *LendingTree*, 2022, https://www.lendingtree.com/home/mortgage/internet-access-study/.

42. Thomas Elisha Jones and Len Edward Necefer, *Identifying Barriers and Pathways for Success for Renewable Energy Development on American Indian Lands* (Department of Energy, November 2016).

43. Margaret Tallmadge, "What Is Holding Back Renewable Energy Development in Indian Country?," *Clean Energy Finance Forum*, December 11, 2019, https://cleanenergyfinanceforum.com/2019/12/11/what-is-holding-back-renewable-energy-development-in-indian-country.

44. Miranda Willson, "Climate Law Boost for Renewables Hits Barrier on Tribal Lands," *E&E News*, December 22, 2022, https://www.eenews.net/articles/climate-law-boost-for-renewables-hits-barrier-on-tribal-lands/.

45. Willson, "Climate Law Boost for Renewables Hits Barrier on Tribal Lands."

46. Willson, "Climate Law Boost for Renewables Hits Barrier on Tribal Lands."

47. Consider, for example, the 250 MW solar project on the Moapa Band of Paiutes land, as completed in 2017, and other solar projects on Choctaw land in Oklahoma and on Navajo land in New Mexico. Sources for additional information include Felicity Barringer, "What It May Take to Harness Solar Energy on Native Lands," Stanford Doerr School of Sustainability, May 6, 2021, https://earth.stanford.edu/news/what-it-may-take-harness-solar-energy-native-lands; Solar Energy Industries Association, "Oklahoma State Solar Overview," *SEIA.org*, 2022, https://www.seia.org/state-solar-policy/oklahoma-solar; Sarabeth Henne, "Navajo Nation Makes History with Solar Powered Plant," *Indianz.com*, August 16, 2018, https://www.indianz.com/News/2018/08/16/navajo-nation-makes-history-with-solar-p.asp.

48. Susan Montoya Bryan, "US Launches Program to Provide Electricity to More Native American Homes," *AP News*, August 15, 2023, https://apnews.com/article/native-american-homes-electricity-infrastructure-energy-6ebeee0b027e5aa1d1204e0c0ddcc7b1.

49. Valerie Volcovici, "Why Native American Tribes Struggle to Tap Billions in Clean Energy Incentives," *Reuters*, September 8, 2023, https://www.reuters.com/sustainability/climate-energy/why-us-tribes-struggle-tap-billions-clean-energy-incentives-2023-09-08/.

50. Congressional Research Service, "The Plug-In Electric Vehicle Tax Credit," *In Focus* 11017, version 3 (updated May 14, 2019), https://sgp.fas.org/crs/misc/IF11017.pdf.

51. Severin Borenstein and Lucas W. Davis, "The Distributional Effects of US Clean Energy Tax Credits," *Tax Policy and the Economy* 30, no. 1 (January 2016): 191–234, https://doi.org/10.1086/685597.

52. Rachel Reed, "What the US Is Getting Right—and Wrong—About the Move to Electric Vehicles," *Harvard Law Today*, June 2023, https://hls.harvard.edu/today/what-the-us-is-getting-right-and-wrong-about-the-move-to-electric-vehicles/.

53. Internal Revenue Service, "Used Clean Vehicle Credit," *IRS.gov*, accessed October 22, 2023, https://www.irs.gov/credits-deductions/used-clean-vehicle-credit.

54. US Department of Agriculture Rural Development, "Energy Efficiency and Conservation Loan Program," *USDA.gov*, January 16, 2015, https://www.rd.usda.gov/programs-services/electric-programs/energy-efficiency-and-conservation-loan-program.

55. Connecticut Green Bank, "Smart-E Loans," *CTGreenBank.com*, July 14, 2022, https://www.ctgreenbank.com/home-solutions/smart-e-loans/.

56. Arthur L. Ku and John D. Graham, "Is California's Electric Vehicle Rebate Regressive? A Distributional Analysis," *Journal of Benefit-Cost Analysis* 13, no. 1 (March 2022): 1–19, https://doi.org/10.1017/bca.2022.2.

57. Darghouth et al., "Characterizing Local Rooftop Solar Adoption Inequity in the US."

58. Larisa Manescu, "Sierra Club Releases Nationwide Investigation into Electric Vehicle Shopping Experience," Sierra Club press release, May 8, 2023, https://www.sierraclub

.org/press-releases/2023/05/sierra-club-releases-nationwide-investigation-electric-vehicle
-shopping.

59. Le Hieu and Andrew Linhardt, *Rev Up Electric Vehicles: A Nationwide Study of the Electric Vehicle Shopping Experience* (Sierra Club, November 2019).

60. Tony G. Reames et al., "An Incandescent Truth: Disparities in Energy-Efficient Lighting Availability and Prices in an Urban U.S. County," *Applied Energy* 218 (May 2018): 95–103, https://doi.org/10.1016/j.apenergy.2018.02.143.

61. Sarah Lehmann et al., *Diversity in the U.S. Energy Workforce: Data Findings to Inform State Energy, Climate, and Workforce Development Policies and Programs* (National Association of State Energy Officials, BW Research Partnership, HBCU CDAC Clean Energy Initiative, April 2021).

62. Lehmann et al., *Diversity in the U.S. Energy Workforce.*

63. Maya Weber, "The Changing Face of Energy," *S&P Global*, 2023, archived at https://web.archive.org/web/20230328020514/https://www.spglobal.com/en/research-insights/featured/the-changing-face-of-energy.

64. Jacqui Patterson et al., *Who Holds the Power: Demystifying and Democratizing Public Utilities Commissions* (Chisholm Legacy Project, December 2022).

65. Interstate Renewable Energy Council, *National Solar Jobs Census 2023* (Interstate Renewable Energy Council, September 2024), https://irecusa.org/census-executive-summary/.

66. David Keyser et al., *United States Energy and Employment Report 2023* (Department of Energy, 2023), https://www.energy.gov/sites/default/files/2023-06/2023%20USEER%20REPORT-v2.pdf.

67. Borenstein, "Private Net Benefits of Residential Solar PV."

68. Borenstein, "Private Net Benefits of Residential Solar PV."

69. This phenomenon is referred to as "peer effects," and its presence has been documented in the literature. Refer to Bryan Bollinger and Kenneth Gillingham, "Peer Effects in the Diffusion of Solar Photovoltaic Panels," *Marketing Science* 31, no. 6 (2012): 900–912.

70. Darghouth et al., "Characterizing Local Rooftop Solar Adoption Inequity in the US." We have observed these trends going in the opposite direction in some places in recent years, as utilities companies have petitioned the utility commissions to reallocate costs so that more are now recovered in fixed charges rather than variable charges.

71. Cherrelle Eid et al., "The Economic Effect of Electricity Net-Metering with Solar PV: Consequences for Network Cost Recovery, Cross Subsidies and Policy Objectives," *Energy Policy* 75 (December 2014): 244–54, https://doi.org/10.1016/j.enpol.2014.09.011.

72. Stranded assets can refer to those assets connected to physical infrastructure, such as a coal power plant, as well as to those assets that are owned but still in the ground. For example, if a company owns land that has significant coal deposits, then that coal—when left in the ground and unsold—is referred to as a stranded asset. A team of energy modelers estimated the total stranded assets of coal operations across the world through 2050 to be worth $1.3 to $2.3 trillion. See Yen-Heng Henry Chen et al., "An Economy-Wide Framework for Assessing the Stranded Assets of Energy Production Sector Under Climate Policies," *Climate Change Economics* 14, no. 1 (2023): 2350003, https://doi.org/10.1142/S2010007823500033.

73. The problem here is not the adoption of solar power, it is the institutional and market arrangements that create and perpetuate this cross-subsidization. For a discussion of how to

minimize the potential problem of a" utility death spiral," including changes in institutional arrangements, see Monica Castaneda et al., "Myths and Facts of the Utility Death Spiral," *Energy Policy* 110 (2017): 105–16.

74. Borenstein, "Private Net Benefits of Residential Solar PV."

75. Lucas W. Davis and Catherine Hausman, "Who Will Pay for Legacy Utility Costs?," *Journal of the Association of Environmental and Resource Economists* 9, no. 6 (November 1, 2022): 1047–85, https://doi.org/10.1086/719793.

CHAPTER 6

1. US Census Bureau, "QuickFacts: Crawford County, Ohio," *Census.gov*, accessed October 31, 2024, https://www.census.gov/quickfacts/fact/table/crawfordcountyohio /PST045223.

2. Honey Creek Wind Farm, "Honey Creek Wind," *HoneyCreekWindPower.com*, accessed October 26, 2023, https://www.honeycreekwindpower.com/about_honey_creek; Gere Goble, "Wind Turbines in Crawford County, Ohio: Here's What We Know About Honey Creek Wind," *Bucyrus Telegraph-Forum*, April 2, 2022, https://www.bucyrustelegraphforum .com/story/news/2022/04/02/wind-turbines-crawford-county-ohio-honey-creek-wind-faq /9382089002/.

3. Peter Krouse, "Republican-Led Effort Singles Out Wind and Solar Power for Local Control," *Cleveland.com*, July 12, 2021, https://www.cleveland.com/news/2021/07/republican -led-effort-singles-out-wind-and-solar-power-for-local-control.html.

4. Robert Zullo, "Across the Country, a Big Backlash to New Renewables Is Mounting," *Virginia Mercury* (blog), February 23, 2023, https://www.virginiamercury.com/2023/02/23 /across-the-country-a-big-backlash-to-new-renewables-is-mounting/.

5. Todd Hill, "Proposed Wind Farm in County Halted," *Bucyrus Telegraph-Forum*, October 22, 2015, https://www.bucyrustelegraphforum.com/story/news/local/2015/10/22 /proposed-wind-farm-county-halted/74386956/.

6. Goble, "Wind Turbines in Crawford County, Ohio."

7. Zullo, "Across the Country, a Big Backlash to New Renewables Is Mounting."

8. Zullo, "Across the Country, a Big Backlash to New Renewables Is Mounting."

9. Zullo, "Across the Country, a Big Backlash to New Renewables Is Mounting."

10. Crawford County, Ohio, Board of Elections, *Official Results Report: 2022 GENERAL ELECTION NOVEMBER 8, 2022* (Crawford County, Ohio, Board of Elections, November 21, 2022).

11. Matthew Eisenson, *Opposition to Renewable Energy Facilities in the United States: May 2023 Edition* (Sabin Center for Climate Change Law, May 2023), https://scholarship .law.columbia.edu/sabin_climate_change/200/.

12. Stephen Ansolabehere and David M. Konisky, *Cheap and Clean: How Americans Think About Energy in the Age of Global Warming* (MIT Press, 2016).

13. European Environment Agency, *Air Pollution Impacts from Carbon Capture and Storage (CCS)* (Publications Office of the European Union, 2011), https://data.europa.eu/doi /10.2800/84208.

14. The other business sectors include restaurants, farming and agriculture, computer, accounting, travel, telephone, automobile, publishing, grocery, retail, movie, banking, inter-

net, sports, airline, television and radio, education, real estate, healthcare, advertising and public relations, legal, federal government, and pharmaceutical. The data are available at Gallup, Inc., "Business and Industry Sector Ratings," *Gallup.com*, accessed October 31, 2024, https://news.gallup.com/poll/12748/business-industry-sector-ratings.aspx.

15. See, for example, Roger Fouquet, "The Slow Search for Solutions: Lessons from Historical Energy Transitions by Sector and Service," *Energy Policy* 38, no. 11 (November 2010): 6586–96, https://doi.org/10.1016/j.enpol.2010.06.029; Vaclav Smil, "Examining Energy Transitions: A Dozen Insights Based on Performance," *Energy Research and Social Science* 22 (December 2016): 194–97, https://doi.org/10.1016/j.erss.2016.08.017.

16. US Energy Information Administration, "Frequently Asked Questions (FAQs): What Is U.S. Electricity Generation by Energy Source?," *EIA.gov*, accessed October 26, 2023, https://www.eia.gov/tools/faqs/faq.php?id=427&t=3#.

17. US Energy Information Administration, "Nuclear Explained: U.S. Nuclear Industry," *EIA.gov*, accessed October 26, 2023, https://www.eia.gov/energyexplained/nuclear/us -nuclear-industry.php.

18. Alliance for Automotive Innovation, "Electric Vehicle Sales Dashboard," *Autosinnovate.org*, accessed November 1, 2024, https://www.autosinnovate.org/EVDashboard.

19. Elena Shao, "As Heat Pumps Go Mainstream, a Big Question: Can They Handle Real Cold?," *New York Times*, February 22, 2023, https://www.nytimes.com/interactive/2023/02 /22/climate/heat-pumps-extreme-cold.html.

20. At the global level, some studies provide a more optimistic set of estimates about the ability of rooftop solar to meet world energy needs. See, for example, Siddharth Joshi et al., "High Resolution Global Spatiotemporal Assessment of Rooftop Solar Photovoltaics Potential for Renewable Electricity Generation," *Nature Communications* 12, no. 1 (October 5, 2021): 5738, https://doi.org/10.1038/s41467-021-25720-2.

21. Pieter Gagnon et al., *Rooftop Solar Photovoltaic Technical Potential in the United States: A Detailed Assessment* (National Renewable Energy Laboratory, January 2016), https://www.nrel.gov/docs/fy16osti/65298.pdf.

22. Sammy Roth, "Solar Sprawl Is Tearing Up the Mojave Desert: Is There a Better Way?," *Los Angeles Times*, June 27, 2023, https://www.latimes.com/environment/story/2023-06-27 /solar-panels-could-save-california-but-they-hurt-the-desert.

23. Kelsey Misbrener, "Electrical Grid Interconnection Backlog Grew 30% in 2023," *Solar Power World*, April 10, 2024, https://www.solarpowerworldonline.com/2024/04/electrical -grid-interconnection-backlog-grew-30-percent-2023/.

24. Brad Plumer, "The U.S. Has Billions for Wind and Solar Projects: Good Luck Plugging Them In," *New York Times*, February 23, 2023, https://www.nytimes.com/2023/02/23 /climate/renewable-energy-us-electrical-grid.html.

25. Although not the topic of this book, the same can be said of other infrastructure, such as bridges that collapse because of wear and tear and overuse.

26. Michael Liedtke, "PG&E Confesses to Killing 84 People in 2018 California Fire," *AP News*, June 16, 2020, https://apnews.com/article/67810cb4d9b6b90e451415b76215d6c9.

27. Kavya Balaraman, "PG&E, SCE Detail Plans to Spend More than $23B Through 2025 to Prevent Wildfires in Their Footprints," *Utility Dive*, March 28, 2023, https://www.utilitydive .com/news/pge-sce-vegetation-management-resilience-california-wildfires/646163/.

28. US Department of Energy, *National Transmission Needs Study: Draft for Public Com-*

ment (US Department of Energy, February 2023), https://www.energy.gov/sites/default/files/2023-02/022423-DRAFTNeedsStudyforPublicComment.pdf.

29. Michelle Solomon, "DOE Study Highlights America's Transmission Needs, but How Do We Accelerate Buildout?," *Utility Dive*, March 31, 2023, https://www.utilitydive.com/news/doe-study-transmission-clean-energy/646589/.

30. Sara Budinis, "Direct Air Capture," International Energy Agency, July 2023, https://www.iea.org/energy-system/carbon-capture-utilisation-and-storage/direct-air-capture.

31. US Department of Transportation Federal Highway Administration, "Bipartisan Infrastructure Law," *FHWA.dot.gov*, accessed October 30, 2023, https://www.fhwa.dot.gov/bipartisan-infrastructure-law/.

32. The private sector owns about 80 percent of US energy infrastructure. See Cybersecurity and Infrastructure Security Agency, "Energy Sector," *CISA.gov*, accessed October 26, 2023, https://www.cisa.gov/topics/critical-infrastructure-security-and-resilience/critical-infrastructure-sectors/energy-sector.

33. Douglas L. Bessette and Sarah B. Mills, "Farmers vs. Lakers: Agriculture, Amenity, and Community in Predicting Opposition to United States Wind Energy Development," *Energy Research and Social Science* 72 (February 2021): 101873, https://doi.org/10.1016/j.erss.2020.101873.

34. Lucas Davis et al., "Transmission Impossible? Prospects for Decarbonizing the US Grid," working paper no. 31377 (National Bureau of Economic Research, June 2023), https://doi.org/10.3386/w31377.

35. Michael J. Graetz, *The End of Energy: The Unmaking of America's Environment, Security, and Independence* (MIT Press, 2011), https://doi.org/10.7551/mitpress/8653.001.0001.

36. As an example, the Federal Energy Regulatory Commission, or FERC, has some jurisdiction over the interstate transmission of electricity, natural gas, and oil, but it is not all encompassing. For example, FERC regulates the transmission and wholesale sale of electricity in interstate commerce and approves the siting of interstate natural-gas pipelines and storage facilities, but it does not approve the construction of electric generation facilities or license the construction of oil pipelines. For more details, see Federal Energy Regulatory Commission, "What FERC Does," *FERC.gov*, accessed October 30, 2023, https://www.ferc.gov/what-ferc-does.

37. Barbara C. Farhar, "Trends: Public Opinion About Energy," *Public Opinion Quarterly* 58, no. 4 (1994): 603–32; Eugene A. Rosa and Riley E. Dunlap, "Poll Trends: Nuclear Power: Three Decades of Public Opinion," *Public Opinion Quarterly* 58, no. 2 (1994): 295, https://doi.org/10.1086/269425; Ansolabehere and Konisky, *Cheap and Clean*.

38. Ansolabehere and Konisky, *Cheap and Clean*; David M. Konisky et al., "Corrigendum to: Proximity, NIMBYism, and Public Support for Energy Infrastructure," *Public Opinion Quarterly* 85, no. 2 (October 1, 2021): 733, https://doi.org/10.1093/poq/nfab008.

39. Ansolabehere and Konisky, *Cheap and Clean*; Deidra Miniard et al., "Shared Vision for a Decarbonized Future Energy System in the United States," *Proceedings of the National Academy of Sciences* 117, no. 13 (March 31, 2020): 7108–14, https://doi.org/10.1073/pnas.1920558117.

40. Parrish Bergquist et al., "Energy Policy and Public Opinion: Patterns, Trends and Future Directions," *Progress in Energy* 2, no. 3 (August 10, 2020): 032003, https://doi.org/10.1088/2516-1083/ab9592.

41. Stephen Ansolabehere and David M. Konisky, "Public Attitudes Toward Construction of New Power Plants," *Public Opinion Quarterly* 73, no. 3 (2009): 566–77; Konisky et al., "Corrigendum to: Proximity, NIMBYism, and Public Support for Energy Infrastructure."

42. US Energy Information Administration, "Electricity Data," *EIA.gov*, accessed October 26, 2023, https://www.eia.gov/electricity/data.php.

43. Matthew Eisenson et al., *Opposition to Renewable Energy Facilities in the United States: June 2024 Edition* (Sabin Center for Climate Change Law, June 2024), https://scholarship.law.columbia.edu/cgi/viewcontent.cgi?article=1227&context=sabin_climate _change.

44. Eisenson, *Opposition to Renewable Energy Facilities in the United States: May 2023 Edition*.

45. Eisenson et al., *Opposition to Renewable Energy Facilities in the United States: June 2024 Edition*.

46. Anthony Lopez et al., "Land Use and Turbine Technology Influences on Wind Potential in the United States," *Energy* 223 (May 2021): 120044, https://doi.org/10.1016/j.energy .2021.120044.

47. Samantha Gross, *Renewables, Land Use, and Local Opposition in the United States* (Brookings Institution, January 2020), https://www.brookings.edu/articles/renewables -land-use-and-local-opposition-in-the-united-states/.

48. It is worth noting that large landowners (i.e., those with the most land to lease) benefit the most, while small landowners may not benefit much at all.

49. David M. Konisky et al., "Proximity, NIMBYism, and Public Support for Energy Infrastructure," *Public Opinion Quarterly* 84, no. 2 (2020): 391–418.

50. Sanya Carley et al., "Energy Infrastructure, NIMBYism, and Public Opinion: A Systematic Literature Review of Three Decades of Empirical Survey Literature," *Environmental Research Letters* 15, no. 9 (September 1, 2020): 093007, https://doi.org/10.1088/1748-9326 /ab875d.

51. Patrick Devine-Wright, "Rethinking NIMBYism: The Role of Place Attachment and Place Identity in Explaining Place-Protective Action," *Journal of Community and Applied Social Psychology* 19, no. 6 (November 2009): 426–41, https://doi.org/10.1002/casp.1004; Patrick Devine-Wright and Yuko Howes, "Disruption to Place Attachment and the Protection of Restorative Environments: A Wind Energy Case Study," *Journal of Environmental Psychology* 30, no. 3 (September 2010): 271–80, https://doi.org/10.1016/j.jenvp.2010.01.008.

52. Devine-Wright and Howes, "Disruption to Place Attachment and the Protection of Restorative Environments."

53. Zullo, "Across the Country, a Big Backlash to New Renewables Is Mounting."

54. "Cape Wind Project, Massachusetts," *Power Technology* (blog), May 11, 2017, https://www.power-technology.com/projects/cape-wind-project-massachusetts/.

55. Bureau of Ocean Energy Management, "Cape Wind Energy Project," fact sheet, last updated September 2015, https://www.boem.gov/sites/default/files/renewable-energy -program/Studies/Cape-Wind-Fact-Sheet---Sept-2015-clean-%281%29.pdf.

56. Katharine Q. Seelye, "After 16 Years, Hopes for Cape Cod Wind Farm Float Away," *New York Times*, December 19, 2017, https://www.nytimes.com/2017/12/19/us/offshore -cape-wind-farm.html.

57. Seelye, "After 16 Years, Hopes for Cape Cod Wind Farm Float Away."

58. Bessette and Mills, "Farmers vs. Lakers"; Martin J. Pasqualetti, "Social Barriers to Renewable Energy Landscapes," *Geographical Review* 101, no. 2 (2011): 201–23; Loren D. Knopper and Christopher A. Ollson, "Health Effects and Wind Turbines: A Review of the Literature," *Environmental Health* 10, no. 1 (December 2011): 78, https://doi.org/10.1186/1476-069X-10-78; Alice Freiberg et al., "Health Effects of Wind Turbines on Humans in Residential Settings: Results of a Scoping Review," *Environmental Research* 169 (February 2019): 446–63, https://doi.org/10.1016/j.envres.2018.11.032; Ben Hoen et al., *The Impact of Wind Power Projects on Residential Property Values in the United States: A Multi-Site Hedonic Analysis* (US Department of Energy Office of Scientific and Technical Information, December 2, 2009), https://doi.org/10.2172/978870.

59. Catherine Brinkley and Andrew Leach, "Energy Next Door: A Meta-Analysis of Energy Infrastructure Impact on Housing Value," *Energy Research and Social Science* 50 (2019): 51–65; Joseph Rand and Ben Hoen, "Thirty Years of North American Wind Energy Acceptance Research: What Have We Learned?," *Energy Research and Social Science* 29 (2017): 135–48; and George Parsons and Martin D. Heintzelman, "The Effect of Wind Power Projects on Property Values: A Decade (2011–2021) of Hedonic Price Analysis," *International Review of Environmental and Resource Economics* 16, no. 1 (2022): 93–170.

60. Eric J. Brunner et al., "Commercial Wind Turbines and Residential Home Values: New Evidence from the Universe of Land-Based Wind Projects in the United States," *Energy Policy* 185 (2024): 113837.

61. Leah Cardamore Stokes, *Short Circuiting Policy: Interest Groups and the Battle over Clean Energy and Climate Policy in the American States* (Oxford University Press, 2020); Miranda Green et al., "An Activist Group Is Spreading Misinformation to Stop Solar Projects in Rural America," *NPR*, February 18, 2023, https://www.npr.org/2023/02/18/1154867064/solar-power-misinformation-activists-rural-america; David Gelles, "The Texas Group Waging a National Crusade Against Climate Action," *New York Times*, December 4, 2022, https://www.nytimes.com/2022/12/04/climate/texas-public-policy-foundation-climate-change.html; Julia Simon, "Misinformation Is Derailing Renewable Energy Projects Across the United States," *NPR*, March 28, 2022, https://www.npr.org/2022/03/28/1086790531/renewable-energy-projects-wind-energy-solar-energy-climate-change-misinformation; Akielly Hu, "The GOP Donors Behind a Growing Misinformation Campaign to Stop Offshore Wind," *Grist*, April 20, 2023, https://grist.org/politics/republicans-fossil-fuels-the-gop-donors-behind-a-growing-misinformation-campaign-to-stop-offshore-wind/.

62. Sarah Banas Mills et al., "Exploring Landowners' Post-Construction Changes in Perceptions of Wind Energy in Michigan," *Land Use Policy* 82 (March 2019): 754–62, https://doi.org/10.1016/j.landusepol.2019.01.010.

63. Simon, "Misinformation Is Derailing Renewable Energy Projects Across the United States."

64. Patrick J. Egan and Megan Mullin, "Climate Change: US Public Opinion," *Annual Review of Political Science* 20, no. 1 (May 11, 2017): 209–27, https://doi.org/10.1146/annurev-polisci-051215-022857.

65. Gallup asked the following question: "I'm going to read you a list of environmental problems. As I read each one, please tell me if you personally worry about this problem a great deal, a fair amount, only a little or not at all. How much do you personally worry about global warming or climate change?" See Lydia Saad, "A Steady Six in 10 Say Global Warm-

ing's Effects Have Begun," *Gallup.com*, April 20, 2023, https://news.gallup.com/poll/474542 /steady-six-say-global-warming-effects-begun.aspx.

66. Pew asked the following question: "How much of a priority should each of the following be for the president and Congress to address this year? Dealing with global climate change?" See Pew Research Center, *Economy Remains the Public's Top Policy Priority; COVID-19 Concerns Decline Again* (Pew Research Center, February 6, 2023), https://www .pewresearch.org/politics/2023/02/06/economy-remains-the-publics-top-policy-priority -covid-19-concerns-decline-again/.

67. Patrick J. Egan et al., "Ascendant Public Opinion," *Public Opinion Quarterly* 86, no. 1 (March 18, 2022): 134–48, https://doi.org/10.1093/poq/nfab071.

68. Bergquist et al., "Energy Policy and Public Opinion."

69. Bessette and Mills, "Farmers vs. Lakers"; Leah C. Stokes et al., "Prevalence and Predictors of Wind Energy Opposition in North America," *Proceedings of the National Academy of Sciences* 120, no. 40 (October 3, 2023): e2302313120, https://doi.org/10.1073/pnas .2302313120.

70. Timothy B. Gravelle and Erick Lachapelle, "Politics, Proximity and the Pipeline: Mapping Public Attitudes Toward Keystone XL," *Energy Policy* 83 (August 1, 2015): 99–108, https:// doi.org/10.1016/j.enpol.2015.04.004.

71. Daniel J. Hopkins, *The Increasingly United States: How and Why American Political Behavior Nationalized* (University of Chicago Press, 2018).

72. Davis et al., "Transmission Impossible?"

73. Miriam Waser, "It's Hard to Build Transmission Lines in the Northeast, so 8 States Are Asking the Feds for Help," *WBUR*, July 5, 2023, https://www.wbur.org/news/2023/07 /05/electrical-transmission-power-lines-massachusetts-new-england-clean-energy-doe.

74. Alex Lawson, "Global Electricity Grid Must Be Upgraded Urgently to Hit Climate Goals, Says IEA," *Guardian*, October 17, 2023, https://www.theguardian.com/environment /2023/oct/17/global-electricity-grid-climate-iea.

75. Sanya Carley et al., "Are All Electrons the Same? Evaluating Support for Local Transmission Lines Through an Experiment," ed. Bing Xue, *PLOS ONE* 14, no. 7 (July 17, 2019): e0219066, https://doi.org/10.1371/journal.pone.0219066.

76. To maintain the confidentiality of the interviewees, we do not name the specific project. More details on the study methodology are available in the original paper: Parrish Bergquist et al., "Backyard Voices: How Sense of Place Shapes Views of Large-Scale Energy Transmission Infrastructure," *Energy Research and Social Science* 63 (May 2020): 101396, https://doi.org/10.1016/j.erss.2019.101396.

77. Bergquist et al., "Backyard Voices."

78. Bergquist et al., "Backyard Voices."

79. Mathias Einberger, "Reality Check: The United States Has the Only Major Power Grid Without a Plan," *RMI*, January 12, 2023, https://rmi.org/the-united-states-has-the-only -major-power-grid-without-a-plan/.

80. Davis, Hausman, and Rose, "Transmission Impossible?"

81. Congressional Research Service, *Federal Land Ownership: Overview and Data*, report no. R42346 (Congressional Research Service, updated February 1, 2020), https://crsreports .congress.gov/product/pdf/r/r42346#:~:text=The%20federal%20government%20owns %20roughly,of%20September%2030%2C%202018).

82. Davis et al., "Transmission Impossible?"

83. Matthew Bandyk, "Largest Planned Wind Farm in US Gets Key Federal Approval," *Utility Dive*, October 25, 2019, https://www.utilitydive.com/news/largest-planned-wind-farm-in-us-gets-key-federal-approval/565795/.

84. Robert Walton, "Feds Approve 1st Phase of Largest US Wind Project in Wyoming," *Utility Dive*, January 19, 2017, https://www.utilitydive.com/news/feds-approve-1st-phase-of-largest-us-wind-project-in-wyoming/434342/.

85. Jason Plautz, "Western Transmission Line Breaks Ground After 18-Year Wait," *E&E News*, June 21, 2023, https://www.eenews.net/articles/western-transmission-line-breaks-ground-after-18-year-wait/.

86. Morgan Browning et al., "Net-Zero CO2 by 2050 Scenarios for the United States in the Energy Modeling Forum 37 Study," *Energy and Climate Change* 4 (December 2023): 100104, https://doi.org/10.1016/j.egycc.2023.100104.

87. Emily Pontecorvo, "Will the Inflation Reduction Act Jumpstart Carbon Capture?," *Grist*, August 22, 2022, https://grist.org/technology/will-the-inflation-reduction-act-jumpstart-carbon-capture-45q/.

88. Carlos Plautz and Jason Anchondo, "CCS 2.0: Company Reboots Bid to Save N.M. Coal Plant," *E&E News*, August 18, 2023, https://www.eenews.net/articles/ccs-2-0-company-reboots-bid-to-save-n-m-coal-plant/.

89. Albert Lin, "Does Geoengineering Present a Moral Hazard?," *Ecology Law Quarterly* 40, no. 3 (2013): 673, https://escholarship.org/uc/item/7th0d0pd.

90. European Environment Agency, *Air Pollution Impacts from Carbon Capture and Storage (CCS)*.

91. Timothy Puko, "Why These Environmentalists Are Resisting Part of Biden's Climate Push," *Washington Post*, June 22, 2023, https://www.washingtonpost.com/nation/2023/06/22/biden-carbon-capture-climate-environmentalists/.

92. Puko, "Why These Environmentalists Are Resisting Part of Biden's Climate Push."

93. Puko, "Why These Environmentalists Are Resisting Part of Biden's Climate Push."

CHAPTER 7

1. Chris Paine, dir., *Who Killed the Electric Car?* (Sony Pictures Home Entertainment, 2006). However, *Who Killed the Electric Car?* is also quick to point out that the failure of 1990s electric cars was not solely an issue of consumer demand but was due primarily to oil and gas companies who sought to remove the cars from the market due to vested interests and concerns about production costs.

2. Nicholas Wallace and Austin Irwin, "Longest-Range Electric Cars We've Ever Tested," *Car and Driver*, June 7, 2022, https://www.caranddriver.com/shopping-advice/g32634624/ev-longest-driving-range/; Drew Dorian, "2024 Tesla Model S," *Car and Driver*, accessed March 8, 2024, https://www.caranddriver.com/tesla/model-s.

3. Renee Valdes, "How Much Are Electric Cars?," *Kelley Blue Book* (blog), July 10, 2023, https://www.kbb.com/car-advice/how-much-electric-car-cost/.

4. As of 2024, Tesla is the most valuable automobile company in the world, with a market capitalization of over $560 billion, comfortably ahead other notable automobile companies, such as Toyota ($319 billion), Mercedes-Benz ($77 billion), Ford ($48 billion), and General

Motors ($465 billion). Tesla holds a dominant share of the EV market, selling 55 percent of all EVs in 2023 and more than 4 percent of all "light-duty" vehicles (i.e., cars, light trucks, SUVs) sold nationwide. See Yahoo! Finance, "Tesla, Inc. (TSLA)," *Finance.Yahoo.com*, accessed March 11, 2024, https://finance.yahoo.com/quote/TSLA?.tsrc=fin-srch; Yahoo! Finance, "Toyota Motor Corporation (TM)," *Finance.Yahoo.com*, accessed March 11, 2024, https://finance.yahoo.com/quote/TM?.tsrc=fin-srch; Yahoo! Finance, "Mercedes-Benz Group AG (MBG.DE)," *Finance.Yahoo.com*, accessed March 11, 2024, https://finance.yahoo.com/quote/MBG.DE?.tsrc=fin-srch; Yahoo! Finance, "Ford Motor Company (F)," *Finance.Yahoo.com*, accessed March 11, 2024, https://finance.yahoo.com/quote/F?.tsrc=fin-srch; Yahoo! Finance, "General Motors Company (GM)," *Finance.Yahoo.com*, accessed March 11, 2024, https://finance.yahoo.com/quote/GM?.tsrc=fin-srch; Cox Automotive, "A Record 1.2 Million EVs Were Sold in the U.S. in 2023, According to Estimates from Kelley Blue Book," *CoxAutoInc.com*, January 9, 2024, https://www.coxautoinc.com/market-insights/q4-2023-ev-sales/; and Kelley Blue Book, *Electric Vehicle Sales Report—Q4 2023* (Kelley Blue Book, January 2024), https://www.coxautoinc.com/wp-content/uploads/2024/01/Q4-2023-Kelley-Blue-Book-Electric-Vehicle-Sales-Report.pdf.

5. Alan N. Hoffman, "Case 26: Tesla Motors, Inc.: The First U.S. Car Company IPO Since 1956," in *Strategic Management and Business Policy: Globalization, Innovation, and Sustainability*, 14th ed., eds. Thomas L. Wheelen et al. (Pearson, 2015), 671–86.

6. Manuel Moritz et al., "Tesla Motors, Inc.: Pioneer Towards a New Strategic Approach in the Automobile Industry Along the Open Source Movement?," in *2015 Portland International Conference on Management of Engineering and Technology (PICMET)* (Institute of Electrical and Electronics Engineers, 2015), 85–92.

7. Dorian, "2024 Tesla Model S."

8. Eric Stafford, "2024 Tesla Cybertruck," *Car and Driver*, accessed March 8, 2024, https://www.caranddriver.com/tesla/cybertruck.

9. Adam Ruggiero, "'Starship Pooper': Hilarious New Names for Tesla's Cybertruck," *GearJunkie*, November 22, 2019, https://gearjunkie.com/motors/tesla-cybertruck-funny-names.

10. Alliance for Automotive Innovation, "Electric Vehicle Sales Dashboard," *Autosinnovate.org*, accessed November 1, 2024, https://www.autosinnovate.org/EVDashboard.

11. Joann Muller, "The World's Car Buyers Are Ready to Go Electric," *Axios*, May 28, 2022, https://www.axios.com/2022/05/23/electric-vehicles-consumer-interest.

12. Jim Motavalli, "Every Automaker's EV Plans Through 2035 and Beyond," *Forbes*, October 4, 2021, https://www.forbes.com/wheels/news/automaker-ev-plans/.

13. Steve Hanley, "31 Countries, States, and Cities Have Gas/Diesel Car Bans In Place," *CleanTechnica*, January 2, 2021, https://cleantechnica.com/2021/01/02/31-countries-states-and-cities-have-ice-bans-in-place/; Maxine Joselow, "Three Electric Vehicle Fights to Watch Today," *Washington Post*, September 14, 2023, https://www.washingtonpost.com/politics/2023/09/14/three-electric-vehicle-fights-watch-today/.

14. International Energy Agency, "Transport," *IEA.org*, accessed November 1, 2024, https://www.iea.org/reports/transport.

15. Based on 2021 data reported in US Environmental Protection Agency, "Sources of Greenhouse Gas Emissions," *EPA.gov*, last updated October 22, 2024, https://www.epa.gov/ghgemissions/sources-greenhouse-gas-emissions.

16. J.-F. Mercure et al., "Integrated Assessment Modelling as a Positive Science: Private Passenger Road Transport Policies to Meet a Climate Target Well Below 2°C," *Climatic Change* 151, no. 2 (November 2018): 109–29, https://doi.org/10.1007/s10584-018-2262-7.

17. Noel Healy et al., "Embodied Energy Injustices: Unveiling and Politicizing the Transboundary Harms of Fossil Fuel Extractivism and Fossil Fuel Supply Chains," *Energy Research and Social Science* 48 (February 1, 2019): 219–34, https://doi.org/10.1016/j.erss.2018.09.016.

18. Troy R. Hawkins et al., "Environmental Impacts of Hybrid and Electric Vehicles—a Review," *International Journal of Life Cycle Assessment* 17, no. 8 (September 2012): 997–1014, https://doi.org/10.1007/s11367-012-0440-9.

19. US Department of Energy Alternative Fuels Data Center, "Batteries for Electric Vehicles," *AFDC.energy.gov*, accessed March 25, 2024, https://afdc.energy.gov/vehicles/electric_batteries.html.

20. John Voelcker, "Electric-Vehicle Battery Basics," *Car and Driver*, updated July 30, 2024, https://www.caranddriver.com/features/a43093875/electric-vehicle-battery/.

21. Battery life can vary based on battery type, manufacturer, climate in which it is used, and driving and charging habits, but an expected lifespan is approximately ten years or a hundred thousand miles, with lifespans in more moderate climates extending to twelve to fifteen years and approaching two hundred thousand miles. Replacement costs also vary widely, but a reported cost range is between $5,000 and $20,000, depending on the vehicle and battery make. See John M. Vincent, "How Long Do Electric Car Batteries Last?," *US News and World Report*, August 16, 2023, archived at https://cars.usnews.com/cars-trucks/advice/how-long-do-ev-batteries-last; Jane Ulitskaya, "How Much Do Electric Car Batteries Cost to Replace?," *Cars.com*, April 24, 2023, https://www.cars.com/articles/how-much-do-electric-car-batteries-cost-to-replace-465308/.

22. Dionne Searcey, Michael Forsythe, and Eric Lipton, "A Power Struggle over Cobalt Rattles the Clean Energy Revolution," *New York Times*, November 20, 2021, https://www.nytimes.com/2021/11/20/world/china-congo-cobalt.html.

23. US Aid and Global Waters, "Democratic Republic of Congo," *Global Waters.org*, 2022, https://www.globalwaters.org/WhereWeWork/Africa/DRC; World Bank Group, "Dem. Rep. Cong Data," *WorldBank.org*, 2023, https://data.worldbank.org/country/congo-dem-rep.

24. Caroline Bersch, "4 Health Care Facts About the Democratic Republic of the Congo," Borgen Project, September 6, 2021, https://borgenproject.org/health-care-in-the-drc/.

25. US Geological Survey, *Mineral Commodity Summaries 2022* (US Geological Survey, 2022), https://www.usgs.gov/publications/mineral-commodity-summaries-2022.

26. Vladimir Basov, "The World's Largest Cobalt Reserves by Country in 2022," *Kitco News*, February 7, 2023, archived at https://web.archive.org/web/20230314082433/https://www.kitco.com/news/2023-02-07/The-world-s-largest-cobalt-reserves-by-country-in-2022.html.

27. When made into a compound, cobalt can also provide a blue color to glass, ceramics, and other media. In fact, archeologists have discovered the presence of cobalt in ceramics and other pieces in Egyptian sculptures from the eighteenth dynasty and in the ruins of Pompeii in 79 CE, among other locations marked by ancient history.

28. Benjamin K. Sovacool, "When Subterranean Slavery Supports Sustainability Transitions? Power, Patriarchy, and Child Labor in Artisanal Congolese Cobalt Mining," *Extractive Industries and Society* 8, no. 1 (March 2021): 271–93, https://doi.org/10.1016/j.exis.2020.11.018.

29. Siddharth Kara, *Cobalt Red: How the Blood of the Congo Powers Our Lives* (St. Martin's Press, 2023).

30. Amnesty International, *"This Is What We Die For": Human Rights Abuses in the Democratic Republic of the Congo Power the Global Trade in Cobalt* (Amnesty International, 2016).

31. Kara, *Cobalt Red*.

32. Amnesty International, *"This Is What We Die For."*

33. Benjamin K. Sovacool et al., "The Decarbonisation Divide: Contextualizing Landscapes of Low-Carbon Exploitation and Toxicity in Africa," *Global Environmental Change* 60 (January 2020): 102028, https://doi.org/10.1016/j.gloenvcha.2019.102028.

34. Lena Mucha et al., "The Hidden Costs of Cobalt Mining," *Washington Post*, February 28, 2018, https://www.washingtonpost.com/news/in-sight/wp/2018/02/28/the-cost-of-cobalt/.

35. Amnesty International, *"This Is What We Die For."*

36. Kara, *Cobalt Red*.

37. Amnesty International, *"This Is What We Die For."*

38. Siddharth Kara, "Our Device-Driven Lives Depend More Than Ever on Tragedy in the Democratic Republic of Congo," *CNN*, December 17, 2021, https://www.cnn.com/2021/12/17/opinions/siddharth-kara-mining-dr-congo/index.html.

39. Mucha et al., "Hidden Costs of Cobalt Mining."

40. Clare Church and Alec Crawford, "Minerals and the Metals for the Energy Transition: Exploring the Conflict Implications for Mineral-Rich, Fragile States," in *The Geopolitics of the Global Energy Transition*, eds. Manfred Hafner and Simone Tagliapietra, Lecture Notes in Energy (Springer International Publishing, 2020), 279–304, https://doi.org/10.1007/978-3-030-39066-2_12; Kara, *Cobalt Red*.

41. Sovacool, "When Subterranean Slavery Supports Sustainability Transitions?"

42. Amnesty International, "Phones, Electric Cars and Human Rights Abuses—5 Things You Need to Know," *Amnesty.org*, May 1, 2018, https://www.amnesty.org/en/latest/news/2018/05/phones-electric-cars-and-human-rights-abuses-5-things-you-need-to-know/.

43. Benjamin Rubbers, "Governing New Mining Projects in D. R. Congo: A View from the HR Department of a Chinese Company," *Extractive Industries and Society* 7, no. 1 (January 2020): 191–98, https://doi.org/10.1016/j.exis.2019.12.006.

44. David Manley et al., *No Time to Waste: Governing Cobalt amid the Energy Transition* (Berkeley Law and Natural Resource Governance Institute, March 2022).

45. Amnesty International, *"This Is What We Die For."*

46. Amnesty International, *"This Is What We Die For."*

47. Amnesty International, *"This Is What We Die For."*

48. Amnesty International, *"This Is What We Die For."*

49. Sovacool, "When Subterranean Slavery Supports Sustainability Transitions?"

50. Kara, *Cobalt Red*.

51. Amnesty International, *"This Is What We Die For."*

52. Sovacool, "When Subterranean Slavery Supports Sustainability Transitions?"

53. Jack Ewing and Eric Lipton, "Carmakers Race to Control Next-Generation Battery Technology," *New York Times*, March 7, 2022, https://www.nytimes.com/2022/03/07/business/energy-environment/next-generation-auto-battery.html.

54. Alexander Beadle, "Cobalt-Free Batteries Could Power the Next Generation of Elec-

tric Vehicles," *Technology Networks*, January 23, 2024, https://www.technologynetworks.com/applied-sciences/news/cobalt-free-batteries-could-power-the-next-generation-of-electric-vehicles-383059.

55. Jim Motavalli, "Closing the Loop on EV Battery Recycling," *SAE International*, October 7, 2022, https://www.sae.org/news/2022/10/ev-battery-recycling.

56. Bipartisan Policy Center, *Inflation Reduction Act Summary: Energy and Climate Provisions* (Bipartisan Policy Center, August 4, 2022), https://bipartisanpolicy.org/blog/inflation-reduction-act-summary-energy-climate-provisions/.

57. White House, "Fact Sheet: Biden-Harris Administration Driving U.S. Battery Manufacturing and Good-Paying Jobs," *WhiteHouse.gov*, October 19, 2022, https://www.whitehouse.gov/briefing-room/statements-releases/2022/10/19/fact-sheet-biden-harris-administration-driving-u-s-battery-manufacturing-and-good-paying-jobs/.

58. Mariana Ambrose et al., "CHIPS and Science Act Summary: Energy, Climate, and Science Provisions," Bipartisan Policy Center, November 14, 2022, https://bipartisanpolicy.org/blog/chips-science-act-summary/.

59. The US Geological Survey has identified fifty critical minerals that are essential to the digital economy. While mineral deposits are situated around the world, China is the unquestioned leader at processing these deposits into usable materials. See US Geological Survey, "U.S. Geological Survey Releases 2022 List of Critical Minerals," *USGS.gov*, February 22, 2022, https://www.usgs.gov/news/national-news-release/us-geological-survey-releases-2022-list-critical-minerals; Jared Cohen, "Resource Realism: The Geopolitics of Critical Mineral Supply Chains," Goldman Sachs, September 13, 2023, https://www.goldmansachs.com/intelligence/pages/resource-realism-the-geopolitics-of-critical-mineral-supply-chains.html.

60. Jack Healy and Mike Baker, "As Miners Chase Clean-Energy Minerals, Tribes Fear a Repeat of the Past," *New York Times*, December 27, 2021, https://www.nytimes.com/2021/12/27/us/mining-clean-energy-antimony-tribes.html.

61. Samuel Block, "Mining Energy-Transition Metals: National Aims, Local Conflicts," MSCI, June 3, 2021, https://www.msci.com/www/blog-posts/mining-energy-transition-metals/02531033947.

62. See the IRMA website, accessed March 24, 2024, https://responsiblemining.net/.

63. Natural Resources Canada, *The Canadian Critical Minerals Strategy—From Exploration to Recycling: Powering the Green and Digital Economy for Canada and the World* (Natural Resources Canada, 2022), https://www.canada.ca/content/dam/nrcan-rncan/site/critical-minerals/Critical-minerals-strategyDec09.pdf.

64. Natural Resources Canada, *Canadian Critical Minerals Strategy*.

65. Matt Hall, "Map of China Manufacturing Distribution," Berkeley Sourcing Group, June 6, 2016, https://www.berkeleysg.com/china-manufacturing-distribution-map/.

66. Paulo Cavallo and Clint Peinhardt, "Foreign Investment and Right-to-Work Laws," *Business and Politics* 23, no. 3 (2021): 406–18. Although GM, Ford, and Chrysler plants have been unionized in these states, foreign automakers traditionally resisted efforts toward unionization of the workforce, most commonly by paying wages that were comparable to those paid to unionized employees. This approach proved successful, as notable attempts to unionize plants in Alabama and Tennessee failed during the 2010s. See Andrea Hsu and

Stephan Bisaha, "Seeking to Defy History, the UAW is Coming Closer to Unionizing the South," *NPR*, April 5, 2024.

67. While the reduction of trade barriers increased US automotive exports to Mexico (increasing 262 percent between 1993 and 2016), reciprocal trade in the form of imports to the US from Mexico increased 765 percent over the same period. Mexico's light-vehicle production grew from 1.1 million units in 1994 to nearly 3.5 million units in 2016, and light-vehicle exports increased from 579,000 to 2.8 million units over the same period. Approximately 90 percent of the increase in North American vehicle production between 1995 and 2016 was produced in Mexican plants. See M. Angeles Villarreal and Ian F. Fergusson, *The North American Free Trade Agreement (NAFTA)*, report R42964 (Congressional Research Service, May 24, 2017), https://sgp.fas.org/crs/row/R42965.pdf; Thomas H. Klier and James M. Rubenstein, "Mexico's Growing Role in the Auto Industry Under NAFTA: Who Makes What and What Goes Where," *Economic Perspectives* 41, no. 6 (2017): 1–29.

According to the US Bureau of Labor Statistics, automotive-sector (parts and assembly) employment dropped from 1,168,000 in 1994 to 820,000 in 2013, although the employment total had rebounded to 1,040,500 in 2023. Over the 1994 to 2013 period, automotive-sector employment in Mexico increased from 122,000 (1994) to 552,000 (2013). See Gary Clyde Hufbauer et al., *NAFTA at 20: Misleading Charges and Positive Achievements*, report PB14-13 (Peterson Institute for International Economics, May 2014), https://www.piie.com/sites/default/files/publications/pb/pb14-13.pdf; US Bureau of Labor Statistics, "Automotive Industry: Employment, Earnings, and Hours," *BLS.gov*, accessed March 20, 2024, https://www.bls.gov/iag/tgs/iagauto.htm#emp_national.

68. Peter Weber, "The Rise and Fall of Detroit: A Timeline," *Week*, January 8, 2015. https://theweek.com/articles/461968/rise-fall-detroit-timeline.

69. Brent D. Ryan and Daniel Campo, "Autopia's End: The Decline and Fall of Detroit's Automotive Manufacturing Landscape," *Journal of Planning History* 12, no. 2 (May 2013): 95–132, https://doi.org/10.1177/1538513212471166.

70. Amy Padnani, "Anatomy of Detroit's Decline," *New York Times*, August 17, 2013, https://www.nytimes.com/interactive/2013/08/17/us/detroit-decline.html.

71. Frank Goeddeke Jr. and Marick F. Masters, "The UAW: Then and Now 50 Years of United Auto Workers Union History," *LERA for Libraries* 24 (2020), https://lerawebillinois.web.illinois.edu/index.php/PFL/article/view/3392.

72. Timothy J. Minchin, "'A Gallant Fight': The UAW and the 1970 General Motors Strike," *International Review of Social History* 68, no. 1 (April 2023), 41–73, https://doi.org/10.1017/S0020859022000293.

73. Weber, "Rise and Fall of Detroit."

74. Ross Eisenbrey, "Management—Bad Management—Crippled the Auto Industry's Big Three, Not the UAW," *Economic Policy Institute Working Economics Blog*, May 24, 2012, https://www.epi.org/blog/bad-management-crippled-auto-industry-big-three/.

75. T. Koshiba et al., "Japanese Automakers and the NAFTA Environment: Global Context," *Environments* 29, no. 3 (2001).

76. Robert B. Reich, "Bailout: A Comparative Study in Law and Industrial Structure." *Yale Journal on Regulation* 2 (1984): 163224.

77. Ryan and Campo, "Autopia's End."

78. Jane Kulik et al., *The Downriver Community Conference Economic Readjustment Program: Final Evaluation Report* (Abt Associates, September 30, 1984).

79. Padnani, "Anatomy of Detroit's Decline."

80. Ryan and Campo, "Autopia's End."

81. Mathieu Dupuis and Ian Greer, "Recentralizing Industrial Relations? Local Unions and the Politics of Insourcing in Three North American Automakers," *ILR Review* 75, no. 4 (2022): 918–42.

82. Goeddeke Jr. and Masters, "UAW: Then and Now."

83. Two of the stand-alone companies formed by the Big Three filed for bankruptcy during the first decade of the 2000s: GM's Delphi (formed by GM) filed in 2005 and Visteon (formed by Ford) filed in 2009. See Dupuis and Greer, "Recentralizing Industrial Relations?"

84. 2009 also represented the end of a thirty-year period of UAW membership contraction: membership shrank from a high of approximately 1.5 million members in 1979 to fewer than four hundred thousand members in 2009. At the end of the decade, the wage and benefit concession agreed upon in the 2007 contract negotiation had reduced the compensation gap between Big Three unionized workers and transplant workers; for Ford, the gap shrank from $20.55 in 2005 to $6.00 in 2010. See Goeddeke Jr. and Masters, "UAW: Then and Now"; and Joel Cutcher-Gershenfeld et al., *The Decline and Resurgence of the U.S. Auto Industry* (Economic Policy Institute, May 6, 2015), https://files.epi.org/2015/the-decline-and -resurgence-of-the-us-auto-industry.pdf.

85. Camilia Domonoske, "Big 3 Autoworkers Vote 'Yes' to Historic UAW Contracts," *NPR*, November 17, 2023, https://www.npr.org/2023/11/16/1212381342/gm-autoworkers-vote-yes -approve-uaw-contract-ford-stellantis; David Shepardson and Ben Klayman, "UAW Ratifies Labor Deal with General Motors," *Reuters*, December 16, 2023, https://www.reuters.com /business/autos-transportation/gms-labor-deal-with-uaw-clinches-ratification-2023-11 -16/. Nonunion workers took note of the concessions gained by UAW members following the 2023 strike. In contrast to their anti-unionization strategy in American plants, where wages had been comparable to those received by unionized workers, Mercedes has kept wages relatively stagnant over the past several years and has adopted the two-tier wage system used by other automakers. In April 2024, workers at a Vance, Alabama, Mercedes plant filed a petition for a UAW union election, following the April 2024 affirmative vote to unionize a Volkswagen plant in Chattanooga, Tennessee. Additionally, emboldened by their 2023 successes in negotiating contracts with the Big Three, the UAW announced plans to unionize approximately 150,000 nonunion employees at thirty-six plants of thirteen automakers—three American companies (Tesla, Rivian, and Lucid) and ten foreign-based companies (BMW, Honda, Hyundai, Mazda, Mercedes, Nissan, Subaru, Toyota, Volkswagen, and Volvo). The 150,000 employees targeted by UAW efforts exceeded the approximate membership in November 2023 among the Big Three automakers (145,000 members). See Hsu and Bisaha, "Seeking to Defy History, the UAW is Coming Closer to Unionizing the South"; Nora Eckert, "UAW Wins Big in Historic Union Vote at Volkswagen Tennessee Factory," *Reuters*, April 22, 2024, https://www.reuters.com/business/autos-transportation/uaw-clinches-watershed -union-victory-volkswagen-tennessee-factory-2024-04-20/; Chris Isidore, "The UAW Just Won Its Battle with the Big Three: Now It's Aiming at 13 Non-Union Automakers," *CNN*, November 29, 2023, https://www.cnn.com/2023/11/29/business/uaw-organize-nonunion -automakers/index.html.

86. Frances Kai Hwa Wang, "What Will the EV Revolution Mean for Detroit?," *PBS News-Hour*, October 11, 2022, https://www.pbs.org/newshour/nation/detroit-leads-the-way-as-the-nation-grapples-with-impending-electric-car-revolution.

87. Environmental Defense Fund and WSP, *U.S. Electric Vehicle Manufacturing Investments and Jobs—Characterizing the Impacts of the Inflation Reduction Act After 1 Year* (Environmental Defense Fund and WSP, August 2023), https://www.edf.org/sites/default/files/2023-08/EDF%20WSP%20EV%20report%208-16-23%20FINAL%20FINAL.pdf.

88. Susan Ager, "Tough, Cheap, and Real, Detroit Is Cool Again," *National Geographic*, January 14, 2016, http://www.nationalgeographic.com/taking-back-detroit/see-detroit.html.

89. Lawrence Ulrich, "GM Bets Big on Batteries: A New $2.3 Billion Plant Cranks Out Ultium Cells to Power a Future Line of Electric Vehicles," *IEEE Spectrum* 57, no. 12 (December 2020): 26–31, https://doi.org/10.1109/MSPEC.2020.9271805.

90. Sherry Linkon and John Russo, "With GM Job Cuts, Youngstown Faces a New 'Black Monday,'" *Bloomberg*, November 27, 2018, https://www.bloomberg.com/news/articles/2018-11-27/gm-s-job-cuts-reopen-old-wounds-in-youngstown-ohio.

91. Jamie L. LaReau, "GM's Former Plant in Lordstown Will Return to Mass Vehicle Production, Thousands of Jobs," *Detroit Free Press*, May 22, 2020, https://www.freep.com/story/money/cars/general-motors/2020/05/22/gm-lordstown-motors-plant-jobs-electric-pickups-vehicles/5229669002.

92. Ian Kullgren and David Welch, "Lordstown's Auto Industry Is Coming Back: The Jobs? Not So Much," *Bloomberg*, November 10, 2021, https://www.bloomberg.com/news/features/2021-11-10/gm-brings-back-lordstown-s-auto-industry-the-jobs-not-so-much.

93. LaToya Ruby Frazier and Dan Kaufman, "What Happens to a Factory Town When the Factory Shuts Down?," *New York Times Magazine*, May 2, 2019, https://www.nytimes.com/interactive/2019/05/01/magazine/lordstown-general-motors-plant.html.

94. Tom Krisher, "GM Rebounds with $8.1B 2018 Profit on Strong Pricing," *Phys*, February 6, 2019, https://phys.org/news/2019-02-gm-strong-profits-good-sales.html.

95. Frazier and Kaufman, "What Happens to a Factory Town When the Factory Shuts Down?"

96. Frazier and Kaufman, "What Happens to a Factory Town When the Factory Shuts Down?"

97. Michael Sainato, "Youngstown's Hopes for Reinvention Fade as Electric Truck Firm Sputters," *Guardian*, June 30, 2021, https://www.theguardian.com/business/2021/jun/30/youngstown-ohio-car-electric-truck.

98. Matthew S. Wood et al., "Fake It 'Til You Make It: Hazards of a Cultural Norm in Entrepreneurship," *Business Horizons* 65, no. 5 (September 2022): 681–96, https://doi.org/10.1016/j.bushor.2021.12.001.

99. Bradley Brownell, "Lordstown Motors Prototype Electric Truck Burned to the Ground in Testing," *Jalopnik*, February 25, 2021, https://jalopnik.com/lordstown-motors-prototype-electric-truck-burned-to-the-1846350243.

100. Wood et al., "Fake It 'Til You Make It."

101. Lora Kolodny, "Foxconn Buys Lordstown Motors' Old GM Factory for $230 Million," *CNBC*, November 10, 2021, https://www.cnbc.com/2021/11/10/foxconn-buys-lordstown-motors-old-gm-factory-for-230-million.html.

102. Mike Spector et al., "Lordstown Motors Files for Bankruptcy, Sues Foxconn," *Reuters*, June 27, 2023, https://www.reuters.com/business/autos-transportation/lordstown-motors -files-bankruptcy-sues-foxconn-2023-06-27/.

103. Ulrich, "GM Bets Big on Batteries."

104. Mark Kane, "Ultium Cells Battery Plant Is Almost Complete: June 1, 2021," *InsideEVs*, June 2, 2021, https://insideevs.com/news/511582/ultium-cells-battery-plant-june2021/.

105. Ulrich, "GM Bets Big on Batteries."

106. Our research team conducted semi-structured focus groups in six locations across the Midwest, including Lordstown and Detroit, in the spring and summer of 2021. In total, we had 150 respondents; 50 were from Detroit and 48 were from Lordstown. Focus groups lasted about seventy-five minutes and were conducted online over Zoom. The focus groups were guided by an instrument.

107. US Bureau of Labor Statistics, "Bureau of Labor Statistics Motor Vehicle Manufacturing Data," *BLS.gov*, accessed February 1, 2023, https://data.bls.gov/timeseries/CEU3133610001 ?amp%25253bdata_tool=XGtable&output_view=data&include_graphs=true.

108. Karla Walter et al., *Electric Vehicles Should Be a Win for American Workers* (Center for American Progress, September 23, 2020), https://www.americanprogress.org/article /electric-vehicles-win-american-workers/.

109. Klaus Barthel et al., *The Future of the German Automotive Industry: Structural Change in the Automotive Industry: Challenges and Perspectives* (Friedrich Ebert Stiftung, 2015).

110. Turner Cotterman et al., "The Transition to Electrified Vehicles: Evaluating the Labor Demand of Manufacturing Conventional Versus Battery Electric Vehicle Powertrains," *Energy Policy* 188 (June 4, 2022): 114064, https://doi.org/10.2139/ssrn.4128130.

111. Cotterman et al., "Transition to Electrified Vehicles."

112. Nuri Onat et al., "Towards Life Cycle Sustainability Assessment of Alternative Passenger Vehicles," *Sustainability* 6, no. 12 (December 16, 2014): 9305–42, https://doi.org/10 .3390/su6129305.

113. Scott Hardman et al., "A Perspective on Equity in the Transition to Electric Vehicles," eds. Bertrand Neyhouse and Yana Petri, *MIT Science Policy Review* 2 (August 30, 2021): 46– 54, https://doi.org/10.38105/spr.e10rdoaoup.

114. Severin Borenstein and Lucas W. Davis, "The Distributional Effects of US Clean Energy Tax Credits," *Tax Policy and the Economy* 30, no. 1 (January 2016): 191–234, https://doi .org/10.1086/685597.

115. Jae Hyun Lee et al., "Who Is Buying Electric Vehicles in California? Characterising Early Adopter Heterogeneity and Forecasting Market Diffusion," *Energy Research and Social Science* 55 (September 2019): 218–26, https://doi.org/10.1016/j.erss.2019.05.011.

116. Fuels Institute Electric Vehicle Council, *EV Consumer Behavior* (Fuels Institute, June 2021), https://www.transportationenergy.org/wp-content/uploads/2022/11/21FI_-EVC _ConsumerBehaviorReport_V07-FINAL.pdf.

117. Sanya Carley et al., "Evolution of Plug-In Electric Vehicle Demand: Assessing Consumer Perceptions and Intent to Purchase over Time," *Transportation Research Part D: Transport and Environment* 70 (May 2019): 94–111, https://doi.org/10.1016/j.trd.2019.04.002.

118. Maddy Yozwiak et al., *Clean and Just: Electric Vehicle Innovation to Accelerate More Equitable Early Adoption* (Information Technology and Innovation Foundation, June 27,

2022), https://itif.org/publications/2022/06/27/electric-vehicle-innovation-to-accelerate-more-equitable-early-adoption/.

119. Katie Bohn, "Misestimating Travel Times May Stop People from Walking or Biking to Work," Penn State, March 28, 2018, https://www.psu.edu/news/research/story/misestimating-travel-times-may-stop-people-walking-or-biking-work/.

120. Notably, this offer only was available for the first two hundred thousand cars sold of each make and model.

121. Internal Revenue Service, "Credits for New Clean Vehicles Purchased in 2023 or After," *IRS.gov*, March 31, 2023, https://www.irs.gov/credits-deductions/credits-for-new-clean-vehicles-purchased-in-2023-or-after; see also Hardman et al., "Perspective on Equity in the Transition to Electric Vehicles."

122. In 2023, the US federal government allowed tax credits to be applied at point of sale.

123. Ellis Davies, "Consumers Three Times More Likely to Buy Used Cars over New," *Motor Trader* (blog), October 28, 2019, https://www.motortrader.com/motor-trader-news/automotive-news/majority-buy-consumers-opt-used-new-cars-28-10-2019.

124. Will Englund, "Without Access to Charging Stations, Black and Hispanic Communities May Be Left Behind in the Era of Electric Vehicles," *Washington Post*, December 9, 2021, https://www.washingtonpost.com/business/2021/12/09/charging-deserts-evs/.

125. Chih-Wei Hsu and Kevin Fingerman, "Public Electric Vehicle Charger Access Disparities Across Race and Income in California," *Transport Policy* 100 (January 2021): 59–67, https://doi.org/10.1016/j.tranpol.2020.10.003.

126. Yozwiak et al., *Clean and Just*, 9.

127. Yozwiak et al., *Clean and Just*, 9.

128. Martin Oteng-Ababio et al., "Building Policy Coherence for Sound Waste Electrical and Electronic Equipment Management in a Developing Country," *Journal of Environment and Development* 29, no. 3 (September 2020): 306–28, https://doi.org/10.1177/1070496519898218.

129. Vanessa Forti et al., *The Global E-Waste Monitor 2020* (United Nations University, Unitar, ITU, and ISWA, November 2020), https://ewastemonitor.info/wp-content/uploads/2020/11/GEM_2020_def_july1_low.pdf.

130. Katie Campbell and Ken Christensen, "Where Does America's E-Waste End Up? GPS Tracker Tells All," *PBS NewsHour*, May 10, 2016, https://www.pbs.org/newshour/science/america-e-waste-gps-tracker-tells-all-earthfix.

131. Tamba S. Lebbie et al., "E-Waste in Africa: A Serious Threat to the Health of Children," *International Journal of Environmental Research and Public Health* 18, no. 16 (August 11, 2021): 8488, https://doi.org/10.3390/ijerph18168488.

132. Roland Kofi Srigboh et al., "Multiple Elemental Exposures Amongst Workers at the Agbogbloshie Electronic Waste (E-Waste) Site in Ghana," *Chemosphere* 164 (December 2016): 68–74, https://doi.org/10.1016/j.chemosphere.2016.08.089.

133. Grace A. Akese and Peter C. Little, "Electronic Waste and the Environmental Justice Challenge in Agbogbloshie," *Environmental Justice* 11, no. 2 (April 2018): 77–83, https://doi.org/10.1089/env.2017.0039.

134. Karoline Owusu-Sekyere et al., "Assessing Data in the Informal E-Waste Sector: The Agbogbloshie Scrapyard," *Waste Management* 139 (February 2022): 158–67, https://doi.org/10.1016/j.wasman.2021.12.026.

135. Florian Weigensamer and Christian Krönes, dirs., *Welcome to Sodom* (Torch Films, 2018).

136. Weigensamer and Krönes, dirs., *Welcome to Sodom.*

137. Lebbie et al., "E-Waste in Africa."

138. Oteng-Ababio et al., "Building Policy Coherence."

139. Neil Shaw, "For Ghana E-Waste Recyclers, a Safer Option amid Toxic Fumes," *AP News*, January 5, 2019, https://apnews.com/article/f9a0d071d1d646edb2b53fd22fd8548c.

140. Srigboh et al., "Multiple Elemental Exposures."

141. Srigboh et al., "Multiple Elemental Exposures."

142. Jacklin Kwan, "Your Old Electronics Are Poisoning People at This Toxic Dump in Ghana," *Wired UK*, November 26, 2020, https://www.wired.co.uk/article/ghana-ewaste-dump-electronics.

143. Lebbie et al., "E-Waste in Africa."

144. World Health Organization, "Dioxins and Their Effects on Human Health," *WHO.int*, October 4, 2016, archived at https://web.archive.org/web/20180427100129/http://www.who.int/news-room/fact-sheets/detail/dioxins-and-their-effects-on-human-health.

145. Tash Morgan, "Agbogbloshie: Welcome to the World's Digital Dumping Ground (Part 1) | Human Impact," *Earth Touch News Network*, March 26, 2014, https://www.earthtouchnews.com/conservation/human-impact/agbogbloshie-welcome-to-the-worlds-digital-dumping-ground-part-1.

146. Kawdwo Ansong Asante et al., "E-Waste Interventions in Ghana," *Reviews on Environmental Health* 31, no. 1 (March 1, 2016): 145–48, https://doi.org/10.1515/reveh-2015-0047.

147. Doreen Andoh, "Phase One of Agbogbloshie E-Waste Project Inaugurated," *Graphic Online*, March 28, 2019, https://www.graphic.com.gh/news/general-news/ghananews-phase-one-of-agbogbloshie-e-waste-project-inaugurated.html.

148. Muntaka Chasant "Agbogbloshie Demolition: The End of an Era or an Injustice?," *Muntaka* (blog), November 23, 2022, https://www.muntaka.com/agbogbloshie-demolition/.

149. Benjamin K. Sovacool, "Who Are the Victims of Low-Carbon Transitions? Towards a Political Ecology of Climate Change Mitigation," *Energy Research and Social Science* 73 (March 2021): 101916, https://doi.org/10.1016/j.erss.2021.101916.

150. Minerals needed for clean-energy technologies will change global demand and production in substantial but to-some-degree-uncertain ways in the future. Energy experts typically assume an increase in the use of graphite, copper, lithium, nickel, cobalt, and rare earth minerals. The major producing countries of these minerals are as follows: graphite (China, Mozambique, Brazil), copper (Chile, Peru, China, DRC), lithium (Australia, Chile, China, Argentina), nickel (Indonesia, Philippines, Russia, Canada), cobalt (DRC, Australia, Russia), and rare earth minerals (China).

151. Sovacool et al., "Decarbonisation Divide."

CHAPTER 8

1. International Energy Agency, *CO2 Emissions in 2023* (International Energy Agency, March 2024), https://www.iea.org/reports/co2-emissions-in-2023.

2. Timothy M. Lenton et al., "Climate Tipping Points—Too Risky to Bet Against," *Nature* 575, no. 7784 (November 2019): 592–95, https://doi.org/10.1038/d41586-019-03595-0.

3. Robert McSweeney, "Explainer: Nine 'Tipping Points' That Could Be Triggered by Climate Change," *CarbonBrief*, October 2020, https://www.carbonbrief.org/explainer-nine -tipping-points-that-could-be-triggered-by-climate-change/.

4. National Snow and Ice Data Center, "Antarctic Melt Season off to Fast Start; Green-land 2023 Melt Season Review," *NSIDC.org*, December 21, 2023, https://nsidc.org/ice-sheets -today/analyses/antarctic-melt-season-fast-start-greenland-2023-melt-season-review.

5. Rebecca Newman and Ilan Noy, "The Global Costs of Extreme Weather That Are At-tributable to Climate Change," *Nature Communications* 14, no. 1 (September 29, 2023): 6103, https://doi.org/10.1038/s41467-023-41888-1.

6. US Energy Information Administration, "Total Energy," *EIA.gov*, 2023, https://www .eia.gov/totalenergy/data/browser/index.php?tbl=T01.03#/?f=A&start=1997&end=2022& charted=1-2-3-5-12.

7. US Energy Information Administration, "Frequently Asked Questions (FAQs)—What Countries Are the Top Producers and Consumers of Oil?," *EIA.gov*, 2023, https://www.eia .gov/tools/faqs/faq.php?id=709&t=6.

8. US Energy Information Administration, "Total Energy."

9. International Energy Agency, *World Energy Outlook 2023* (International Energy Agency, October 2023), https://www.iea.org/reports/world-energy-outlook-2023.

10. US Energy Information Administration, *Monthly Energy Review* (US Energy Infor-mation Administration, November 2024), https://www.eia.gov/totalenergy/data/monthly /pdf/mer.pdf.

11. US Environmental Protection Agency, "Sources of Greenhouse Gas Emissions," *EPA.gov*, last updated October 22, 2024, https://www.epa.gov/ghgemissions/sources -greenhouse-gas-emissions.

12. US Environmental Protection Agency, "Highlights of the Automotive Trends Re-port," *EPA.gov*, last updated December 20, 2023, https://www.epa.gov/automotive-trends /highlights-automotive-trends-report.

13. Ronald Montoya and Steven Ewing, "How Many Electric Cars Are There in the U.S.?" *Edmunds*, July 17, 2024, https://www.edmunds.com/electric-car/articles/how-many -electric-cars-in-us.html; Alexa St. John, "US Electric Vehicle Sales to Hit Record This Year, but Still Lag Behind China and Germany," *AP News*, November 24, 2023, https://apnews.com /article/automakers-electric-vehicles-us-china-sales-d121c09a61f50e7357f5675af4b6056b.

14. US Environmental Protection Agency, "Sources of Greenhouse Gas Emissions."

15. US Energy Information Administration, "Total Energy."

16. US Bureau of Labor Statistics, "All Employees, Coal Mining," retrieved from FRED, Federal Reserve Bank of St. Louis, 2024, https://fred.stlouisfed.org/series/CES1021210001.

17. US Energy Information Administration, "Table 4.1: Count of Electric Power Industry Power Plants, by Sector, by Predominant Energy Sources within Plant, 2013 through 2023," *EIA.gov*, 2023, https://www.eia.gov/electricity/annual/html/epa_04_01.html.

18. Seth Feaster, *U.S. on Track to Close Half of Coal Capacity by 2026* (Institute for Energy Economics and Financial Analysis, April 3, 2023), https://ieefa.org/resources/us-track-close -half-coal-capacity-2026; US Energy Information Administration, "Nearly a Quarter of the

Operating U.S. Coal-Fired Fleet Scheduled to Retire by 2029," *EIA.gov*, November 7, 2022, https://www.eia.gov/todayinenergy/detail.php?id=54559.

19. Jamie LaReau, "GM's Former Plant in Lordstown Will Return to Mass Vehicle Production, Thousands of Jobs," *Detroit Free Press*, May 22, 2020, https://www.freep.com/story/money/cars/general-motors/2020/05/22/gm-lordstown-motors-plant-jobs-electric-pickups-vehicles/5229669002/.

20. Phil LeBeau, "GM's Lordstown Factory Goes Dark as Automaker Idles Underused Plants," *CNBC*, March 6, 2019, https://www.cnbc.com/2019/03/06/gms-lordstown-factory-goes-dark-as-automaker-closes-underused-plants.html.

21. LaToya Ruby Frazier and Dan Kaufman, "What Happens to a Factory Town When the Factory Shuts Down?," *New York Times Magazine*, May 1, 2019, https://www.nytimes.com/interactive/2019/05/01/magazine/lordstown-general-motors-plant.html.

22. Alexander Gazmararian and Dustin Tingley, *Uncertain Futures: How to Unlock the Climate Impasse* (Cambridge University Press, 2023).

23. US Energy Information Administration, *Table HC11.1: Household Energy Insecurity, 2020* (US Energy Information Administration, March 2023), https://www.eia.gov/consumption/residential/data/2020/hc/pdf/HC%2011.1.pdf.

24. David Wallace-Wells, *The Uninhabitable Earth: Life After Warming* (Penguin UK, 2019).

REFERENCES

AES Indiana. "Harding Street Station." *AESIndiana.com*, March 11, 2020. https://www
.aesindiana.com/harding-street-station.

AES Indiana. "IPL Burns Coal for the Last Time at Harding Street Station." Press release,
March 25, 2016. https://www.aesindiana.com/press-release/ipl-burns-coal-last-time
-harding-street-station.

Ager, Susan. "Tough, Cheap, and Real, Detroit Is Cool Again." *National Geographic*, January 14, 2016. http://www.nationalgeographic.com/taking-back-detroit/see-detroit.html.

Akese, Grace A., and Peter C. Little. "Electronic Waste and the Environmental Justice Challenge in Agbogbloshie." *Environmental Justice* 11, no. 2 (April 2018): 77–83. https://doi
.org/10.1089/env.2017.0039.

Aklin, Michael, and Johannes Urpelainen. "Enable a Just Transition for American Fossil Fuel
Workers Through Federal Action." *Brookings*, August 2022. https://www.brookings.edu
/articles/enable-a-just-transition-for-american-fossil-fuel-workers-through-federal
-action/.

Aldhous, Peter, Stephanie Lee, and Zahra Hirji. "The Texas Winter Storm and Power Outages Killed Hundreds More People Than the State Says." *BuzzFeed News*, May 26, 2021.
https://www.buzzfeednews.com/article/peteraldhous/texas-winter-storm-power
-outage-death-toll.

Alliance for Automotive Innovation. "Electric Vehicle Sales Dashboard." *Autosinnovate.org*,
accessed November 1, 2024. https://www.autosinnovate.org/EVDashboard.

Ambrose, Mariana, John Jacobs, and Natalie Tham. "CHIPS and Science Act Summary: Energy, Climate, and Science Provisions." Bipartisan Policy Center, November 14, 2022.
https://bipartisanpolicy.org/blog/chips-science-act-summary/.

American Lung Association. "Key Findings." American Lung Association *State of the Air* report, 2023. https://www.lung.org/research/sota/key-findings.

Amnesty International. "Phones, Electric Cars and Human Rights Abuses—5 Things
You Need to Know." *Amnesty.org*, May 1, 2018. https://www.amnesty.org/en/latest

/news/2018/05/phones-electric-cars-and-human-rights-abuses-5-things-you-need
-to-know/.

Amnesty International. *"This Is What We Die For": Human Rights Abuses in the Democratic Republic of the Congo Power the Global Trade in Cobalt.* Amnesty International, 2016.

Andoh, Doreen. "Phase One of Agbogbloshie E-Waste Project Inaugurated." *Graphic Online*, March 28, 2019. https://www.graphic.com.gh/news/general-news/ghananews-phase -one-of-agbogbloshie-e-waste-project-inaugurated.html.

Ansolabehere, Stephen, and David M. Konisky. *Cheap and Clean: How Americans Think About Energy in the Age of Global Warming.* MIT Press, 2016.

Ansolabehere, Stephen, and David M. Konisky. "Public Attitudes Toward Construction of New Power Plants." *Public Opinion Quarterly* 73, no. 3 (2009): 566–77.

"Arrowhead Landfill: Proven, Efficient and Environmentally Superior." Arrowhead Environmental Partners, 2024. https://arrowheadenvironmentalpartners.com/.

Asante, Kwadwo Ansong, John A. Pwamang, Yaw Amoyaw-Osei, and Joseph Addo Ampofo. "E-Waste Interventions in Ghana." *Reviews on Environmental Health* 31, no. 1 (March 1, 2016): 145–48. https://doi.org/10.1515/reveh-2015-0047.

Attari, Shahzeen Z., Michael L. DeKay, Cliff I. Davidson, and Wändi Bruine de Bruin. "Public Perceptions of Energy Consumption and Savings." *Proceedings of the National Academy of Sciences* 107, no. 37 (September 14, 2010): 16054–59. https://doi.org/10.1073/pnas .1001509107.

Avila, Eric. *The Folklore of the Freeway: Race and Revolt in the Modernist City.* University of Minnesota Press, 2014.

Balaraman, Kavya. "PG&E, SCE Detail Plans to Spend More than $23B Through 2025 to Prevent Wildfires in Their Footprints." *Utility Dive*, March 28, 2023. https://www.utilitydive .com/news/pge-sce-vegetation-management-resilience-california-wildfires/646163/.

Baldauf, R., E. Thoma, A. Khlystov, V. Isakov, G. Bowker, T. Long, et al. "Impacts of Noise Barriers on Near-Road Air Quality." *Atmospheric Environment* 42, no. 32 (October 1, 2008): 7502–7. https://doi.org/10.1016/j.atmosenv.2008.05.051.

Bandyk, Matthew. "Largest Planned Wind Farm in US Gets Key Federal Approval." *Utility Dive*, October 25, 2019. https://www.utilitydive.com/news/largest-planned-wind-farm -in-us-gets-key-federal-approval/565795/.

Barringer, Felicity. "What It May Take to Harness Solar Energy on Native Lands." Stanford Doerr School of Sustainability, May 6, 2021. https://earth.stanford.edu/news/what-it -may-take-harness-solar-energy-native-lands.

Barthel, Klaus, Susanne Bohler-Baedeker, Rene Bormann, Jurgen Dispan, Philipp Fink, Thorsten Koska, et al. *The Future of the German Automotive Industry: Structural Change in the Automotive Industry: Challenges and Perspectives.* Friedrich Ebert Stiftung, 2015.

Basov, Vladimir. "The World's Largest Cobalt Reserves by Country in 2022." *Kitco News*, February 7, 2023. Archived at https://web.archive.org/web/20230314082433/https://www.kitco .com/news/2023-02-07/The-world-s-largest-cobalt-reserves-by-country-in-2022.html.

Beadle, Alexander. "Cobalt-Free Batteries Could Power the Next Generation of Electric Vehicles." *Technology Networks*, January 23, 2024. https://www.technologynetworks.com /applied-sciences/news/cobalt-free-batteries-could-power-the-next-generation-of -electric-vehicles-383059.

Beckfield, Jason, Devin Booker, Kerry Bowie, Brianna Castro, Christine DeMyers, Daniel Alain Evrard, et al. *How the Gulf Coast Can Lead the Energy Transition.* Roosevelt Project, MIT Center for Energy and Environmental Policy Research, Harvard University, April 2022.

Beckfield, Jason, and Daniel Alain Evrard. "The Social Impacts of Supply-Side Decarbonization." *Annual Review of Sociology* 49, no. 1 (July 31, 2023): 155–75. https://doi.org/10.1146/annurev-soc-031021-012201.

Bell, Shannon. *Fighting King Coal: The Challenges to Micromobilization in Central Appalachia.* MIT Press, 2016.

Bell, Shannon Elizabeth, and Richard York. "Community Economic Identity: The Coal Industry and Ideology Construction in West Virginia: Community Economic Identity." *Rural Sociology* 75, no. 1 (January 19, 2010): 111–43. https://doi.org/10.1111/j.1549-0831.2009.00004.x.

Bergquist, Parrish, Stephen Ansolabehere, Sanya Carley, and David Konisky. "Backyard Voices: How Sense of Place Shapes Views of Large-Scale Energy Transmission Infrastructure." *Energy Research and Social Science* 63 (May 2020): 101396. https://doi.org/10.1016/j.erss.2019.101396.

Bergquist, Parrish, David M. Konisky, and John Kotcher. "Energy Policy and Public Opinion: Patterns, Trends and Future Directions." *Progress in Energy* 2, no. 3 (August 10, 2020): 032003. https://doi.org/10.1088/2516-1083/ab9592.

Bersch, Caroline. "4 Health Care Facts About the Democratic Republic of the Congo." Borgen Project, September 6, 2021. https://borgenproject.org/health-care-in-the-drc/.

Bessette, Douglas L., and Sarah B. Mills. "Farmers vs. Lakers: Agriculture, Amenity, and Community in Predicting Opposition to United States Wind Energy Development." *Energy Research and Social Science* 72 (February 2021): 101873. https://doi.org/10.1016/j.erss.2020.101873.

Betz, Michael R., Mark D. Partridge, Michael Farren, and Linda Lobao. "Coal Mining, Economic Development, and the Natural Resources Curse." *Energy Economics* 50 (July 2015): 105–16. https://doi.org/10.1016/j.eneco.2015.04.005.

Bhattacharya, Jayanta, Thomas DeLeire, Steven Haider, and Janet Currie. "Heat or Eat? Cold-Weather Shocks and Nutrition in Poor American Families." *American Journal of Public Health* 93, no. 7 (July 2003): 1149–54. https://doi.org/10.2105/ajph.93.7.1149.

Bhutada, Govind. "Charted: Energy Consumption by Source and Country (1969–2018)." *Visual Capitalist*, September 9, 2020. https://www.visualcapitalist.com/energy-consumption-by-source-and-country-1969-2018/.

Bipartisan Policy Center. *Inflation Reduction Act Summary: Energy and Climate Provisions.* Bipartisan Policy Center, August 4, 2022. https://bipartisanpolicy.org/blog/inflation-reduction-act-summary-energy-climate-provisions/.

Black, Katie Jo, Andrew J. Boslett, Elaine L. Hill, Lala Ma, and Shawn J. McCoy. "Economic, Environmental, and Health Impacts of the Fracking Boom." *Annual Review of Resource Economics* 13, no. 1 (2021): 311–34. https://doi.org/10.1146/annurev-resource-110320-092648.

Blackley, David J., Cara N. Halldin, and A. Scott Laney. "Continued Increase in Prevalence of Coal Workers' Pneumoconiosis in the United States, 1970–2017." *American Journal of*

Public Health 108, no. 9 (September 2018): 1220–22. https://doi.org/10.2105/AJPH.2018
.304517.

Block, Samuel. "Mining Energy-Transition Metals: National Aims, Local Conflicts." MSCI,
June 3, 2021. https://www.msci.com/www/blog-posts/mining-energy-transition-metals
/02531033947.

Bodenhamer, Aysha. "King Coal: A Study of Mountaintop Removal, Public Discourse, and
Power in Appalachia." *Society and Natural Resources* 29, no. 10 (2016): 1139–53. https://
doi.org/10.1080/08941920.2016.1138561.

Boehm, Jessica. "Phoenix Experienced Its Hottest and Driest Summer on Record." *Axios*, October 10, 2023. https://www.axios.com/local/phoenix/2023/10/10/2023-hottest-driest
-summer-monsoon.

Bohn, Katie. "Misestimating Travel Times May Stop People from Walking or Biking to Work."
Penn State, March 28, 2018. https://www.psu.edu/news/research/story/misestimating
-travel-times-may-stop-people-walking-or-biking-work/.

Bollinger, Bryan, and Kenneth Gillingham. "Peer Effects in the Diffusion of Solar Photovoltaic Panels." *Marketing Science* 31, no. 6 (2012): 900–12.

Borenstein, Severin. "Private Net Benefits of Residential Solar PV: The Role of Electricity
Tariffs, Tax Incentives, and Rebates." *Journal of the Association of Environmental and
Resource Economists* 4, no. S1 (September 2017): S85–122. https://doi.org/10.1086/691978.

Borenstein, Severin, and Lucas W. Davis. "The Distributional Effects of US Clean Energy
Tax Credits." *Tax Policy and the Economy* 30, no. 1 (January 2016): 191–234. https://doi
.org/10.1086/685597.

Bowman, Jennifer, and Kelly Johnson. *2015 Stream Health Report: An Evaluation of Wa-
ter Quality, Biology, and Acid Mine Drainage Reclamation in Five Watersheds: Raccoon
Creek, Monday Creek, Sunday Creek, Huff Run, and Leading Creek*. Voinovich School of
Leadership and Public Affairs at Ohio University, June 30, 2016.

Bowman, Sarah. "Indiana Supreme Court: Duke Can't Make Customers Pay $212 Million to
Clean up Coal Ash." *Indianapolis Star*, March 11, 2022. https://www.indystar.com/story
/news/environment/2022/03/11/court-duke-energy-cant-make-customers-pay-212-m
-clean-up-coal-ash/7001908001/.

Bridge, Gavin, Stefan Bouzarovski, Michael Bradshaw, and Nick Eyre. "Geographies of Energy Transition: Space, Place and the Low-Carbon Economy." *Energy Policy* 53 (2013):
331–40.

Brinkley, Catherine, and Andrew Leach. "Energy Next Door: A Meta-Analysis of Energy Infrastructure Impact on Housing Value." *Energy Research and Social Science* 50 (2019):
51–65.

Brown, David P. "Socioeconomic and Demographic Disparities in Residential Battery Storage Adoption: Evidence from California." *Energy Policy* 164 (May 2022): 112877. https://
doi.org/10.1016/j.enpol.2022.112877.

Brown, Kate Pride. "In the Pocket: Energy Regulation, Industry Capture, and Campaign
Spending." *Sustainability: Science, Practice and Policy* 12, no. 2 (2016): 1–15. https://doi
.org/10.1080/2052546.2016.11949232.

Brownell, Bradley. "Lordstown Motors Prototype Electric Truck Burned to the Ground in
Testing." *Jalopnik*, February 25, 2021. https://jalopnik.com/lordstown-motors-prototype
-electric-truck-burned-to-the-1846350243.

Browning, Morgan, James McFarland, John Bistline, Gale Boyd, Matteo Muratori, Matthew Binsted, et al. "Net-Zero CO2 by 2050 Scenarios for the United States in the Energy Modeling Forum 37 Study." *Energy and Climate Change* 4 (December 2023): 100104. https://doi.org/10.1016/j.egycc.2023.100104.

Brugge, Doug, John L. Durant, and Christine Rioux. "Near-Highway Pollutants in Motor Vehicle Exhaust: A Review of Epidemiologic Evidence of Cardiac and Pulmonary Health Risks." *Environmental Health* 6, no. 1 (August 9, 2007): 23. https://doi.org/10.1186/1476-069X-6-23.

Brugge, Doug, and Rob Goble. "The History of Uranium Mining and the Navajo People." *American Journal of Public Health* 92, no. 9 (September 2002): 1410–19. https://doi.org/10.2105/AJPH.92.9.1410.

Brugger, Kelsey. "'It Should Not Have Taken This Long': Regan Confronts EJ Vows." *E&E News*, November 23, 2021. https://www.eenews.net/articles/it-should-not-have-taken-this-long-regan-confronts-ej-vows/.

Brulle, Robert, and David Pellow. "Environmental Justice: Human Health and Environmental Inequalities." *Annual Review of Public Health* 27 (April 21, 2006): 103–24.

Brunner, Eric J., Ben Hoen, Joe Rand, and David Schwegman. "Commercial Wind Turbines and Residential Home Values: New Evidence from the Universe of Land-Based Wind Projects in the United States." *Energy Policy* 185 (2024): 113837.

Bryan, Susan Montoya. "US Launches Program to Provide Electricity to More Native American Homes." *AP News*, August 15, 2023. https://apnews.com/article/native-american-homes-electricity-infrastructure-energy-6ebeee0b027e5aa1d1204e0c0ddcc7b1.

Budinis, Sara. "Direct Air Capture." International Energy Agency, July 2023. https://www.iea.org/energy-system/carbon-capture-utilisation-and-storage/direct-air-capture.

Budryk, Zack. "Almost 5 Million Without Power as Winter Storm Stresses Grid in Texas, 13 Other States." *Hill*, February 16, 2021. https://thehill.com/regulation/energy-environment/538953-almost-5-million-without-power-as-winter-storm-stresses-grid/.

Bullard, Nat. "Decarbonization: Stocks and Flows, Abundance and Scarcity, Net Zero." *Nathanielbullard.com*, January 31, 2024. https://www.nathanielbullard.com/presentations.

Bullard, Robert D. *Dumping in Dixie: Race, Class, and Environmental Quality*. 3rd ed. Westview Press, 2008.

Bullard, Robert Doyle, Glenn Steve Johnson, and Angel O. Torres, eds. *Highway Robbery: Transportation Racism and New Routes to Equity*. South End Press, 2004.

Bureau of Ocean Energy Management. "Cape Wind Energy Project." Fact sheet, last updated September 2015. https://www.boem.gov/sites/default/files/renewable-energy-program/Studies/Cape-Wind-Fact-Sheet---Sept-2015-clean-%281%29.pdf.

Burnett, Richard, Hong Chen, Mieczysław Szyszkowicz, Neal Fann, Bryan Hubbell, C. Arden Pope, et al. "Global Estimates of Mortality Associated with Long-Term Exposure to Outdoor Fine Particulate Matter." *Proceedings of the National Academy of Sciences* 115, no. 38 (September 18, 2018): 9592–97. https://doi.org/10.1073/pnas.1803222115.

Campbell, Katie, and Ken Christensen. "Where Does America's E-Waste End Up? GPS Tracker Tells All." *PBS NewsHour*, May 10, 2016. https://www.pbs.org/newshour/science/america-e-waste-gps-tracker-tells-all-earthfix.

"Cape Wind Project, Massachusetts." *Power Technology* (blog). May 11, 2017. https://www.power-technology.com/projects/cape-wind-project-massachusetts/.

Carley, Sanya. "Normative Dimensions of Sustainable Energy Policy." *Ethics, Policy and Environment* 14, no. 2 (June 1, 2011): 211–29.

Carley, Sanya, Stephen Ansolabehere, and David M. Konisky. "Are All Electrons the Same? Evaluating Support for Local Transmission Lines Through an Experiment." Edited by Bing Xue. *PLOS ONE* 14, no. 7 (July 17, 2019): e0219066. https://doi.org/10.1371/journal.pone.0219066.

Carley, Sanya, Tom P. Evans, and David M. Konisky. "Adaptation, Culture, and the Energy Transition in American Coal Country." *Energy Research and Social Science* 37 (March 2018): 133–39. https://doi.org/10.1016/j.erss.2017.10.007.

Carley, Sanya, Michelle Graff, David M. Konisky, and Trevor Memmott. "Behavioral and Financial Coping Strategies Among Energy-Insecure Households." *Proceedings of the National Academy of Sciences* 119, no. 36 (September 6, 2022): e2205356119. https://doi.org/10.1073/pnas.2205356119.

Carley, Sanya, and David Konisky. *Survey of Household Energy Insecurity in Time of COVID: Preliminary Results of Wave-4, and Waves 1 Through 4 Combined.* Indiana University Energy Justice Lab, July 5, 2021. https://energyjustice.indiana.edu/doc/wave_4_report.pdf.

Carley, Sanya, and David Konisky. "Utility Disconnections Dashboard." Energy Justice Lab, 2023. https://utilitydisconnections.org/.

Carley, Sanya, David M. Konisky, Zoya Atiq, and Nick Land. "Energy Infrastructure, NIMBY-ism, and Public Opinion: A Systematic Literature Review of Three Decades of Empirical Survey Literature." *Environmental Research Letters* 15, no. 9 (September 1, 2020): 093007. https://doi.org/10.1088/1748-9326/ab875d.

Carley, Sanya, David Konisky, and Emily Nash. *Electric Utility Disconnections: Legal Protections and Policy Recommendations.* Indiana University and University of Pennsylvania Energy Justice Lab, June 2023. https://utilitydisconnections.org/doc/electric-utility-disconnections-legal-protections-and-policy-recommendations.pdf.

Carley, Sanya, Saba Siddiki, and Sean Nicholson-Crotty. "Evolution of Plug-In Electric Vehicle Demand: Assessing Consumer Perceptions and Intent to Purchase over Time." *Transportation Research Part D: Transport and Environment* 70 (May 2019): 94–111. https://doi.org/10.1016/j.trd.2019.04.002.

Castaneda, Monica, Maritza Jimenez, Sebastian Zapata, Carlos J. Franco, and Isaac Dyner. "Myths and Facts of the Utility Death Spiral." *Energy Policy* 110 (2017): 105–16.

Cavallo, Paulo, and Clint Peinhardt. "Foreign Investment and Right-to-Work Laws." *Business and Politics* 23, no. 3 (2021): 406–18.

Centers for Disease Control and Prevention. "2020 Adult Asthma Data: Prevalence Tables and Maps." *CDC.gov*, March 22, 2022. https://www.cdc.gov/asthma/brfss/2020/tableL1.html.

Centers for Disease Control and Prevention. "Injury Data Visualization Tools." WISQARS (Web-Based Injury Statistics Query and Reporting System), 2020. https://wisqars.cdc.gov/data/non-fatal/home.

Centers for Disease Control and Prevention. "PLACES: Local Data for Better Health." *CDC.gov*, July 13, 2023. https://www.cdc.gov/places/index.html.

Centers for Disease Control and Prevention. "Suicide Prevention: Facts About Suicide." *CDC.gov*, April 6, 2023. https://www.cdc.gov/suicide/facts/index.html.

Cha, J. Mijin. "A Just Transition for Whom? Politics, Contestation, and Social Identity in the

Disruption of Coal in the Powder River Basin." *Energy Research and Social Science* 69 (November 1, 2020): 101657. https://doi.org/10.1016/j.erss.2020.101657.

Cha, J. Mijin, Caroline Farrell, and Dimitris Stevis. "Climate and Environmental Justice Policies in the First Year of the Biden Administration." *Publius: The Journal of Federalism* 52, no. 3 (July 1, 2022): 408–27. https://doi.org/10.1093/publius/pjac017.

Cha, J. Mijin, and Manuel Pastor. "Just Transition: Framing, Organizing, and Power-Building for Decarbonization." *Energy Research and Social Science* 90 (August 1, 2022): 102588. https://doi.org/10.1016/j.erss.2022.102588.

Channel, Jacob, Dan Shepard, and Pearly Huang. "More Than 14 Million Households Across U.S. Don't Have Internet." *LendingTree*, 2022. https://www.lendingtree.com/home /mortgage/internet-access-study/.

Chapman, Isabelle. "Gambling 'America's Amazon.'" *CNN*, December 5, 2021. https://www .cnn.com/interactive/2021/12/us/coal-ash-ponds-plant-barry/.

Chasant, Muntaka. "Agbogbloshie Demolition: The End of an Era or an Injustice?" *Muntaka* (blog), August 22, 2021. https://www.muntaka.com/agbogbloshie-demolition/.

Chen, Yen-Heng Henry, Erik Landry, and John M. Reilly. "An Economy-Wide Framework for Assessing the Stranded Assets of Energy Production Sector Under Climate Policies." *Climate Change Economics* 14, no. 1 (2023): 2350003. https://doi.org/10.1142 /S2010007823500033.

Cheng-Chun Lee, Mikel Maron, and Ali Mostafavi, "Community-Scale Big Data Reveals Disparate Impacts of the Texas Winter Storm of 2021 and Its Managed Power Outage," *Humanities and Social Sciences Communications* 9, no. 1 (September 24, 2022): 1–12. https:// www.nature.com/articles/s41599-022-01353-8.

Church, Clare, and Alec Crawford. "Minerals and the Metals for the Energy Transition: Exploring the Conflict Implications for Mineral-Rich, Fragile States." In *The Geopolitics of the Global Energy Transition*, edited by Manfred Hafner and Simone Tagliapietra, 279–304. Lecture Notes in Energy. Springer International Publishing, 2020. https://doi.org /10.1007/978-3-030-39066-2_12.

City-Data. "Fifth Ward Neighborhood in Houston, Texas (TX), 77020, 77026 Subdivision Profile—Real Estate, Apartments, Condos, Homes, Community, Population, Jobs, Income, Streets." *City-Data.com*, accessed April 17, 2023. http://www.city-data.com /neighborhood/Fifth-Ward-Houston-TX.html.

Climate Action Tracker. "Net Zero Targets." *ClimateActionTracker.org*, 2021. https:// climateactiontracker.org/countries/usa/net-zero-targets/.

Cohen, Jared. "Resource Realism: The Geopolitics of Critical Mineral Supply Chains." Goldman Sachs, September 13, 2023. https://www.goldmansachs.com/intelligence/pages /resource-realism-the-geopolitics-of-critical-mineral-supply-chains.html.

Cohn, Michael Bennett. "Dakota Access Pipeline Standing Rock Standoff: Behind the Front Lines." *Newsweek*, November 28, 2016. https://www.newsweek.com/2016/12/09/dakota -access-pipeline-protest-standing-rock-sioux-525894.html.

Colmer, Jonathan, Ian Hardman, Jay Shimshack, and John Voorheis. "Disparities in PM2.5 Air Pollution in the United States." *Science* 369, no. 6503 (July 31, 2020): 575–78. https:// doi.org/10.1126/science.aaz9353.

Community Action Partnership. "Weatherization—Apply for Services." Community Council of South Central Texas, 2023, https://www.ccsct.org/weatherization-apply-for-services/.

Cong, Shuchen, Destenie Nock, Yueming Lucy Qiu, and Bo Xing. "Unveiling Hidden Energy Poverty Using the Energy Equity Gap." *Nature Communications* 13, no. 1 (May 4, 2022): 2456. https://doi.org/10.1038/s41467-022-30146-5.

Congressional Research Service. *Federal Land Ownership: Overview and Data*. Report no. R42346. Congressional Research Service, updated February 1, 2020. https://crsreports .congress.gov/product/pdf/r/r42346#:~:text=The%20federal%20government%20owns %20roughly,of%20September%2030%2C%202018).

Congressional Research Service. *LIHEAP: Program and Funding*. Congressional Research Service, updated June 22, 2018. https://www.everycrsreport.com/files/20180622 _RL31865_85805bac2287a504f2a4eb05e4637a3cd21eaa2e.pdf.

Congressional Research Service. "The Plug-In Electric Vehicle Tax Credit." *In Focus* 11017, version 3 (updated May 14, 2019). https://sgp.fas.org/crs/misc/IF11017.pdf.

Connecticut Green Bank. "Smart-E Loans." *CTGreenBank.com*, July 14, 2022. https://www .ctgreenbank.com/home-solutions/smart-e-loans/.

Consortium for Energy Efficiency. *CEE Annual Industry Report: 2019 State of the Efficiency Program Industry: Budgets, Expenditures, and Impacts*. Consortium for Energy Efficiency, August 2020. https://cee1.org/images/pdf/2019_AIR_Report_1.pdf.

Consortium for Energy Efficiency. *CEE Annual Industry Report: 2020 State of the Efficiency Program Industry: Budgets, Expenditures, and Impacts*. Consortium for Energy Efficiency, September 2021. https://cee1.org/images/uploads/2020_AIR_Final.pdf.

Cook, John T., Deborah A. Frank, Patrick H. Casey, Ruth Rose-Jacobs, Maureen M. Black, Mariana Chilton, et al. "A Brief Indicator of Household Energy Security: Associations with Food Security, Child Health, and Child Development in US Infants and Toddlers." *Pediatrics* 122, no. 4 (October 2008): e867–75. https://doi.org/10.1542/peds.2008-0286.

Cotterman, Turner, Erica R. H. Fuchs, and Kate Whitefoot. "The Transition to Electrified Vehicles: Evaluating the Labor Demand of Manufacturing Conventional Versus Battery Electric Vehicle Powertrains." *Energy Policy* 188 (June 4, 2022): 114064. https://doi.org /10.2139/ssrn.4128130.

Cox Automotive. "A Record 1.2 Million EVs Were Sold in the U.S. in 2023, According to Estimates from Kelley Blue Book." *CoxAutoInc.com*, January 9, 2024. https://www.coxautoinc .com/market-insights/q4-2023-ev-sales/.

Cox, Colin. "EPA Agrees to Investigate Texas for Alleged Civil Rights Violations Caused by Air Pollution from Port Arthur Plant." *Environmental Integrity Project* (blog), October 15, 2021. https://environmentalintegrity.org/news/epa-agrees-to-investigate-texas -for-alleged-civil-rights-violations-caused-by-air-pollution-from-port-arthur-plant/.

Crawford County, Ohio, Board of Elections. *Official Results Report: 2022 GENERAL ELECTION NOVEMBER 8, 2022*. Crawford County, Ohio, Board of Elections, November 21, 2022. https://crawfordcountyohioboe.gov/wp-content/uploads/Official-Summary -Report-11-8-2022.pdf.

Crimmins, A. R., C. W. Avery, D. R. Easterling, K. E. Kunkel, B. C. Stewart, and T. K. Maycock, eds. *Fifth National Climate Assessment*. US Global Change Research Program, 2023. https://doi.org/10.7930/NCA5.2023.

Crippa, M., D. Guizzardi, M. Banja, E. Solazzo, M. Muntean, E. Schaaf, et al. *CO2 Emissions of All World Countries: JRC/IEA/PBL 2022 Report*. Publications Office of the European

Union, 2022. https://op.europa.eu/en/publication-detail/-/publication/6c10e2bd-3892
-11ed-9c68-01aa75ed71a1/language-en.

Culhane, Trevor, Galen Hall, and J. Timmons Roberts. "Who Delays Climate Action? Interest Groups and Coalitions in State Legislative Struggles in the United States." *Energy
Research and Social Science* 79 (September 2021): 102114. https://doi.org/10.1016/j.erss
.2021.102114.

Currie, Janet, Michael Greenstone, and Katherine Meckel. "Hydraulic Fracturing and Infant
Health: New Evidence from Pennsylvania." *Science Advances* 3, no. 12 (December 13,
2017): e1603021. https://doi.org/10.1126/sciadv.1603021.

Curtis, E. Mark, Layla O'Kane, and R. Jisung Park. "Workers and the Green-Energy Transition: Evidence from 300 Million Job Transitions." Working paper no. 31539. National Bureau of Economic Research, August 2023. https://doi.org/10.3386/w31539.

Cushing, Lara J., Shiwen Li, Benjamin B. Steiger, and Joan A. Casey. "Historical Red-Lining
Is Associated with Fossil Fuel Power Plant Siting and Present-Day Inequalities in Air
Pollutant Emissions." *Nature Energy* 8, no. 1 (January 2023): 52–61. https://doi.org/10
.1038/s41560-022-01162-y.

Cutcher-Gershenfeld, Joel, Dan Brooks, and Martin Mulloy. *The Decline and Resurgence of
the U.S. Auto Industry.* Economic Policy Institute, May 6, 2015. https://files.epi.org/2015
/the-decline-and-resurgence-of-the-us-auto-industry.pdf.

Cybersecurity and Infrastructure Security Agency. "Energy Sector." *CISA.gov*, accessed
October 26, 2023. https://www.cisa.gov/topics/critical-infrastructure-security-and
-resilience/critical-infrastructure-sectors/energy-sector.

Darghouth, Naïm R, Eric O'Shaughnessy, Sydney Forrester, and Galen Barbose. "Characterizing Local Rooftop Solar Adoption Inequity in the US." *Environmental Research Letters*
17, no. 3 (March 1, 2022): 034028. https://doi.org/10.1088/1748-9326/ac4fdc.

Davies, Ellis. "Consumers Three Times More Likely to Buy Used Cars over New." *Motor
Trader* (blog), October 28, 2019. https://www.motortrader.com/motor-trader-news
/automotive-news/majority-buy-consumers-opt-used-new-cars-28-10-2019.

Davies, Thom. "Slow Violence and Toxic Geographies: 'Out of Sight' to Whom?" *Environment and Planning C: Politics and Space* 40, no. 2 (2022): 409–27.

Davis, Lucas W., and Catherine Hausman. "Who Will Pay for Legacy Utility Costs?" *Journal
of the Association of Environmental and Resource Economists* 9, no. 6 (November 1, 2022):
1047–85. https://doi.org/10.1086/719793.

Davis, Lucas, Catherine Hausman, and Nancy Rose. "Transmission Impossible? Prospects
for Decarbonizing the US Grid." Working paper no. 31377. National Bureau of Economic
Research, June 2023. https://doi.org/10.3386/w31377.

Devine-Wright, Patrick. "Rethinking NIMBYism: The Role of Place Attachment and Place
Identity in Explaining Place-Protective Action." *Journal of Community and Applied Social Psychology* 19, no. 6 (November 2009): 426–41. https://doi.org/10.1002/casp.1004.

Devine-Wright, Patrick, and Yuko Howes. "Disruption to Place Attachment and the Protection of Restorative Environments: A Wind Energy Case Study." *Journal of Environmental Psychology* 30, no. 3 (September 2010): 271–80. https://doi.org/10.1016/j.jenvp
.2010.01.008.

Deziel, Nicole C., Eran Brokovich, Itamar Grotto, Cassandra J. Clark, Zohar Barnett-Itzhaki,

David Broday, et al. "Unconventional Oil and Gas Development and Health Outcomes: A Scoping Review of the Epidemiological Research." *Environmental Research* 182 (March 1, 2020): 109124. https://doi.org/10.1016/j.envres.2020.109124.

Diep, Francie. "Abandoned Uranium Mines: An 'Overwhelming Problem' in the Navajo Nation." *Scientific American*, December 30, 2010. https://www.scientificamerican.com /article/abandoned-uranium-mines-a/.

Dillon, Liam, and Ben Poston. "The Racist History of America's Interstate Highway Boom." *Los Angeles Times*, November 11, 2021. https://www.latimes.com/homeless-housing /story/2021-11-11/the-racist-history-of-americas-interstate-highway-boom.

Domonoske, Camila. "Big 3 Autoworkers Vote 'Yes' to Historic UAW Contracts." *NPR*, November 17, 2023. https://www.npr.org/2023/11/16/1212381342/gm-autoworkers-vote-yes -approve-uaw-contract-ford-stellantis.

Dorian, Drew. "2024 Tesla Model S." *Car and Driver*, accessed March 8, 2024. https://www .caranddriver.com/tesla/model-s.

Dottle, Rachael, Laura Bliss, and Pablo Robles. "What It Looks Like to Reconnect Black Communities Torn Apart by Highways." *Bloomberg.com*, July 28, 2021. https://www .bloomberg.com/graphics/2021-urban-highways-infrastructure-racism/.

Douglas, Stratford, and Anne Walker. "Coal Mining and the Resource Curse in the Eastern United States." *Journal of Regional Science* 57, no. 4 (September 2017): 568–90. https:// doi.org/10.1111/jors.12310.

Drehobl, Ariel, and Lauren Ross. *Lifting the High Energy Burden in America's Largest Cities: How Energy Efficiency Can Improve Low-Income and Underserved Communities.* American Council for an Energy-Efficient Economy, April 2016. https://www.aceee.org /research-report/u1602.

Drehobl, Ariel, Lauren Ross, and Roxana Ayala. *How High Are Household Energy Burdens? An Assessment of National and Metropolitan Energy Burden across the United States.* American Council for an Energy-Efficient Economy, September 2020. https://www .aceee.org/sites/default/files/pdfs/u2006.pdf.

Dubois, Ghislain, Benjamin Sovacool, Carlo Aall, Maria Nilsson, Carine Barbier, Alina Herrmann, et al. "It Starts at Home? Climate Policies Targeting Household Consumption and Behavioral Decisions Are Key to Low-Carbon Futures." *Energy Research and Social Science* 52 (June 2019): 144–58. https://doi.org/10.1016/j.erss.2019.02.001.

Dupuis, Mathieu, and Ian Greer, "Recentralizing Industrial Relations? Local Unions and the Politics of Insourcing in Three North American Automakers." *ILR Review* 75, no. 4 (2022): 918–42.

Eartheasy. "LED Light Bulbs: Comparison Charts." *Eartheasy.com*, accessed October 22, 2023. https://learn.eartheasy.com/guides/led-light-bulbs-comparison-charts/.

Earthjustice. "Ashes: A Community's Toxic Inheritance." *Earthjustice.org*, September 1, 2014. https://earthjustice.org/feature/photos-a-toxic-inheritance.

Eckert, Nora. "UAW Wins Big in Historic Union Vote at Volkswagen Tennessee Factory." *Reuters*, April 22, 2024. https://www.reuters.com/business/autos-transportation/uaw -clinches-watershed-union-victory-volkswagen-tennessee-factory-2024-04-20.

Egan, Patrick J., David M. Konisky, and Megan Mullin. "Ascendant Public Opinion." *Public Opinion Quarterly* 86, no. 1 (March 18, 2022): 134–48. https://doi.org/10.1093/poq /nfab071.

Egan, Patrick J., and Megan Mullin. "Climate Change: US Public Opinion." *Annual Review of Political Science* 20, no. 1 (May 11, 2017): 209–27. https://doi.org/10.1146/annurev-polisci -051215-022857.

Eid, Cherrelle, Javier Reneses Guillén, Pablo Frías Marín, and Rudi Hakvoort. "The Economic Effect of Electricity Net-Metering with Solar PV: Consequences for Network Cost Recovery, Cross Subsidies and Policy Objectives." *Energy Policy* 75 (December 2014): 244–54. https://doi.org/10.1016/j.enpol.2014.09.011.

Einberger, Mathias. "Reality Check: The United States Has the Only Major Power Grid Without a Plan." *RMI*, January 12, 2023. https://rmi.org/the-united-states-has-the-only-major -power-grid-without-a-plan/.

Eisenbrey, Ross. "Management—Bad Management—Crippled the Auto Industry's Big Three, Not the UAW." *Economic Policy Institute Working Economics Blog*, May 24, 2012. https://www.epi.org/blog/bad-management-crippled-auto-industry-big-three/.

Eisenson, Matthew. *Opposition to Renewable Energy Facilities in the United States: May 2023 Edition.* Sabin Center for Climate Change Law, May 2023. https://scholarship.law .columbia.edu/cgi/viewcontent.cgi?article=1201&context=sabin_climate_change.

Eisenson, Matthew, Jacob Elkin, Harmukh Singh, and Noah Schaffir. *Opposition to Renewable Energy Facilities in the United States: June 2024 Edition.* Sabin Center for Climate Change Law, June 2024. https://scholarship.law.columbia.edu/cgi/viewcontent.cgi ?article=1227&context=sabin_climate_change.

Emanuel, Ryan E., Martina A. Caretta, Louie Rivers III, and Pavithra Vasudevan. "Natural Gas Gathering and Transmission Pipelines and Social Vulnerability in the United States." *GeoHealth* 5, no. 6 (2021): e2021GH000442. https://doi.org/10.1029/2021GH000442.

Englund, Will. "Without Access to Charging Stations, Black and Hispanic Communities May Be Left Behind in the Era of Electric Vehicles." *Washington Post*, December 9, 2021. https://www.washingtonpost.com/business/2021/12/09/charging-deserts-evs/.

Environmental Defense Fund and WSP. *U.S. Electric Vehicle Manufacturing Investments and Jobs—Characterizing the Impacts of the Inflation Reduction Act After 1 Year.* Environmental Defense Fund and WSP, August 2023. https://www.edf.org/sites/default/files/2023 -08/EDF%20WSP%20EV%20report%208-16-23%20FINAL%20FINAL.pdf.

European Environment Agency. *Air Pollution Impacts from Carbon Capture and Storage (CCS).* Publications Office of the European Union, 2011. https://data.europa.eu/doi/10 .2800/84208.

Evans, Farrell. "How Interstate Highways Gutted Communities—and Reinforced Segregation." *History*, September 21, 2023. https://www.history.com/news/interstate-highway -system-infrastructure-construction-segregation.

Evans, Gareth. "A Toxic Crisis in America's Coal Country." *BBC News*, February 11, 2019. https://www.bbc.com/news/world-us-canada-47165522.

Evensen, Darrick, and Rich Stedman. "'Fracking': Promoter and Destroyer of 'the Good Life.'" *Journal of Rural Studies* 59 (April 2018): 142–52. https://doi.org/10.1016/j.jrurstud .2017.02.020.

Evergreen Action. "How States Are Centering Workers in the Clean Energy Transition." *Evergreen Action* (blog), January 24, 2024. https://www.evergreenaction.com/blog/how -states-are-centering-workers-in-the-clean-energy-transition.

Ewing, Jack, and Eric Lipton. "Carmakers Race to Control Next-Generation Battery Technol-

ogy." *New York Times*, March 7, 2022. https://www.nytimes.com/2022/03/07/business
/energy-environment/next-generation-auto-battery.html.

Farhar, Barbara C. "Trends: Public Opinion About Energy." *Public Opinion Quarterly* 58,
no. 4 (1994): 603–32.

Fears, Darryl. "'Cancer Has Decimated Our Community': EPA's Regan Vows to Help Hard-
Hit Areas, but Residents Have Doubts." *Washington Post*, January 4, 2022. https://www
.washingtonpost.com/climate-environment/2021/11/28/epa-regan-environmental
-justice-tour/.

Feaster, Seth. *U.S. on Track to Close Half of Coal Capacity by 2026.* Institute for Energy Economics and Financial Analysis, April 3, 2023. https://ieefa.org/resources/us-track-close
-half-coal-capacity-2026.

Federal Energy Regulatory Commission (FERC). *Final Report on February 2021 Freeze Underscores Winterization Recommendations.* FERC, November 16, 2021. https://www.ferc
.gov/news-events/news/final-report-february-2021-freeze-underscores-winterization
-recommendations.

Fenwick, Tyler. "How I-65/70 Shattered Black Neighborhoods." *Indianapolis Recorder*,
December 27, 2018. https://indianapolisrecorder.com/c57d7e9c-09e4-11e9-b276
-83369364ae4e/.

Federal Energy Regulatory Commission. "What FERC Does." *FERC.gov*, accessed October 30, 2023. https://www.ferc.gov/what-ferc-does.

Forrester, Sydney, Galen L. Barbose, Eric O'Shaughnessy, Naïm R. Darghouth, and Cristina
Crespo Montañés. *Residential Solar-Adopter Income and Demographic Trends: November 2022 Update.* Lawrence Berkeley National Laboratory, November 2022. https://emp
.lbl.gov/publications/residential-solar-adopter-income-1.

Forrester, Sydney, Galen L. Barbose, Eric O'Shaughnessy, Naïm Darghouth, and Cristina
Crespo Montañés. *Residential Solar-Adopter Income and Demographic Trends: 2023
Update.* Lawrence Berkeley National Laboratory, December 2023. https://emp.lbl.gov
/publications/residential-solar-adopter-income-2.

Forti, Vanessa, Cornelis Peter Baldé, Ruediger Kuehr, and Garam Bel. *The Global E-Waste
Monitor 2020.* United Nations University, UNITAR, ITU, and ISWA, November 2020.
https://ewastemonitor.info/wp-content/uploads/2020/11/GEM_2020_def_july1
_low.pdf.

Fouquet, Roger. "The Slow Search for Solutions: Lessons from Historical Energy Transitions
by Sector and Service." *Energy Policy* 38, no. 11 (November 2010): 6586–96. https://doi
.org/10.1016/j.enpol.2010.06.029.

Foxhall, Emily, and Erin Douglas. "Appeals Court Says State Agency Set Electricity Prices
Too High During 2021 Winter Storm." *Texas Tribune*, March 17, 2023. https://www
.texastribune.org/2023/03/17/puc-appeals-court-uri-prices/.

Frazier, LaToya Ruby, and Dan Kaufman. "What Happens to a Factory Town When the Factory Shuts Down?" *New York Times Magazine*, May 1, 2019, https://www.nytimes.com
/interactive/2019/05/01/magazine/lordstown-general-motors-plant.html.

Freiberg, Alice, Christiane Schefter, Maria Girbig, Vanise C. Murta, and Andreas Seidler.
"Health Effects of Wind Turbines on Humans in Residential Settings: Results of a Scoping
Review." *Environmental Research* 169 (February 2019): 446–63. https://doi.org/10.1016/j
.envres.2018.11.032.

Freudenburg, William R. "Addictive Economies: Extractive Industries and Vulnerable Localities in a Changing World Economy." *Rural Sociology* 57, no. 3 (September 1992): 305–32. https://doi.org/10.1111/j.1549-0831.1992.tb00467.x.

Fuels Institute Electric Vehicle Council. *EV Consumer Behavior*. Fuels Institute, June 2021. https://www.transportationenergy.org/wp-content/uploads/2022/11/21FI_-EVC _ConsumerBehaviorReport_V07-FINAL.pdf.

Gagnon, Pieter, Robert Margolis, Jennifer Melius, Caleb Phillips, and Ryan Elmore. *Rooftop Solar Photovoltaic Technical Potential in the United States: A Detailed Assessment*. National Renewable Energy Laboratory, January 2016. https://www.nrel.gov/docs/fy16osti /65298.pdf.

Gallup, Inc. "Business and Industry Sector Ratings." *Gallup.com*, accessed October 31, 2024. https://news.gallup.com/poll/12748/business-industry-sector-ratings.aspx.

Gazmararian, Alexander, and Dustin Tingley. *Uncertain Futures: How to Unlock the Climate Impasse*. Cambridge University Press, 2023.

Gelles, David. "The Texas Group Waging a National Crusade Against Climate Action." *New York Times*, December 4, 2022. https://www.nytimes.com/2022/12/04/climate/texas -public-policy-foundation-climate-change.html.

Genoways, Ted. "Port Arthur, Texas: American Sacrifice Zone." Natural Resources Defense Council, November 13, 2014. https://www.nrdc.org/stories/port-arthur-texas-american -sacrifice-zone.

Giles, Cynthia. *Next Generation Compliance: Environmental Regulation for the Modern Era*. Oxford University Press, 2022.

Gilio-Whitaker, Dina. *As Long as Grass Grows: The Indigenous Fight for Environmental Justice, from Colonization to Standing Rock*. Beacon Press, 2019.

Goble, Gere. "Wind Turbines in Crawford County, Ohio: Here's What We Know About Honey Creek Wind." *Bucyrus Telegraph-Forum*, April 2, 2022. https://www .bucyrustelegraphforum.com/story/news/2022/04/02/wind-turbines-crawford-county -ohio-honey-creek-wind-faq/9382089002/.

Goeddeke, Frank, Jr., and Marick F. Masters. "The UAW: Then and Now 50 Years of United Auto Workers Union History." *LERA For Libraries* 24 (2020). https://lerawebillinois.web .illinois.edu/index.php/PFL/article/view/3392.

Global Energy Monitor. "Global Coal Plant Tracker." *GlobalEnergyMonitor.org*, April 25, 2022. https://globalenergymonitor.org/projects/global-coal-plant-tracker/.

Gorski, Sam. "What's the Average Cost of Utilities in West Virginia?" *12WBOY*, April 3, 2023. https://www.wboy.com/news/west-virginia/whats-the-average-cost-of-utilities-in -west-virginia/.

Graetz, Michael J. *The End of Energy: The Unmaking of America's Environment, Security, and Independence*. MIT Press, 2011. https://doi.org/10.7551/mitpress/8653.001.0001.

Graff, Michelle. "Reducing Administrative Burdens in an Energy Bill Assistance Program." *Public Management Review* (2024): 1–26.

Graff, Michelle. "Unpacking the Determinants and Burdens of Energy Assistance in the United States." PhD diss., Indiana University, 2021. https://www.proquest.com/docview /2566108736/abstract/D68EC01092434CE8PQ/1.

Graff, Michelle, Sanya Carley, and David M. Konisky. "Stakeholder Perceptions of the United States Energy Transition: Local-Level Dynamics and Community Responses to National

Politics and Policy." *Energy Research and Social Science* 43 (September 2018): 144–57. https://doi.org/10.1016/j.erss.2018.05.017.

Graff, Michelle, Sanya Carley, David M. Konisky, and Trevor Memmott. "Opportunities to Advance Research on Energy Insecurity." *Nature Energy* 8, no. 6 (June 2023): 550–53. https://doi.org/10.1038/s41560-023-01265-0.

Graff, Michelle, David M. Konisky, Sanya Carley, and Trevor Memmott. "Climate Change and Energy Insecurity: A Growing Need for Policy Intervention." *Environmental Justice* 15, no. 2 (April 2022): 76–82. https://doi.org/10.1089/env.2021.0032.

Gravelle, Timothy B., and Erick Lachapelle. "Politics, Proximity and the Pipeline: Mapping Public Attitudes Toward Keystone XL." *Energy Policy* 83 (August 1, 2015): 99–108. https://doi.org/10.1016/j.enpol.2015.04.004.

Green, Miranda, Michael Copley, and Ryan Kellman. "An Activist Group Is Spreading Misinformation to Stop Solar Projects in Rural America." *NPR*, February 18, 2023. https://www.npr.org/2023/02/18/1154867064/solar-power-misinformation-activists-rural-america.

Greene, Mary, and Keene Kelderman. "Port Arthur, Texas: The End of the Line for an Economic Myth." *Environmental Integrity Project* (blog), August 2017. https://environmentalintegrity.org/what-we-do/oil-and-gas/the-human-cost-of-energy-production/port-arthur-texas/.

Grogan, Maura, Rebecca Morse, and April Youpee-Roll. *Native American Lands and Natural Resource Development.* Revenue Watch Institute, 2011.

Gross, Samantha. *Renewables, Land Use, and Local Opposition in the United States.* Brookings Institution, January 2020. https://www.brookings.edu/articles/renewables-land-use-and-local-opposition-in-the-united-states/.

Guevara, Tom, Tim Slaper, Sanya Carley, Matt Kinghorn, Drew Klacik, Jamie Palmer, et al. *Economic, Fiscal, and Social Impacts of the Transition of Electricity Generation Resources in Indiana.* Indiana University Public Policy Institute, August 2020.

Guterres, António. "Secretary-General's Video Message to Powering Past Coal Alliance Summit." *United Nations, March 2,* 2021. https://www.un.org/sg/en/content/sg/statement/2021-03-02/secretary-generals-video-message-powering-past-coal-alliance-summit.

Habeeb, Dana, Jason Vargo, and Brian Stone. "Rising Heat Wave Trends in Large US Cities." *Natural Hazards* 76, no. 3 (April 1, 2015): 1651–65. https://doi.org/10.1007/s11069-014-1563-z.

Hale, Zack. "World Needs $14 Trillion in Grid Spending by 2050 to Support Renewables—Report." *S&P Global Market Intelligence*, February 23, 2021. https://www.spglobal.com/marketintelligence/en/news-insights/latest-news-headlines/world-needs-14-trillion-in-grid-spending-by-2050-to-support-renewables-8211-report-62816721.

Hall, Matt. "Map of China Manufacturing Distribution." Berkeley Sourcing Group, June 6, 2016. https://www.berkeleysg.com/china-manufacturing-distribution-map/.

Hanley, Steve. "31 Countries, States, and Cities Have Gas/Diesel Car Bans In Place." *CleanTechnica*, January 2, 2021. https://cleantechnica.com/2021/01/02/31-countries-states-and-cities-have-ice-bans-in-place/.

Hardman, Scott, Kelly Fleming, Eesha Kare, and Mahmoud Ramadan. "A Perspective on Equity in the Transition to Electric Vehicles." Edited by Bertrand Neyhouse and Yana Petri. *MIT Science Policy Review* 2 (August 30, 2021): 46–54. https://doi.org/10.38105/spr.e10rdoaoup.

Hawkins, Daniel A. "Two Miles to Hell: A Miner's Story." *Appalachian Voice* (blog), April 3, 2010. https://appvoices.org/2010/04/03/two-miles-to-hell-a-miners-story/.

Hawkins, Troy R., Ola Moa Gausen, and Anders Hammer Strømman. "Environmental Impacts of Hybrid and Electric Vehicles—a Review." *International Journal of Life Cycle Assessment* 17, no. 8 (September 2012): 997–1014. https://doi.org/10.1007/s11367-012 -0440-9.

Hazardous Substance Research Centers/South and Southwest Outreach Program. *Environmental Update #12: Environmental Impact of the Petroleum Industry.* June 2003. https:// cfpub.epa.gov/ncer_abstracts/index.cfm/fuseaction/display.files/fileID/14522.

Healy, Jack, and Mike Baker. "As Miners Chase Clean-Energy Minerals, Tribes Fear a Repeat of the Past." *New York Times*, December 27, 2021. https://www.nytimes.com/2021/12/27 /us/mining-clean-energy-antimony-tribes.html.

Healy, Noel, Jennie C. Stephens, and Stephanie A. Malin. "Embodied Energy Injustices: Unveiling and Politicizing the Transboundary Harms of Fossil Fuel Extractivism and Fossil Fuel Supply Chains." *Energy Research and Social Science* 48 (February 1, 2019): 219–34. https://doi.org/10.1016/j.erss.2018.09.016.

Henne, Sarabeth. "Navajo Nation Makes History with Solar Powered Plant." *Indianz.com*, August 16, 2018. https://www.indianz.com/News/2018/08/16/navajo-nation-makes -history-with-solar-p.asp.

Hendryx, Michael, and Benjamin Holland. "Unintended Consequences of the Clean Air Act: Mortality Rates in Appalachian Coal Mining Communities." *Environmental Science and Policy* 63 (2016): 1–6. https://doi.org/10.1016/j.envsci.2016.04.021.

Herd, Pamela, and Donald P. Moynihan. *Administrative Burden: Policymaking by Other Means.* Russell Sage Foundation, 2018. https://doi.org/10.7758/9781610448789.

Hernández, Diana. "Understanding 'Energy Insecurity' and Why It Matters to Health." *Social Science and Medicine* 167 (October 1, 2016): 1–10. https://doi.org/10.1016/j.socscimed .2016.08.029.

Hersher, Rebecca. "Key Moments in the Dakota Access Pipeline Fight." *NPR*, February 22, 2017. https://www.npr.org/sections/thetwo-way/2017/02/22/514988040/key-moments -in-the-dakota-access-pipeline-fight.

Hertwich, Edgar G., and Glen P. Peters. "Carbon Footprint of Nations: A Global, Trade-Linked Analysis." *Environmental Science and Technology* 43, no. 16 (August 15, 2009): 6414–20. https://doi.org/10.1021/es803496a.

Hieu, Le, and Andrew Linhardt. *Rev Up Electric Vehicles: A Nationwide Study of the Electric Vehicle Shopping Experience.* Sierra Club, November 2019.

Hill, Todd. "Proposed Wind Farm in County Halted." *Bucyrus Telegraph-Forum*, October 22, 2015. https://www.bucyrustelegraphforum.com/story/news/local/2015/10/22 /proposed-wind-farm-county-halted/74386956/.

Hill, Elaine L., and Lala Ma. "Drinking Water, Fracking, and Infant Health." *Journal of Health Economics* 82 (March 1, 2022): 102595. https://doi.org/10.1016/j.jhealeco.2022.102595.

Hirsch, Jameson K., K. Bryant Smalley, Emily M. Selby-Nelson, Jane M. Hamel-Lambert, Michael R. Rosmann, Tammy A. Barnes, et al. "Psychosocial Impact of Fracking: A Review of the Literature on the Mental Health Consequences of Hydraulic Fracturing." *International Journal of Mental Health and Addiction* 16, no. 1 (February 2018): 1–15. https://doi .org/10.1007/s11469-017-9792-5.

Hoen, Ben, Ryan Wiser, Peter Cappers, Mark Thayer, and Gautam Sethi. *The Impact of Wind Power Projects on Residential Property Values in the United States: A Multi-Site Hedonic Analysis*. US Department of Energy Office of Scientific and Technical Information, December 2, 2009. https://doi.org/10.2172/978870.

Hoffman, Alan N. "Case 26: Tesla Motors, Inc.: The First U.S. Car Company IPO Since 1956." In *Strategic Management and Business Policy: Globalization, Innovation, and Sustainability*, 14th ed., eds. Thomas L. Wheelen, J. David Hunger, Alan N. Hoffman, and Charles E. Bamford, 671–86. Pearson, 2015.

Holzhauer, Brett. "Here's the Average Net Worth of Homeowners and Renters." *CNBC*, February 27, 2023. https://www.cnbc.com/select/average-net-worth-homeowners-renters/.

Honey Creek Wind Farm. "Honey Creek Wind." *HoneyCreekWindPower.com*, accessed October 26, 2023. https://www.honeycreekwindpower.com/about_honey_creek.

Hopkins, Daniel J. *The Increasingly United States: How and Why American Political Behavior Nationalized*. University of Chicago Press, 2018.

Hsu, Andrea, and Stephan Bisaha. "Seeking to Defy History, the UAW is Coming Closer to Unionizing the South." *NPR*, April 5, 2024.

Hsu, Chih-Wei, and Kevin Fingerman. "Public Electric Vehicle Charger Access Disparities Across Race and Income in California." *Transport Policy* 100 (January 2021): 59–67. https://doi.org/10.1016/j.tranpol.2020.10.003.

Hsu, Jeremy. "Solar Power's Benefits Don't Shine Equally on Everyone." *Scientific American*, April 4, 2019. https://www.scientificamerican.com/article/solar-powers-benefits-dont-shine-equally-on-everyone/.

Hu, Akielly. "The GOP Donors Behind a Growing Misinformation Campaign to Stop Offshore Wind." *Grist*, April 20, 2023. https://grist.org/politics/republicans-fossil-fuels-the-gop-donors-behind-a-growing-misinformation-campaign-to-stop-offshore-wind/.

Huang, Luling, Destenie Nock, Shuchen Cong, and Yueming Lucy Qiu. "Inequalities Across Cooling and Heating in Households: Energy Equity Gaps." *Energy Policy* 182 (November 1, 2023): 113748. https://doi.org/10.1016/j.enpol.2023.113748.

Hufbauer, Gary Clyde, Cathleen Cimino, and Tyler Moran. *NAFTA at 20: Misleading Charges and Positive Achievements*. Report no, PB14–13. Peterson Institute for International Economics, May 2014.

Hunnicutt, Trevor. "Schwarzenegger Tours Calif. Gas Line Blast Site." *Seattle Times*, September 16, 2010. https://www.seattletimes.com/business/schwarzenegger-tours-calif-gas-line-blast-site/.

Indiana Housing and Community Development Authority. "Weatherization/Energy Conservation." *Indiana.gov*, March 30, 2021. https://www.in.gov/ihcda/homeowners-and-renters/weatherizationenergy-conservation/.

Intergovernmental Panel on Climate Change. *Climate Change 2021: The Physical Science Basis: Contribution of Working Group 1 to the Sixth Assessment Report of the Intergovernmental Panel on Climate Change*. Cambridge University Press, 2021. https://www.ipcc.ch/report/ar6/wg1/chapter/summary-for-policymakers/.

Intergovernmental Panel on Climate Change. *Climate Change 2022—Impacts, Adaptation and Vulnerability: Working Group 2 Contribution to the Sixth Assessment Report of the Intergovernmental Panel on Climate Change*. 1st ed. Cambridge University Press, June 2023. https://doi.org/10.1017/9781009325844.

Intergovernmental Panel on Climate Change (IPCC). *Global Warming of 1.5°C: IPCC Special Report on Impacts of Global Warming of 1.5°C Above Pre-Industrial Levels in Context of Strengthening Response to Climate Change, Sustainable Development, and Efforts to Eradicate Poverty.* 1st ed. Cambridge University Press, 2022. https://doi.org/10.1017/9781009157940.

Internal Revenue Service. "Credits for New Clean Vehicles Purchased in 2023 or After." *IRS.gov*, March 31, 2023. https://www.irs.gov/credits-deductions/credits-for-new-clean-vehicles-purchased-in-2023-or-after.

Internal Revenue Service. "Used Clean Vehicle Credit." *IRS.gov*, accessed October 22, 2023. https://www.irs.gov/credits-deductions/used-clean-vehicle-credit.

International Energy Agency. *CO2 Emissions in 2023.* International Energy Agency, March 2024. https://www.iea.org/reports/co2-emissions-in-2023.

International Energy Agency. "Global Coal Demand Expected to Decline in Coming Years." *IEA.org*, December 15, 2023. https://www.iea.org/news/global-coal-demand-expected-to-decline-in-coming-years.

International Energy Agency. *Global EV Outlook 2020.* International Energy Agency, June 2020. https://www.iea.org/reports/global-ev-outlook-2020.

International Energy Agency. "Transport." *IEA.org*, accessed November 1, 2024. https://www.iea.org/reports/transport.

International Energy Agency. *World Energy Outlook 2021.* International Energy Agency, October 2021. https://www.iea.org/reports/world-energy-outlook-2021.

International Energy Agency. *World Energy Outlook 2023.* International Energy Agency, October 2023. https://www.iea.org/reports/world-energy-outlook-2023.

International Energy Agency. "World Total Coal Production, 1971–2020." *IEA.org*, updated July 29, 2021. https://www.iea.org/data-and-statistics/charts/world-total-coal-production-1971-2020.

International Renewable Energy Agency (IRENA). *Renewable Power Generation Costs in 2022.* IRENA, August 29, 2023. https://www.irena.org/Publications/2023/Aug/Renewable-Power-Generation-Costs-in-2022.

Interstate Renewable Energy Council. *National Solar Jobs Census 2023.* Interstate Renewable Energy Council, September 2024. https://irecusa.org/census-executive-summary/.

Isidore, Chris. "The UAW Just Won Its Battle with the Big Three: Now It's Aiming at 13 Non-Wnion Automakers." *CNN*, November 29, 2023. https://www.cnn.com/2023/11/29/business/uaw-organize-nonunion-automakers/index.html.

Joint Research Centre (European Commission), M. Crippa, D. Guizzardi, E. Schaaf, F. Monforti-Ferrario, R. Quadrelli, et al. *GHG Emissions of All World Countries: 2023.* Publications Office of the European Union, 2023. https://data.europa.eu/doi/10.2760/953322.

Jones, Andrew, Destenie Nock, Constantine Samaras, Yueming Lucy Qiu, and Bo Xing. "Climate Change Impacts on Future Residential Electricity Consumption and Energy Burden: A Case Study in Phoenix, Arizona." *Energy Policy* 183 (2023): 113811. https://doi.org/10.1016/j.enpol.2023.113811.

Jones, Thomas Elisha, and Len Edward Necefer. *Identifying Barriers and Pathways for Success for Renewable Energy Development on American Indian Lands.* Department of Energy, November 2016.

Joselow, Maxine. "Three Electric Vehicle Fights to Watch Today." *Washington Post*, September 14, 2023. https://www.washingtonpost.com/politics/2023/09/14/three-electric-vehicle-fights-watch-today/.

Josey, Kevin P., Scott W. Delaney, Xiao Wu, Rachel C. Nethery, Priyanka DeSouza, Danielle Braun, et al. "Air Pollution and Mortality at the Intersection of Race and Social Class." *New England Journal of Medicine* 388, no. 15 (April 13, 2023): 1396–404. https://doi.org/10.1056/NEJMsa2300523.

Joshi, Siddharth, Shivika Mittal, Paul Holloway, Priyadarshi Ramprasad Shukla, Brian Ó Gallachóir, and James Glynn. "High Resolution Global Spatiotemporal Assessment of Rooftop Solar Photovoltaics Potential for Renewable Electricity Generation." *Nature Communications* 12, no. 1 (October 5, 2021): 5738. https://doi.org/10.1038/s41467-021-25720-2.

Juskus, Ryan. "Sacrifice Zones: A Genealogy and Analysis of an Environmental Justice Concept." *Environmental Humanities* 15, no. 1 (March 1, 2023): 3–24.

Kahn, Peter A., Christopher M. Worsham, and Gretchen Berland. "Characterization of Prescription Patterns and Estimated Costs for Use of Oxygen Concentrators for Home Oxygen Therapy in the US." *JAMA Network Open* 4, no. 10 (October 19, 2021): e2129967–e2129967. https://doi.org/10.1001/jamanetworkopen.2021.29967.

Kai Hwa Wang, Frances. "What Will the EV Revolution Mean for Detroit?" *PBS NewsHour*, October 11, 2022. https://www.pbs.org/newshour/nation/detroit-leads-the-way-as-the-nation-grapples-with-impending-electric-car-revolution.

Kane, Mark. "Ultium Cells Battery Plant Is Almost Complete: June 1, 2021." *InsideEVs*, June 2, 2021. https://insideevs.com/news/511582/ultium-cells-battery-plant-june2021/.

Kara, Siddharth. *Cobalt Red: How the Blood of the Congo Powers Our Lives*. St. Martin's Press, 2023.

Kara, Siddharth. "Our Device-Driven Lives Depend More Than Ever on Tragedy in the Democratic Republic of Congo." *CNN*, December 17, 2021. https://www.cnn.com/2021/12/17/opinions/siddharth-kara-mining-dr-congo/index.html.

Ke, Bryan. "Devastating House Fire Claims Lives of 3 Children, Grandmother During Texas Winter Storm." *Yahoo News*, February 22, 2021. https://news.yahoo.com/devastating-house-fire-claims-lives-170707185.html.

Kelley Blue Book. *Electric Vehicle Sales Report—Q4 2023*. Kelley Blue Book, January 2024. https://www.coxautoinc.com/wp-content/uploads/2024/01/Q4-2023-Kelley-Blue-Book-Electric-Vehicle-Sales-Report.pdf.

Keyser, David, Marie Fiori, Betony Jones, Hugh Ho, Shrayas Jatkar, Kate Gordon, et al. *United States Energy and Employment Report 2023*. Department of Energy, 2023. https://www.energy.gov/sites/default/files/2023-06/2023%20USEER%20REPORT-v2.pdf.

KFF. "State Health Facts: Distribution of the Total Population by Federal Poverty Level (Above and Below 200% FPL)." *KFF.org*, 2022, https://www.kff.org/other/state-indicator/population-up-to-200-fpl/.

Kier, Laura. "How Much Is 1° Worth?" *EnergyHub* (blog), May 16, 2012. https://www.energyhub.com/blog/how-much-is-one-degree-worth/.

Kleeman, Hanzelle, Baruch Fischhoff, and Daniel Erian Armanios, "Effects of Redesigning the Communication of Low-Income Residential Energy Efficiency Programs in the U.S." *Energy Policy* 178 (July 1, 2023): 113568. https://doi.org/10.1016/j.enpol.2023.113568.

Klier, Thomas H., and James M. Rubenstein. "Mexico's Growing Role in the Auto Industry Under NAFTA: Who Makes What and What Goes Where." *Economic Perspectives* 41, no. 6 (2017): 1–29.

Knopper, Loren D., and Christopher A. Ollson. "Health Effects and Wind Turbines: A Review of the Literature." *Environmental Health* 10, no. 1 (December 2011): 78. https://doi.org/10.1186/1476-069X-10-78.

Kohli, Anisha. "The U.S. Government Is Requiring Washers, Refrigerators and Freezers to Be More Efficient: Here's What It Means for You." *TIME*, February 11, 2023. https://time.com/6254877/energy-department-efficiency-home-appliances/.

Kojola, Erik, and David N. Pellow. "New Directions in Environmental Justice Studies: Examining the State and Violence." *Environmental Politics* 30, no. 1–2 (2021): 100–18.

Kolodny, Lora. "Foxconn Buys Lordstown Motors' Old GM Factory for $230 Million." *CNBC*, November 10, 2021. https://www.cnbc.com/2021/11/10/foxconn-buys-lordstown-motors-old-gm-factory-for-230-million.html.

Komarek, Timothy M. "Labor Market Dynamics and the Unconventional Natural Gas Boom: Evidence from the Marcellus Region." *Resource and Energy Economics* 45 (August 2016): 1–17. https://doi.org/10.1016/j.reseneeco.2016.03.004.

Konisky, David M., ed. *Failed Promises: Evaluating the Federal Government's Response to Environmental Justice.* MIT Press, 2015.

Konisky, David M. "Inequities in Enforcement? Environmental Justice and Government Performance." *Journal of Policy Analysis and Management* 28, no. 1 (2009): 102–21.

Konisky, David M., Stephen Ansolabehere, and Sanya Carley. "Corrigendum to: Proximity, NIMBYism, and Public Support for Energy Infrastructure." *Public Opinion Quarterly* 85, no. 2 (October 1, 2021): 733. https://doi.org/10.1093/poq/nfab008.

Konisky, David M., Stephen Ansolabehere, and Sanya Carley. "Proximity, NIMBYism, and Public Support for Energy Infrastructure." *Public Opinion Quarterly* 84, no. 2 (2020): 391–418.

Konisky, David M., Sanya Carley, Michelle Graff, and Trevor Memmott. "The Persistence of Household Energy Insecurity During the COVID-19 Pandemic." *Environmental Research Letters* 17, no. 10 (September 2022): 104017. https://doi.org/10.1088/1748-9326/ac90d7.

Konisky, David M., and Christopher Reenock. "Regulatory Enforcement, Riskscapes, and Environmental Justice." *Policy Studies Journal* 46, no. 1 (2018): 7–36.

Koshiba, T., P. Parker, T. Rutherford, D. Sanford, and R. Olson. "Japanese Automakers and the NAFTA Environment: Global Context." *Environments* 29, no. 3 (2001).

Kovats, Sari, Michael Depledge, Andy Haines, Lora E. Fleming, Paul Wilkinson, Seth B. Shonkoff, et al. "The Health Implications of Fracking." *Lancet* 383, no. 9919 (March 2014): 757–58. https://doi.org/10.1016/S0140-6736(13)62700-2.

Krisher, Tom. "GM Rebounds with $8.1B 2018 Profit on Strong Pricing." *Phys*, February 6, 2019. https://phys.org/news/2019-02-gm-strong-profits-good-sales.html.

Krouse, Peter. "Republican-Led Effort Singles Out Wind and Solar Power for Local Control." *Cleveland.com*, July 12, 2021. https://www.cleveland.com/news/2021/07/republican-led-effort-singles-out-wind-and-solar-power-for-local-control.html.

Ku, Arthur L., and John D. Graham. "Is California's Electric Vehicle Rebate Regressive? A Distributional Analysis." *Journal of Benefit-Cost Analysis* 13, no. 1 (March 2022): 1–19. https://doi.org/10.1017/bca.2022.2.

Kulik, Jane, Alton D. Smith, and Ernst W. Stromsdorfer. *The Downriver Community Confer-*

ence Economic Readjustment Program: Final Evaluation Report. Abt Associates, September 30, 1984.

Kullgren, Ian, and David Welch. "Lordstown's Auto Industry Is Coming Back: The Jobs? Not So Much." *Bloomberg*, November 10, 2021. https://www.bloomberg.com/news/features /2021-11-10/gm-brings-back-lordstown-s-auto-industry-the-jobs-not-so-much.

Kwan, Jacklin. "Your Old Electronics Are Poisoning People at This Toxic Dump in Ghana." *Wired UK*, November 26, 2020. https://www.wired.co.uk/article/ghana-ewaste-dump -electronics.

Laney, A. Scott, and David N. Weissman. "Respiratory Diseases Caused by Coal Mine Dust." *Journal of Occupational and Environmental Medicine / American College of Occupational and Environmental Medicine* 56, supplement 10 (October 2014): S18–22. https:// doi.org/10.1097/JOM.0000000000000260.

LaReau, Jamie. "GM's Former Plant in Lordstown Will Return to Mass Vehicle Production, Thousands of Jobs." *Detroit Free Press*, May 22, 2020. https://www.freep.com/story /money/cars/general-motors/2020/05/22/gm-lordstown-motors-plant-jobs-electric -pickups-vehicles/5229669002/.

Larson, Lance N. *The Front End of the Nuclear Fuel Cycle: Current Issues*. Congressional Research Service, July 29, 2019. https://sgp.fas.org/crs/nuke/R45753.pdf.

Lawson, Alex. "Global Electricity Grid Must Be Upgraded Urgently to Hit Climate Goals, Says IEA." *Guardian*, October 17, 2023. https://www.theguardian.com/environment/2023/oct /17/global-electricity-grid-climate-iea.

Lebbie, Tamba S., Omosehin D. Moyebi, Kwadwo Ansong Asante, Julius Fobil, Marie Noel Brune-Drisse, William A. Suk, et al. "E-Waste in Africa: A Serious Threat to the Health of Children." *International Journal of Environmental Research and Public Health* 18, no. 16 (August 11, 2021): 8488. https://doi.org/10.3390/ijerph18168488.

LeBeau, Phil. "GM's Lordstown Factory Goes Dark as Automaker Idles Underused Plants." *CNBC*, March 6, 2019. https://www.cnbc.com/2019/03/06/gms-lordstown-factory-goes -dark-as-automaker-closes-underused-plants.html.

Lee, Jae Hyun, Scott J. Hardman, and Gil Tal. "Who Is Buying Electric Vehicles in California? Characterising Early Adopter Heterogeneity and Forecasting Market Diffusion." *Energy Research and Social Science* 55 (September 2019): 218–26. https://doi.org/10.1016/j.erss .2019.05.011.

Lehmann, Sarah, Nate Hunt, Cobi Frongillo, and Philip Jordan. *Diversity in the U.S. Energy Workforce: Data Findings to Inform State Energy, Climate, and Workforce Development Policies and Programs*. National Association of State Energy Officials, BW Research Partnership, HBCU CDAC Clean Energy Initiative, April 2021.

Lenton, Timothy M., Johan Rockström, Owen Gaffney, Stefan Rahmstorf, Katherine Richardson, Will Steffen, et al. "Climate Tipping Points—Too Risky to Bet Against." *Nature* 575, no. 7784 (November 2019): 592–95. https://doi.org/10.1038/d41586-019-03595-0.

Lerner, Steve. *Sacrifice Zones: The Front Lines of Toxic Chemical Exposure in the United States*. MIT Press, 2012.

Liedtke, Michael. "PG&E Confesses to Killing 84 People in 2018 California Fire." *AP News*, June 16, 2020. https://apnews.com/article/67810cb4d9b6b90e451415b76215d6c9.

LIHEAP Clearinghouse. "LIHEAP and WAP Funding." *Liheapch.acf.hhs.gov*, 2023. https:// liheapch.acf.hhs.gov/Funding/funding.htm.

Lile, Samantha. "How Can Efficient HVAC Save You Money?" *Motili*, June 17, 2023. https://www.motili.com/blog/how-can-efficient-hvac-save-you-money/.

Lin, Albert. "Does Geoengineering Present a Moral Hazard?" *Ecology Law Quarterly* 40, no. 3 (2013): 673.

Linkon, Sherry, and John Russo. "With GM Job Cuts, Youngstown Faces a New 'Black Monday.'" *Bloomberg*, November 27, 2018. https://www.bloomberg.com/news/articles/2018-11-27/gm-s-job-cuts-reopen-old-wounds-in-youngstown-ohio.

Lobao, Linda, Minyu Zhou, Mark Partridge, and Michael Betz. "Poverty, Place, and Coal Employment Across Appalachia and the United States in a New Economic Era." *Rural Sociology* 81, no. 3 (September 2016): 343–86. https://doi.org/10.1111/ruso.12098.

Locke, Katherine. "Navajo Generating Station Shuts Down Permanently." *Navajo-Hopi Observer News*. November 18, 2019. https://www.nhonews.com/news/2019/nov/18/navajo-generating-station-shuts-down-permanently/.

Lopez, Anthony, Trieu Mai, Eric Lantz, Dylan Harrison-Atlas, Travis Williams, and Galen Maclaurin. "Land Use and Turbine Technology Influences on Wind Potential in the United States." *Energy* 223 (May 2021): 120044. https://doi.org/10.1016/j.energy.2021.120044.

Lou, Robert, Kevin P. Hallinan, Kefan Huang, and Timothy Reissman. "Smart Wifi Thermostat-Enabled Thermal Comfort Control in Residences." *Sustainability* 12, no. 5 (March 3, 2020): 1919. https://doi.org/10.3390/su12051919.

Louie, Edward P., and Joshua M. Pearce. "Retraining Investment for U.S. Transition from Coal to Solar Photovoltaic Employment." *Energy Economics* 57 (June 2016): 295–302. https://doi.org/10.1016/j.eneco.2016.05.016.

Lozanova, Sarah. "Solar Payback Period." *GreenLancer*, May 18, 2022. https://www.greenlancer.com/post/solar-payback-period.

Manescu, Larisa. "Sierra Club Releases Nationwide Investigation into Electric Vehicle Shopping Experience." Sierra Club press release, May 8, 2023. https://www.sierraclub.org/press-releases/2023/05/sierra-club-releases-nationwide-investigation-electric-vehicle-shopping.

Manley, David, Patrick R. P. Heller, and William Davis. *No Time to Waste: Governing Cobalt amid the Energy Transition*. Berkeley Law and Natural Resource Governance Institute, March 2022.

Marvel, K., W. Su, R. Delgado, S. Aarons, A. Chatterjee, M. E. Garcia, et al. "Chapter 2: Climate Trends." In *Fifth National Climate Assessment*, edited by A. R. Crimmins, C. W. Avery, D. R. Easterling, K. E. Kunkel, B. C. Stewart, and T. K. Maycock. US Global Change Research Program, 2023. https://doi.org/10.7930/NCA5.2023.CH2.

Masterson, Kathleen. "Q&A: Examining the Tennessee Coal Ash Spill." *NPR*, January 8, 2009. https://www.npr.org/2009/01/08/99134153/q-a-examining-the-tennessee-coal-ash-spill.

Mayer, Adam. "Quality of Life and Unconventional Oil and Gas Development: Towards a Comprehensive Impact Model for Host Communities." *Extractive Industries and Society* 4 (2017): 923–30.

Mayer, Adam, Shawn K. Olson-Hazboun, and Stephanie Malin. "Fracking Fortunes: Economic Well-Being and Oil and Gas Development Along the Urban-Rural Continuum." *Rural Sociology* 83, no. 3 (September 2018): 532–67. https://doi.org/10.1111/ruso.12198.

Mayfield, Debra. "Five Years Later and the Story of the TVA Spill Continues." *Earthjustice. org*, December 23, 2013. https://earthjustice.org/article/five-years-later-and-the-story-of-the-tva-spill-continues.

Mazurek, Jacek M., John Wood, David Blackley, and David Weissman. "Coal Workers' Pneumoconiosis: Attributable Years of Potential Life Lost to Life Expectancy and Potential Life Lost Before Age 65 Years—United States, 1999–2016." *Morbidity and Mortality Weekly Report* 67 (August 3, 2018). https://doi.org/10.15585/mmwr.mm6730a3.

McCleary, Kelly, Amir Vera, and Liam Reilly. "All 17 Victims of Bronx Apartment Fire, Including 2-Year-Old, Died of Smoke Inhalation, NYC Medical Examiner Rules." *CNN*, January 13, 2022. https://www.cnn.com/2022/01/11/us/new-york-bronx-apartment-fire-tuesday/index.html.

McFarland, Victor, and Jeff D. Colgan. "Oil and Power: The Effectiveness of State Threats on Markets." *Review of International Political Economy* 30, no. 2 (2023): 487–510. https://doi.org/10.1080/09692290.2021.2014931.

McGlade, Christophe, and Paul Ekins. "The Geographical Distribution of Fossil Fuels Unused When Limiting Global Warming to 2 °C." *Nature* 517, no. 7533 (January 2015): 187–90.

McKibben, Bill. "With the Keystone Pipeline, Drawing a Line in the Tar Sands." *YaleEnvironment360*, October 2011. https://e360.yale.edu/features/with_the_keystone_pipeline_drawing_a_line_in_the_tar_sands.

McSweeney, Robert. "Explainer: Nine 'Tipping Points' That Could Be Triggered by Climate Change." *CarbonBrief*, October 2020. https://www.carbonbrief.org/explainer-nine-tipping-points-that-could-be-triggered-by-climate-change/.

Meckling, Jonas, Thomas Sterner, and Gernot Wagner. "Policy Sequencing Toward Decarbonization." *Nature Energy* 2, no. 12 (November 13, 2017): 918–22. https://doi.org/10.1038/s41560-017-0025-8.

Memmott, Trevor, Sanya Carley, Michelle Graff, and David M. Konisky. "Sociodemographic Disparities in Energy Insecurity Among Low-Income Households Before and During the COVID-19 Pandemic." *Nature Energy* 6, no. 2 (February 2021): 186–93. https://doi.org/10.1038/s41560-020-00763-9.

Memmott, Trevor, Sanya Carley, Michelle Graff, and David M. Konisky. "Utility Disconnection Protections and the Incidence of Energy Insecurity in the United States." *Iscience* 26, no. 3 (2023): 106244.

Mendell, Mark J., Anna G. Mirer, Kerry Cheung, My Tong, and Jeroen Douwes. "Respiratory and Allergic Health Effects of Dampness, Mold, and Dampness-Related Agents: A Review of the Epidemiologic Evidence." *Environmental Health Perspectives* 119, no. 6 (June 2011): 748–56. https://doi.org/10.1289/ehp.1002410.

Mercure, J.-F., A. Lam, S. Billington, and H. Pollitt. "Integrated Assessment Modelling as a Positive Science: Private Passenger Road Transport Policies to Meet a Climate Target Well Below 2°C." *Climatic Change* 151, no. 2 (November 2018): 109–29. https://doi.org/10.1007/s10584-018-2262-7.

Meyer, Alex. "Appalachia's Orange Stain." *Post*, accessed October 27, 2023. https://projects.thepostathens.com/SpecialProjects/AcidMines-AMeyer/.

Mills, Sarah Banas, Douglas Bessette, and Hannah Smith. "Exploring Landowners' Post-Construction Changes in Perceptions of Wind Energy in Michigan." *Land Use Policy* 82 (March 2019): 754–62. https://doi.org/10.1016/j.landusepol.2019.01.010.

Minchin, Timothy J. "'A Gallant Fight': The UAW and the 1970 General Motors Strike." *International Review of Social History* 68, no. 1 (April 2023): 41–73. https://doi.org/10.1017/S0020859022000293.

Miniard, Deidra, Joseph Kantenbacher, and Shahzeen Z. Attari. "Shared Vision for a Decarbonized Future Energy System in the United States." *Proceedings of the National Academy of Sciences* 117, no. 13 (March 31, 2020): 7108–14. https://doi.org/10.1073/pnas.1920558117.

Misbrener, Kelsey. "Electrical Grid Interconnection Backlog Grew 30% in 2023." *Solar Power World*, April 10, 2024. https://www.solarpowerworldonline.com/2024/04/electrical-grid-interconnection-backlog-grew-30-percent-2023/.

Mohai, Paul, and Bunyan Bryant. "Environmental Injustice: Weighing Race and Class as Factors in the Distribution of Environmental Hazards." *University of Colorado Law Review* 63, no. 4 (1992): 921–32.

Morgan, Tash. "Agbogbloshie: Welcome to the World's Digital Dumping Ground (Part 1) | Human Impact." *Earth Touch News Network*, March 26, 2014. https://www.earthtouchnews.com/conservation/human-impact/agbogbloshie-welcome-to-the-worlds-digital-dumping-ground-part-1.

Moritz, Manuel, Tobias Redlich, Pascal Krenz, Sonja Buxbaum-Conradi, and Jens P. Wulfsberg. "Tesla Motors, Inc.: Pioneer Towards a New Strategic Approach in the Automobile Industry Along the Open Source Movement?" In *2015 Portland International Conference on Management of Engineering and Technology (PICMET)*, 85–92. Institute of Electrical and Electronics Engineers, 2015.

Motavalli, Jim. "Closing the Loop on EV Battery Recycling." *SAE International*, October 7, 2022. https://www.sae.org/news/2022/10/ev-battery-recycling.

Motavalli, Jim. "Every Automaker's EV Plans Through 2035 and Beyond." *Forbes*, October 4, 2021. https://www.forbes.com/wheels/news/automaker-ev-plans/.

Mucha, Lena, Todd C. Frankel, and Karly Domb Sadof. "The Hidden Costs of Cobalt Mining." *Washington Post*, February 28, 2018. https://www.washingtonpost.com/news/in-sight/wp/2018/02/28/the-cost-of-cobalt/.

Mueller, Rose M. "Surface Coal Mining and Public Health Disparities: Evidence from Appalachia." *Resources Policy* 76 (2022): 102567, https://doi.org/10.1016/j.resourpol.2022.102567.

Muller, Joann. "The World's Car Buyers Are Ready to Go Electric." *Axios*, May 28, 2022. https://www.axios.com/2022/05/23/electric-vehicles-consumer-interest.

National Academies of Sciences, Engineering, and Medicine. *Monitoring and Sampling Approaches to Assess Underground Coal Mine Dust Exposures*. National Academies Press, 2018. https://doi.org/10.17226/25111.

National Council on Aging. "Energy and Utility Assistance: How Do I Apply for LIHEAP?" *NCOA.org*, November 11, 2022. https://www.ncoa.org/article/how-do-i-apply-for-liheap.

National Highway Traffic Safety Administration. "Newly Released Estimates Show Traffic Fatalities Reached a 16-Year High in 2021." United States Department of Transportation press release, May 17, 2022. https://www.nhtsa.gov/press-releases/early-estimate-2021-traffic-fatalities.

National Snow and Ice Data Center. "Antarctic Melt Season off to Fast Start; Greenland 2023

Melt Season Review." *NSIDC.org*, December 21, 2023. https://nsidc.org/ice-sheets-today
/analyses/antarctic-melt-season-fast-start-greenland-2023-melt-season-review.

Natural Resources Canada. *The Canadian Critical Minerals Strategy—From Exploration
to Recycling: Powering the Green and Digital Economy for Canada and the World*. Natural Resources Canada, 2022. https://www.canada.ca/content/dam/nrcan-rncan/site
/critical-minerals/Critical-minerals-strategyDec09.pdf.

Neamt, Ioana. "We Are the Ones Who Knock—on Walter White's Fictional Door in Breaking
Bad." *FancyPantsHomes.com* (blog), February 2, 2024. https://www.fancypantshomes
.com/movie-homes/walter-white-house-in-breaking-bad/.

Neavling, Steve. "Struggling to Breathe in 48217, Michigan's Most Toxic ZIP Code." *Detroit
Metro Times*, January 2020. https://www.metrotimes.com/news/struggling-to-breathe
-in-48217-michigans-most-toxic-zip-code-23542211.

Nelson, Robert, Justin Madron, and Nathaniel Ayers. "Mapping Inequality: Redlining in New
Deal America." University of Richmond, accessed April 22, 2023. https://dsl.richmond
.edu/panorama/redlining/.

Newman, Rebecca, and Ilan Noy. "The Global Costs of Extreme Weather That Are Attributable to Climate Change." *Nature Communications* 14, no. 1 (September 29, 2023): 6103.
https://doi.org/10.1038/s41467-023-41888-1.

Nieto del Rio, Giulia McDonnell, Nicholas Bogel-Burroughs, and Ivan Penn. "His Lights
Stayed On During Texas' Storm. Now He Owes $16,752." *New York Times*, February 21,
2021. https://www.nytimes.com/2021/02/20/us/texas-storm-electric-bills.html.

Nieto del Rio, Giulia McDonnell, Richard Fausset, and Johnny Diaz. "Extreme Cold Killed
Texans in Their Bedrooms, Vehicles and Backyards." *New York Times*, February 19, 2021,
https://www.nytimes.com/2021/02/19/us/texas-deaths-winter-storm.html.

Nixon, Rob. *Slow Violence and the Environmentalism of the Poor*. Harvard University Press,
2011.

NOAA National Centers for Environmental Information. "U.S. Billion-Dollar Weather
and Climate Disasters." NOAA National Centers for Environmental Information, 2023.
https://www.ncei.noaa.gov/access/billions/.

Obama, Barack. "Remarks by the President in the State of the Union Address." White House
Office of the Press Secretary, February 12, 2013. https://obamawhitehouse.archives.gov
/the-press-office/2013/02/12/remarks-president-state-union-address.

O'Leary, Sean. *The Natural Gas Fracking Boom and Appalachia's Lost Economic Decade*.
Ohio River Valley Institute, February 12, 2021.

Onat, Nuri, Murat Kucukvar, and Omer Tatari. "Towards Life Cycle Sustainability Assessment
of Alternative Passenger Vehicles." *Sustainability* 6, no. 12 (December 16, 2014): 9305–42.
https://doi.org/10.3390/su6129305.

Oteng-Ababio, Martin, Maja van der Velden, and Mark B. Taylor. "Building Policy Coherence for Sound Waste Electrical and Electronic Equipment Management in a Developing
Country." *Journal of Environment and Development* 29, no. 3 (September 2020): 306–28.
https://doi.org/10.1177/1070496519898218.

Owusu-Sekyere, Karoline, Alexander Batteiger, Richard Afoblikame, Gerold Hafner, and
Martin Kranert. "Assessing Data in the Informal E-Waste Sector: The Agbogbloshie
Scrapyard." *Waste Management* 139 (February 2022): 158–67. https://doi.org/10.1016/j
.wasman.2021.12.026.

Padnani, Amy. "Anatomy of Detroit's Decline." *New York Times*, August 17, 2013. https://www.nytimes.com/interactive/2013/08/17/us/detroit-decline.html.

Parsons, George, and Martin D. Heintzelman. "The Effect of Wind Power Projects on Property Values: A Decade (2011–2021) of Hedonic Price Analysis." *International Review of Environmental and Resource Economics* 16, no. 1 (2022): 93–170.

Partridge, Mark D., Michael R. Betz, and Linda Lobao. "Natural Resource Curse and Poverty in Appalachian America." *American Journal of Agricultural Economics* 95, no. 2 (2013): 449–56.

Pasqualetti, Martin J. "Social Barriers to Renewable Energy Landscapes." *Geographical Review* 101, no. 2 (2011): 201–23.

Pasternak, Judy. *Yellow Dirt: An American Story of a Poisoned Land and a People Betrayed.* Simon and Schuster, 2010.

Patterson, Jacqui, Charles Hua, Vikky Angelico, Josie Karout, Jackson Koeppel, Elizabeth Mathis, et al. *Who Holds the Power: Demystifying and Democratizing Public Utilities Commissions.* Chisholm Legacy Project, December 2022.

Penney, Brad, and Phil Kloer. *Shelter Report 2015: Less Is More: Transforming Low-Income Communities Through Energy Efficiency.* Habitat for Humanity, 2015. https://www.habitat.org/sites/default/files/2015-habitat-for-humanity-shelter-report.pdf.

Perl, Libby. *LIHEAP: Program and Funding.* Congressional Research Service, May 22, 2015. https://digital.library.unt.edu/ark:/67531/metadc807073/.

Pew Research Center. *Economy Remains the Public's Top Policy Priority; COVID-19 Concerns Decline Again.* Pew Research Center, February 6, 2023. https://www.pewresearch.org/politics/2023/02/06/economy-remains-the-publics-top-policy-priority-covid-19-concerns-decline-again/.

Philbrick, Ian Prasad. "Why Isn't Biden's Expanded Child Tax Credit More Popular?" *New York Times*, January 5, 2022. https://www.nytimes.com/2022/01/05/upshot/biden-child-tax-credit.html.

Pierre-Louis, Kendra. "This Land Is (Still) Their Land: Meet the Nebraskan Farmers Fighting Keystone XL." *Popular Science*, September 15, 2017. https://www.popsci.com/keystone-xl-pipeline-nebraska-farmers/.

Plautz, Carlos, and Jason Anchondo. "CCS 2.0: Company Reboots Bid to Save N.M. Coal Plant." *E&E News*, August 18, 2023. https://www.eenews.net/articles/ccs-2-0-company-reboots-bid-to-save-n-m-coal-plant/.

Plautz, Jason. "Western Transmission Line Breaks Ground After 18-Year Wait." *E&E News*, June 21, 2023. https://www.eenews.net/articles/western-transmission-line-breaks-ground-after-18-year-wait/.

Plumer, Brad. "The U.S. Has Billions for Wind and Solar Projects: Good Luck Plugging Them In." *New York Times*, February 23, 2023. https://www.nytimes.com/2023/02/23/climate/renewable-energy-us-electrical-grid.html.

Plumer, Brad, Nadja Popovich, and Blacki Migliozzi. "Electric Cars Are Coming: How Long Until They Rule the Road?" *New York Times*, March 11, 2021. https://www.nytimes.com/interactive/2021/03/10/climate/electric-vehicle-fleet-turnover.html.

Political Economy Research Institute at the University of Massachusetts Amherst. "Top 100 Polluter Indexes." *Peri.UMASS.edu*, 2024. https://peri.umass.edu/top-100-polluter-indexes.

Pollin, Robert, and Brian Callaci. "The Economics of Just Transition: A Framework for Supporting Fossil Fuel–Dependent Workers and Communities in the United States." *Labor Studies Journal* 44, no. 2 (June 2019): 93–138. https://doi.org/10.1177/0160449X18787051.

Pollin, Robert, Heidi Garrett-Peltier, James Heintz, and Bracken Hendricks. *Green Growth: A U.S. Program for Controlling Climate Change and Expanding Job Opportunities.* Center for American Progress and Policy Economy Research Institute, September 2014. https://cdn.americanprogress.org/wp-content/uploads/2014/09/PERI.pdf.

Pontecorvo, Emily. "Will the Inflation Reduction Act Jumpstart Carbon Capture?" *Grist*, August 22, 2022. https://grist.org/technology/will-the-inflation-reduction-act-jumpstart-carbon-capture-45q/.

Popovich, Nadja. "Black Lung Disease Comes Storming Back in Coal Country." *New York Times*, February 2018. https://www.nytimes.com/interactive/2018/02/22/climate/black-lung-resurgence.html.

Power, Dan. "Here's What We Know About Energy Efficiency Access in Low-Income Communities." Alliance to Save Energy, June 15, 2021. https://www.ase.org/blog/heres-what-we-know-about-energy-efficiency-access-low-income-communities.

Puko, Timothy. "Why These Environmentalists Are Resisting Part of Biden's Climate Push." *Washington Post*, June 22, 2023. https://www.washingtonpost.com/nation/2023/06/22/biden-carbon-capture-climate-environmentalists/.

Raimi, Daniel. "The Greenhouse Gas Effects of Increased US Oil and Gas Production." *Energy Transitions* 4, no. 1 (2020): 45–56.

Raimi, Daniel, Sanya Carley, and David Konisky. "Mapping County-Level Vulnerability to the Energy Transition in US Fossil Fuel Communities." *Scientific Reports* 12 (September 21, 2022): 15748. https://doi.org/10.1038/s41598-022-19927-6.

Rand, Joseph, and Ben Hoen. "Thirty Years of North American Wind Energy Acceptance Research: What Have We Learned?" *Energy Research and Social Science* 29 (2017): 135–48.

Reames, Tony Gerard. "A Community-Based Approach to Low-Income Residential Energy Efficiency Participation Barriers." *Local Environment* 21, no. 12 (December 1, 2016): 1449–66. https://doi.org/10.1080/13549839.2015.1136995.

Reames, Tony G., Dorothy M. Daley, and John C. Pierce. "Exploring the Nexus of Energy Burden, Social Capital, and Environmental Quality in Shaping Health in US Counties." *International Journal of Environmental Research and Public Health* 18, no. 2 (January 13, 2021): 620. https://doi.org/10.3390/ijerph18020620.

Reames, Tony G., Michael A. Reiner, and M. Ben Stacey. "An Incandescent Truth: Disparities in Energy-Efficient Lighting Availability and Prices in an Urban U.S. County." *Applied Energy* 218 (May 2018): 95–103. https://doi.org/10.1016/j.apenergy.2018.02.143.

Reed, Rachel. "What the US Is Getting Right—and Wrong—About the Move to Electric Vehicles." *Harvard Law Today*, June 2023. https://hls.harvard.edu/today/what-the-us-is-getting-right-and-wrong-about-the-move-to-electric-vehicles/.

Reich, Robert B. "Bailout: A Comparative Study in Law and Industrial Structure." *Yale Journal on Regulation* 2 (1984): 163–224.

Reuters. "Factbox: U.S. Coal-Fired Power Plants Scheduled to Shut." *Reuters.org*, October 28, 2021. https://www.reuters.com/business/energy/us-coal-fired-power-plants-scheduled-shut-2021-10-28/.

Ringquist, Evan J. "Assessing Evidence of Environmental Inequities: A Meta-Analysis." *Journal of Policy Analysis and Management* 24, no. 2 (2005): 223–47. https://doi.org/10.1002/pam.20088.

Reagan, Ronald. *United States Uranium Mining and Milling Industry: A Comprehensive Review: A Report to the Congress.* US Department of Energy, 1984.

Ritter, Stephen. "A New Life for Coal Ash." *Chemical and Engineering News,* February 15, 2016. https://cen.acs.org/articles/94/i7/New-Life-Coal-Ash.html.

Roemer, Kelli F., and Julia H. Haggerty. "The Energy Transition as Fiscal Rupture: Public Services and Resilience Pathways in a Coal Company Town." *Energy Research and Social Science* 91 (September 2022).

Rogers, Everett M. *Diffusion of Innovations.* 3rd ed. Free Press, 1983.

Rosa, Eugene A., and Riley E. Dunlap. "Poll Trends: Nuclear Power: Three Decades of Public Opinion." *Public Opinion Quarterly* 58, no. 2 (1994): 295. https://doi.org/10.1086/269425.

Roth, Sammy. "Solar Sprawl Is Tearing Up the Mojave Desert: Is There a Better Way?" *Los Angeles Times,* June 27, 2023. https://www.latimes.com/environment/story/2023-06-27/solar-panels-could-save-california-but-they-hurt-the-desert.

Rothstein, Richard. *The Color of Law: A Forgotten History of How Our Government Segregated America.* Liveright Publishing, 2017.

Rubbers, Benjamin. "Governing New Mining Projects in D. R. Congo: A View from the HR Department of a Chinese Company." *Extractive Industries and Society* 7, no. 1 (January 2020): 191–98. https://doi.org/10.1016/j.exis.2019.12.006.

Ruggiero, Adam. "'Starship Pooper': Hilarious New Names for Tesla's Cybertruck." *GearJunkie,* November 22, 2019. https://gearjunkie.com/motors/tesla-cybertruck-funny-names.

Russ, Abel, Keene Kelderman, Lisa Evans, and Caroline Weinberg. *Poisonous Coverup: The Widespread Failure of the Power Industry to Clean Up Coal Ash Dumps.* Environmental Integrity Project and EarthJustice, November 3, 2022. https://environmentalintegrity.org/wp-content/uploads/2022/11/Poisonous-Coverup-Final.pdf.

Russell, John. "Getting Ready to Kick the Coal Habit at IPL's Harding Street Station." *Indy Star,* May 29, 2015. https://www.indystar.com/story/money/2015/05/29/getting-ready-kick-coal-habit-ipls-harding-street-station/28085729/.

Ryan, Brent D., and Daniel Campo. "Autopia's End: The Decline and Fall of Detroit's Automotive Manufacturing Landscape." *Journal of Planning History* 12, no. 2 (May 2013): 95–132. https://doi.org/10.1177/1538513212471166.

Saad, Lydia. "A Steady Six in 10 Say Global Warming's Effects Have Begun." *Gallup.com,* April 20, 2023. https://news.gallup.com/poll/474542/steady-six-say-global-warming-effects-begun.aspx.

Sainato, Michael. "Youngstown's Hopes for Reinvention Fade as Electric Truck Firm Sputters." *Guardian,* June 30, 2021. https://www.theguardian.com/business/2021/jun/30/youngstown-ohio-car-electric-truck.

Salcedo, Andrea. "An 11-Year-Old Boy Died in an Unheated Texas Mobile Home: Authorities Suspect Hypothermia." *Washington Post,* February 19, 2021. https://www.washingtonpost.com/nation/2021/02/19/texas-boy-death-winterstorm-pavon/.

Samuels, Gabe, and Yonah Freemark. *The Polluted Life Near the Highway.* Urban Institute, November 2022.

Save on Energy Team. "Electricity Bill Report: March 2024." *SaveOnEnergy.com*, March 2024. https://www.saveonenergy.com/resources/electricity-bills-by-state/.

Searcey, Dionne, Michael Forsythe, and Eric Lipton. "A Power Struggle over Cobalt Rattles the Clean Energy Revolution." *New York Times*, November 20, 2021. https://www.nytimes.com/2021/11/20/world/china-congo-cobalt.html.

Seelye, Katharine Q. "After 16 Years, Hopes for Cape Cod Wind Farm Float Away." *New York Times*, December 19, 2017. https://www.nytimes.com/2017/12/19/us/offshore-cape-wind-farm.html.

Shah, Zeal, Juan Pablo Carvallo, Feng-Chi Hsu, and Jay Taneja. "The Inequitable Distribution of Power Interruptions During the 2021 Texas Winter Storm Uri." *Environmental Research: Infrastructure and Sustainability* 3, no. 2 (2023): 025011.

Shao, Elena. "As Heat Pumps Go Mainstream, a Big Question: Can They Handle Real Cold?" *New York Times*, February 22, 2023. https://www.nytimes.com/interactive/2023/02/22/climate/heat-pumps-extreme-cold.html.

Shaw, Neil. "For Ghana E-Waste Recyclers, a Safer Option amid Toxic Fumes." *AP News*, January 5, 2019. https://apnews.com/article/f9a0d071d1d646edb2b53fd22fd8548c.

Shenassa, Edmond D., Constantine Daskalakis, Allison Liebhaber, Matthias Braubach, and MaryJean Brown. "Dampness and Mold in the Home and Depression: An Examination of Mold-Related Illness and Perceived Control of One's Home as Possible Depression Pathways." *American Journal of Public Health* 97, no. 10 (October 2007): 1893–99. https://doi.org/10.2105/AJPH.2006.093773.

Shepardson, David, and Ben Klayman. "UAW Ratifies Labor Deal with General Motors." *Reuters*, December 16, 2023. https://www.reuters.com/business/autos-transportation/gms-labor-deal-with-uaw-clinches-ratification-2023-11-16.

Sherlock, John. "The Price of Coal: The Coal Mine Health and Safety Act of 1969." *Catholic University Law Review* 20, no. 3 (January 1, 1971): 496–510.

Simon, Julia. "Misinformation Is Derailing Renewable Energy Projects across the United States." *NPR*, March 28, 2022. https://www.npr.org/2022/03/28/1086790531/renewable-energy-projects-wind-energy-solar-energy-climate-change-misinformation.

Skadowski, Suzanne, and Margot Perez-Sullivan. "$2 Billion in Funds Headed for Cleanups in Nevada and on the Navajo Nation from Historic Anadarko Settlement with U.S. EPA, States." EPA news release, April 15, 2016. https://www.epa.gov/archive/epa/newsreleases/2-billion-funds-headed-cleanups-nevada-and-navajo-nation-historic-anadarko-settlement.html.

Smil, Vaclav. "Examining Energy Transitions: A Dozen Insights Based on Performance." *Energy Research and Social Science* 22 (December 2016): 194–97. https://doi.org/10.1016/j.erss.2016.08.017.

Solar Energy Industries Association. "Oklahoma State Solar Overview." *SEIA.org*, 2022. https://www.seia.org/state-solar-policy/oklahoma-solar.

Solomon, Michelle. "DOE Study Highlights America's Transmission Needs, but How Do We Accelerate Buildout?" *Utility Dive*, March 31, 2023. https://www.utilitydive.com/news/doe-study-transmission-clean-energy/646589/.

Sovacool, Benjamin K. "When Subterranean Slavery Supports Sustainability Transitions? Power, Patriarchy, and Child Labor in Artisanal Congolese Cobalt Mining." *Extractive

Industries and Society 8, no. 1 (March 2021): 271–93. https://doi.org/10.1016/j.exis.2020 .11.018.

Sovacool, Benjamin K. "Who Are the Victims of Low-Carbon Transitions? Towards a Political Ecology of Climate Change Mitigation." *Energy Research and Social Science* 73 (March 2021): 101916. https://doi.org/10.1016/j.erss.2021.101916.

Sovacool, Benjamin K., Andrew Hook, Mari Martiskainen, Andrea Brock, and Bruno Turnheim. "The Decarbonisation Divide: Contextualizing Landscapes of Low-Carbon Exploitation and Toxicity in Africa." *Global Environmental Change* 60 (January 2020): 102028. https://doi.org/10.1016/j.gloenvcha.2019.102028.

Sovacool, Benjamin K., Peter Newell, Sanya Carley, and Jessica Fanzo. "Equity, Technological Innovation and Sustainable Behaviour in a Low-Carbon Future." *Nature Human Behaviour* 6, no. 3 (January 31, 2022): 326–37. https://doi.org/10.1038/s41562-021-01257-8.

Specian, Mike, Weston Berg, Sagarika Subramanian, and Kristin Campbell. *2023 Utility Energy Efficiency Scorecard.* American Council for an Energy-Efficient Economy, August 24, 2023. https://www.aceee.org/research-report/u2304.

Spector, Mike, Joseph White, and Dietrich Knauth. "Lordstown Motors Files for Bankruptcy, Sues Foxconn." *Reuters*, June 27, 2023. https://www.reuters.com/business/autos -transportation/lordstown-motors-files-bankruptcy-sues-foxconn-2023-06-27/.

Srigboh, Roland Kofi, Niladri Basu, Judith Stephens, Emmanuel Asampong, Marie Perkins, Richard L. Neitzel, et al. "Multiple Elemental Exposures Amongst Workers at the Agbogbloshie Electronic Waste (E-Waste) Site in Ghana." *Chemosphere* 164 (December 2016): 68–74. https://doi.org/10.1016/j.chemosphere.2016.08.089.

St. John, Alexa. "US Electric Vehicle Sales to Hit Record This Year, but Still Lag Behind China and Germany." *AP News*, November 24, 2023. https://apnews.com/article/automakers -electric-vehicles-us-china-sales-d121c09a61f50e7357f5675af4b6056b.

Stafford, Eric. "2024 Tesla Cybertruck." *Car and Driver*, accessed March 8, 2024. https://www .caranddriver.com/tesla/cybertruck.

Stavins, Robert N. "What Can We Learn from the Grand Policy Experiment? Lessons from SO2 Allowance Trading." *Journal of Economic Perspectives* 12, no. 3 (August 1, 1998): 69–88. https://doi.org/10.1257/jep.12.3.69.

Stephens, Jennie C. "Beyond Climate Isolationism: A Necessary Shift for Climate Justice." *Current Climate Change Reports* 8, no. 4 (December 1, 2022): 83–90. https://doi.org/10 .1007/s4064122-00186-6.

Stevis, Dimitris, Edouard Morena, and Dunja Krause, "Introduction: The Genealogy and Contemporary Politics of Just Transitions," in *Just Transitions: Social Justice in the Shift Towards a Low-Carbon World*, eds. Dimitris Stevis, Edouard Morena, and Dunja Krause, 1–31. Pluto Press, 2020. https://doi.org/10.2307/j.ctvs09qrx.6.

Stipes, Chris. "New Report Details Impact of Winter Storm Uri on Texans." University of Houston, March 29, 2021. https://www.uh.edu/news-events/stories/2021/march-2021 /03292021-hobby-winter-storm.

Stokes, Leah Cardamore. *Short Circuiting Policy: Interest Groups and the Battle over Clean Energy and Climate Policy in the American States.* Oxford University Press, 2020.

Stokes, Leah C., Emma Franzblau, Jessica R. Lovering, and Chris Miljanich. "Prevalence and Predictors of Wind Energy Opposition in North America." *Proceedings of the National*

Academy of Sciences 120, no. 40 (October 3, 2023): e2302313120. https://doi.org/10.1073/pnas.2302313120.

Su, Jean, and Christopher Kuveke. *Powerless in the Pandemic 2.0: After Bailouts, Electric Utilities Chose Profits over People.* Center for Biological Diversity, April 2022. https://bailout.cdn.prismic.io/bailout/ddebd6e2-b136-4dc8-a1da-f6d4583b4c24_Powerless_Report2022_final.pdf.

Sunter, Deborah A., Sergio Castellanos, and Daniel M. Kammen. "Disparities in Rooftop Photovoltaics Deployment in the United States by Race and Ethnicity." *Nature Sustainability* 2, no. 1 (2019): 71–76.

Surber, Sarah J., and D. Scott Simonton. "Disparate Impacts of Coal Mining and Reclamation Concerns for West Virginia and Central Appalachia." *Resources Policy* 54 (2017): 1–8. https://doi.org/10.1016/j.resourpol.2017.08.004.

Svitek, Patrick. "Texas Puts Final Estimate of Winter Storm Death Toll at 246." *Texas Tribune*, January 2, 2022. https://www.texastribune.org/2022/01/02/texas-winter-storm-final-death-toll-246/.

Tallmadge, Margaret. "What Is Holding Back Renewable Energy Development in Indian Country?" Clean Energy Finance Forum, December 11, 2019. https://cleanenergyfinanceforum.com/2019/12/11/what-is-holding-back-renewable-energy-development-in-indian-country.

Tarekegne, Bethel W., Kamila Kazimierczuk, and Rebecca S. O'Neil. *Coal-Dependent Communities in Transition: Identifying Best Practices to Ensure Equitable Outcomes.* Pacific Northwest National Laboratory, September 2021.

Taylor, Dorceta. *Toxic Communities: Environmental Racism, Industrial Pollution, and Residential Mobility.* New York University Press, 2014. https://doi.org/10.18574/nyu/9781479805150.001.0001.

Teller-Elsberg, Jonathan, Benjamin Sovacool, Taylor Smith, and Emily Laine. "Fuel Poverty, Excess Winter Deaths, and Energy Costs in Vermont: Burdensome for Whom?" *Energy Policy* 90 (March 1, 2016): 81–91. https://doi.org/10.1016/j.enpol.2015.12.009.

Tessum, Christopher W., Joshua S. Apte, Andrew L. Goodkind, Nicholas Z. Muller, Kimberley A. Mullins, David A. Paolella, et al. "Inequity in Consumption of Goods and Services Adds to Racial–Ethnic Disparities in Air Pollution Exposure." *Proceedings of the National Academy of Sciences* 116, no. 13 (March 26, 2019): 6001–6. https://doi.org/10.1073/pnas.1818859116.

Texas Commission on Environmental Quality. "TCEQ Air Emission Event Reports Database." Web application. Texas Commission on Environmental Quality, 2023. https://www2.tceq.texas.gov/oce/eer/index.cfm?fuseaction=main.searchForm&newsearch=yes.

Thakrar, Sumil K., Srinidhi Balasubramanian, Peter J. Adams, Inês M. L. Azevedo, Nicholas Z. Muller, Spyros N. Pandis, et al. "Reducing Mortality from Air Pollution in the United States by Targeting Specific Emission Sources." *Environmental Science and Technology Letters* 7, no. 9 (September 8, 2020): 639–45. https://doi.org/10.1021/acs.estlett.0c00424.

Thomas, Michael. "How Much Money Do Heat Pumps Save?" *Carbon Switch*, December 2022. https://carbonswitch.com/heat-pump-savings/.

Tierney, Susan F. *The U.S. Coal Industry: Challenging Transitions in the 21st Century.* Analysis Group, Inc., September 26, 2016. https://www.analysisgroup.com/globalassets/insights/publishing/2016-tierney-coal-industry-21st-century-challenges.pdf.

Ulitskaya, Jane. "How Much Do Electric Car Batteries Cost to Replace?" *Cars.com*, April 24, 2023. https://www.cars.com/articles/how-much-do-electric-car-batteries-cost -to-replace-465308/.

Ulrich, Lawrence. "GM Bets Big on Batteries: A New $2.3 Billion Plant Cranks Out Ultium Cells to Power a Future Line of Electric Vehicles." *IEEE Spectrum* 57, no. 12 (December 2020): 26–31. https://doi.org/10.1109/MSPEC.2020.9271805.

US Aid and Global Waters. "Democratic Republic of Congo." *Global Waters.org*, 2022. https://www.globalwaters.org/WhereWeWork/Africa/DRC.

US Bureau of Labor Statistics. "All Employees, Coal Mining." Retrieved from FRED, Federal Reserve Bank of St. Louis, February 2024. https://fred.stlouisfed.org/series /CES1021210001.

US Bureau of Labor Statistics. "Automotive Industry: Employment, Earnings, and Hours." *BLS.gov*, accessed March 20, 2024. https://www.bls.gov/iag/tgs/iagauto.htm#emp _national.

US Bureau of Labor Statistics. "Bureau of Labor Statistics Motor Vehicle Manufacturing Data." *BLS.gov*, accessed February 1, 2023. https://data.bls.gov/timeseries/CEU3133610001 ?amp%25253bdata_tool=XGtable&output_view=data&include_graphs=true.

US Bureau of Labor Statistics. "Databases, Tables and Calculators by Subject: Employment, Hours, and Earnings from the Current Employment Statistics Survey (National)." Series ID CES1021210001. *Data.BLS.gov*, accessed February 23, 2023. https://data.bls.gov /timeseries/CES1021210001.

US Census Bureau. "2019: American Community Survey, ACS 1-Year Estimates Subject Tables, Table S2502: Demographic Characteristics for Occupied Housing Units." *Data. census.gov*, accessed October 23, 2023. https://data.census.gov/table/ACSST1Y2019 .S2502?q=Owner/Renter+(Householder)+Characteristics&t=Housing.

US Census Bureau. "American Community Survey 5-Year Estimates." *Census Reporter* profile page for Uniontown, AL, 2021. http://censusreporter.org/profiles/16000US0177904 -uniontown-al/.

US Census Bureau. "Highlights of 2023 Characteristics of New Housing." *Census.gov*, June 1, 2023. https://www.census.gov/construction/chars/highlights.html.

US Census Bureau. "QuickFacts: Crawford County, Ohio." *Census.gov*, accessed October 31, 2024. https://www.census.gov/quickfacts/fact/table/crawfordcountyohio/PST045223.

US Census Bureau. "QuickFacts: Port Arthur City, Texas." *Census.gov*, July 2022. https://www .census.gov/quickfacts/portarthurcitytexas.

US Consumer Product Safety Commission. "Seasons Change, but Fire and Carbon Monoxide Safety Is Year-Round; Warm Up to CPSC's Tips for Staying Safe During Colder Weather." News release, December 14, 2021. https://www.cpsc.gov/Newsroom/News-Releases /2022/Seasons-Change-but-Fire-and-Carbon-Monoxide-Safety-Is-Year-Round-Warm -Up-to-CPSCs-Tips-for-Staying-Safe-During-Colder-Weather.

US Department of Agriculture Economic Research Service. "Poverty Area Measures." *USDA. gov*, November 2022. https://www.ers.usda.gov/data-products/poverty-area-measures.

US Department of Agriculture Rural Development. "Energy Efficiency and Conservation Loan Program." *USDA.gov*, January 16, 2015. https://www.rd.usda.gov/programs -services/electric-programs/energy-efficiency-and-conservation-loan-program.

US Department of Commerce. *1950 Census of Population*. Advance report no. PC-14. Bureau of the Census, July 1953. https://www2.census.gov/library/publications/decennial/1950/pc-14/pc-14-18.pdf.

US Department of Energy. "Communities LEAP (Local Energy Action Program)." *Energy.gov*, 2024. https://www.energy.gov/communitiesLEAP/communities-leap.

US Department of Energy. *National Transmission Needs Study: Draft for Public Comment*. US Department of Energy, February 2023. https://www.energy.gov/sites/default/files/2023-02/022423-DRAFTNeedsStudyforPublicComment.pdf.

US Department of Energy. "Weatherization Assistance Program." *Energy.gov*, December 2019. https://www.energy.gov/scep/wap/weatherization-assistance-program.

US Department of Energy and Environmental Protection Agency. "Benefits of ENERGY STAR Qualified Windows, Doors, and Skylights." *EnergyStar.gov*, accessed January 17, 2023. Archived at https://web.archive.org/web/20240125115518/https://www.energystar.gov/products/building_products/residential_windows_doors_and_skylights/benefits.

US Department of Energy and Environmental Protection Agency. *Consumer Messaging Guide for ENERGY STAR Certified Appliances*. US Environmental Protection Agency, December 2019. https://www.energystar.gov/sites/default/files/asset/document/ES_Consumer_Messaging_Guide_19-20-508.pdf.

US Department of Energy and Energy Efficiency and Renewable Energy. *Weatherization Energy Auditor Single Family: WAP Health and Safety Guidance*. Weatherization Assistance Program standardized curriculum, December 2012. https://www.energy.gov/sites/default/files/2016/07/f33/0_9_wap_health_safety_guidance_v2.0.pptx.

US Department of Energy and Environmental Protection Agency. "Super-Efficient Water Heater." *EnergyStar.gov*, accessed September 6, 2023. https://www.energystar.gov/products/energy_star_home_upgrade/super_efficient_water_heater.

US Department of Energy Alternative Fuels Data Center. "Batteries for Electric Vehicles." *AFDC.energy.gov*, accessed March 25, 2024. https://afdc.energy.gov/vehicles/electric_batteries.html.

US Department of the Interior, Indian Affairs. "Atlas of Oil and Gas Plays on American Indian Lands." *Bia.gov*, accessed October 29, 2024. https://www.bia.gov/bia/ots/demd/oil-gas-plays.

US Department of Transportation. "Pipeline Incident 20 Year Trends." Pipeline and Hazardous Materials Safety and Administration, August 21, 2023. https://www.phmsa.dot.gov/data-and-statistics/pipeline/pipeline-incident-20-year-trends.

US Department of Transportation Federal Highway Administration. "Bipartisan Infrastructure Law." *FHWA.dot.gov*, accessed October 30, 2023. https://www.fhwa.dot.gov/bipartisan-infrastructure-law/.

US Department of Transportation, Federal Highway Administration. "Interstate Frequently Asked Questions." *Highways.dot.gov*, updated June 30, 2023. https://highways.dot.gov/highway-history/interstate-system/50th-anniversary/interstate-frequently-asked-questions#question4.

US Energy Information Administration. "Electricity Data." *EIA.gov*, accessed October 26, 2023. https://www.eia.gov/electricity/data.php.

US Energy Information Administration. "Electricity: Form EIA-860 Detailed Data with Previous Form Data (EIA-860A/860B)." *EIA.gov*, September 22, 2023. https://www.eia.gov/electricity/data/eia860/.

US Energy Information Administration. "Frequently Asked Questions (FAQs)—How Many Gallons of Gasoline and Diesel Fuel Are Made from One Barrel of Oil?" *EIA.gov*, May 2023. https://www.eia.gov/tools/faqs/faq.php?id=327&t=9#:~:text=Petroleum%20refineries %20in%20the%20United,gallon%20barrel%20of%20crude%20oil.

US Energy Information Administration. "Frequently Asked Questions (FAQs)—What Countries Are the Top Producers and Consumers of Oil?" *EIA.gov*, 2023. https://www.eia.gov /tools/faqs/faq.php?id=709&t=6.

US Energy Information Administration. "Frequently Asked Questions (FAQs): What Is U.S. Electricity Generation by Energy Source?" *EIA.gov*, accessed October 26, 2023. https:// www.eia.gov/tools/faqs/faq.php?id=427&t=3#.

US Energy Information Administration. "Frequently Asked Questions (FAQs)—When Was the Last Refinery Built in the United States?" *EIA.gov*, July 8, 2022. https://www.eia.gov /tools/faqs/faq.php?id=29&t=6.

US Energy Information Administration. "How Much of U.S. Carbon Dioxide Emissions Are Associated with Electricity Generation?" *EIA.gov*, updated May 1, 2023. https://www.eia .gov/tools/faqs/faq.php?id=77&t=11.

US Energy Information Administration. *Monthly Energy Review*. US Energy Information Administration, November 2024. https://www.eia.gov/totalenergy/data/monthly/pdf /mer.pdf.

US Energy Information Administration. "Natural Gas Explained: Natural Gas Pipelines." *EIA. gov*, updated March 19, 2024. https://www.eia.gov/energyexplained/natural-gas/natural -gas-pipelines.php.

US Energy Information Administration. "Natural Gas Explained: Where Our Natural Gas Comes From." *EIA.gov*, updated December 21, 2023. https://www.eia.gov /energyexplained/natural-gas/where-our-natural-gas-comes-from.php.

US Energy Information Administration. "Natural Gas Gross Withdrawals and Production." *EIA.gov*, accessed October 16, 2023. https://www.eia.gov/dnav/ng/ng_prod_sum_dc _NUS_mmcf_a.htm.

US Energy Information Administration. "Nearly a Quarter of the Operating U.S. Coal-Fired Fleet Scheduled to Retire by 2029." *EIA.gov*, November 7, 2022. https://www.eia.gov /todayinenergy/detail.php?id=54559.

US Energy Information Administration. "Nonfossil Fuel Energy Sources Accounted for 21% of U.S. Energy Consumption in 2022." *EIA.gov*, June 29, 2023. https://www.eia.gov /todayinenergy/detail.php?id=56980.

US Energy Information Administration. "Nuclear Explained: Where Our Uranium Comes From." *EIA.gov*, accessed April 17, 2023. https://www.eia.gov/energyexplained/nuclear /where-our-uranium-comes-from.php.

US Energy Information Administration. "Nuclear Explained: U.S. Nuclear Industry." *EIA. gov*, accessed October 26, 2023. https://www.eia.gov/energyexplained/nuclear/us -nuclear-industry.php.

US Energy Information Administration. "Oil and Petroleum Products Explained—Refining Crude Oil—Refinery Rankings." *EIA.gov*, June 2022. https://www.eia.gov/energyexplained /oil-and-petroleum-products/refining-crude-oil-refinery-rankings.php.

US Energy Information Administration. "Residential Energy Consumption Survey (RECS)." *EIA.gov*, 2023. https://www.eia.gov/consumption/residential/.

US Energy Information Administration. "Table 4.1: Count of Electric Power Industry Power Plants, by Sector, by Predominant Energy Sources within Plant, 2013 through 2023." *EIA.gov*, 2023. https://www.eia.gov/electricity/annual/html/epa_04_01.html.

US Energy Information Administration. *Table HC11.1: Household Energy Insecurity, 2020.* US Energy Information Administration, March 2023. https://www.eia.gov/consumption /residential/data/2020/hc/pdf/HC%2011.1.pdf.

US Energy Information Administration. "Total Energy." *EIA.gov*, 2023. https://www.eia .gov/totalenergy/data/browser/index.php?tbl=T01.03#/?f=A&start=1997&end=2022& charted=1-2-3-5-12.

US Energy Information Administration. "U.S. Energy Facts Explained—Imports and Exports." *EIA.gov*, August 9, 2023. https://www.eia.gov/energyexplained/us-energy-facts /imports-and-exports.php.

US Energy Information Administration. "U.S. Field Production of Crude Oil (Thousand Barrels per Day)." *EIA.gov*, accessed April 17, 2023. https://www.eia.gov/dnav/pet/hist /LeafHandler.ashx?n=pet&s=mcrfpus2&f=a.

US Energy Information Administration. "U.S. Production of Petroleum and Other Liquids to Be Driven by International Demand." *EIA.gov*, April 4, 2023. https://www.eia.gov /todayinenergy/detail.php?id=56041.

US Environmental Protection Agency. "Air Emission Inventory: Air Pollutant Emissions Trends Data." *EPA.gov*, March 31, 2023. https://www.epa.gov/air-emissions-inventories /air-pollutant-emissions-trends-data.

US Environmental Protection Agency. "Clean Air Power Sector Programs: Progress Report— Affected Communities." *EPA.gov*, March 15, 2023. https://www.epa.gov/power-sector /progress-report-affected-communities.

US Environmental Protection Agency. "Coal Ash Basics." *EPA.gov*, February 27, 2023. https:// www.epa.gov/coalash/coal-ash-basics.

US Environmental Protection Agency. "EPA Administrator." *EPA.gov*, March 9, 2021. https:// www.epa.gov/aboutepa/epa-administrator.

US Environmental Protection Agency. "EPA Administrator Michael S. Regan to Embark on 'Journey to Justice' Tour Through Mississippi, Louisiana, and Texas." News release, November 6, 2021. https://www.epa.gov/newsreleases/epa-administrator-michael-s-regan -embark-journey-justice-tour-through-mississippi.

US Environmental Protection Agency. "Highlights of the Automotive Trends Report." *EPA.gov*, last updated December 20, 2023. https://www.epa.gov/automotive-trends /highlights-automotive-trends-report.

US Environmental Protection Agency. "National Enforcement and Compliance History Online Data Downloads." Enforcement and Compliance History Online, last updated October 2, 2024. https://echo.epa.gov/tools/data-downloads#downloads.

US Environmental Protection Agency. "National Priorities List (NPL) Sites—by State." *EPA.gov*, August 14, 2015. https://www.epa.gov/superfund/national-priorities-list-npl-sites -state.

US Environmental Protection Agency. "Petroleum Refinery National Case Results." *EPA.gov*, December 2022. https://www.epa.gov/enforcement/petroleum-refinery-national -case-results.

US Environmental Protection Agency. "Sources of Greenhouse Gas Emissions." *EPA.gov*,

last updated October 22, 2024. https://www.epa.gov/ghgemissions/sources-greenhouse
-gas-emissions.

US Environmental Protection Agency. "Where You Live." *EPA.gov*, March 2023. https://www
.epa.gov/trinationalanalysis/where-you-live.

US Environmental Protection Agency and Centers for Disease Control and Prevention. *Climate Change and Extreme Heat: What You Can Do to Prepare*. US Environmental Protection Agency, October 2016. Archived at https://web.archive.org/web/20161216180653
/https://www.cdc.gov/climateandhealth/pubs/extreme-heat-guidebook.pdf.

US Geological Survey. "Fifty Years of Glacier Change Research in Alaska." *USGS.gov*, September 28, 2016. https://www.usgs.gov/news/national-news-release/fifty-years-glacier
-change-research-alaska.

US Geological Survey. *Mineral Commodity Summaries 2022*. US Geological Survey, 2022.
https://www.usgs.gov/publications/mineral-commodity-summaries-2022.

US Geological Survey. "U.S. Geological Survey Releases 2022 List of Critical Minerals."
USGS.gov, February 22, 2022. https://www.usgs.gov/news/national-news-release/us
-geological-survey-releases-2022-list-critical-minerals.

US Global Change Research Program. *Climate Science Special Report: Fourth National
Climate Assessment, Volume 2: Impacts, Risks, and Adaptation in the United States*. US
Global Change Research Program, 2018. https://nca2018.globalchange.gov.

US Mine Safety and Health Administration. "Coal Fatalities for 1900 Through 2023." US Department of Labor, 2024. https://arlweb.msha.gov/stats/centurystats/coalstats.asp.

US National Research Council, Ford Foundation, and National Academy of Engineering
Study Committee on the Potential for Rehabilitating Lands Surface Mined for Coal in the
Western United States, eds. *Rehabilitation Potential of Western Coal Lands: A Report to
the Energy Policy Project of the Ford Foundation*. Ballinger Pub. Co., 1974.

Valdes, Renee. "How Much Are Electric Cars?" *Kelley Blue Book* (blog), July 10, 2023. https://
www.kbb.com/car-advice/how-much-electric-car-cost/.

Van Roosbroeck, Sofie, José Jacobs, Nicole A. H. Janssen, Marieke Oldenwening, Gerard
Hoek, and Bert Brunekreef. "Long-Term Personal Exposure to PM2.5, Soot and NOx in
Children Attending Schools Located near Busy Roads, a Validation Study." *Atmospheric
Environment* 41, no. 16 (May 1, 2007): 3381–94. https://doi.org/10.1016/j.atmosenv.2006
.12.023.

Vanatta, Max, Michael T. Craig, Bhavesh Rathod, Julian Florez, Isaac Bromley-Dulfano, and
Dylan Smith. "The Costs of Replacing Coal Plant Jobs with Local Instead of Distant Wind
and Solar Jobs Across the United States." *iScience* 25, no. 8 (August 2022): 104817. https://
doi.org/10.1016/j.isci.2022.104817.

Villarreal, M. Angeles, and Ian F. Fergusson. *The North American Free Trade Agreement
(NAFTA)*. Report no. R42965. Congressional Research Service, May 24, 2017. https://sgp
.fas.org/crs/row/R42965.pdf.

Vincent, John M. "How Long Do Electric Car Batteries Last?" *US News and World Report*,
August 16, 2023. Archived at https://web.archive.org/web/20231102202423/https://cars
.usnews.com/cars-trucks/advice/how-long-do-ev-batteries-last.

Voelcker, John. "Electric-Vehicle Battery Basics." *Car and Driver*, updated July 30, 2024,
https://www.caranddriver.com/features/a43093875/electric-vehicle-battery/.

Volcovici, Valerie. "Why Native American Tribes Struggle to Tap Billions in Clean Energy

Incentives." *Reuters*, September 8, 2023. https://www.reuters.com/sustainability/climate
-energy/why-us-tribes-struggle-tap-billions-clean-energy-incentives-2023-09-08/.

Voyles, Traci Brynne. *Wastelanding: Legacies of Uranium Mining in Navajo Country*. University of Minnesota Press, 2015.

Wallace, Nicholas, and Austin Irwin. "Longest-Range Electric Cars We've Ever Tested." *Car and Driver*, June 7, 2022. https://www.caranddriver.com/shopping-advice/g32634624
/ev-longest-driving-range/.

Wallace-Wells, David. *The Uninhabitable Earth: Life After Warming*. Penguin UK, 2019.

Walsh, Bryan. "Exclusive: How the Sierra Club Took Millions from the Natural Gas Industry—
and Why They Stopped." *Time*, February 2, 2012. https://science.time.com/2012/02/02
/exclusive-how-the-sierra-club-took-millions-from-the-natural-gas-industry-and-why
-they-stopped/.

Walsh, Mary Williams. "Congress Saves Coal Miner Pensions, but What About Others?" *New York Times*, December 24, 2019. https://www.nytimes.com/2019/12/24/business/coal
-miner-pensions-bailout.html.

Walter, Karla, Trevor Higgins, and Bidisha Bhattacharyya. *Electric Vehicles Should Be a Win for American Workers*. Center for American Progress, September 23, 2020. https://www
.americanprogress.org/article/electric-vehicles-win-american-workers/.

Walton, Robert. "Feds Approve 1st Phase of Largest US Wind Project in Wyoming." *Utility Dive*, January 19, 2017. https://www.utilitydive.com/news/feds-approve-1st-phase-of
-largest-us-wind-project-in-wyoming/434342/.

Waser, Miriam. "It's Hard to Build Transmission Lines in the Northeast, so 8 States Are Asking the Feds for Help." *WBUR*, July 5, 2023. https://www.wbur.org/news/2023/07/05
/electrical-transmission-power-lines-massachusetts-new-england-clean-energy-doe.

Watkins, Katie. "As Winter Storm Death Toll Exceeds Hurricane Harvey, Scope of Loss Becomes Clearer." *Houston Public Media*, March 30, 2021. https://www.houstonpublicmedia
.org/articles/news/energy-environment/2021/03/30/394592/remembering-the-lives
-lost-in-the-winter-freeze-as-death-toll-surpasses-that-of-hurricane-harvey/.

Watkins, Katie. "Port Arthur Residents Call for Civil Rights Probe into How Texas Has Handled Air Pollution in Their Neighborhood." *Houston Public Media*, August 18, 2021. https://www.houstonpublicmedia.org/articles/news/energy-environment/2021/08/18
/406282/port-arthur-residents-ask-epa-to-open-a-civil-rights-investigation-into-how
-texas-has-handled-air-pollution-in-their-neighborhood/.

WDTV. "Ex-Coal CEO Don Blankenship at End of Prison Term." *WDTV.com*, May 10, 2017. https://www.wdtv.com/content/news/Ex-coal-CEO-Don-Blankenship-at-end-of
-prison-term-421849674.html.

Weber, Maya. "The Changing Face of Energy." *S&P Global*, 2023. Archived at https://web
.archive.org/web/20230328020514/https://www.spglobal.com/en/research-insights
/featured/the-changing-face-of-energy.

Weber, Peter. "The Rise and Fall of Detroit: A Timeline." *Week*, January 8, 2015. https://
theweek.com/articles/461968/rise-fall-detroit-timeline.

Weigensamer, Florian, and Christian Krönes, dirs. *Welcome to Sodom*. Torch Films, 2018. https://www.torchfilms.com/products/welcome-to-sodom.

Weinberger, Kate R., Daniel Harris, Keith R. Spangler, Antonella Zanobetti, and Gregory A.

Wellenius. "Estimating the Number of Excess Deaths Attributable to Heat in 297 United States Counties." *Environmental Epidemiology* 4, no. 3 (April 23, 2020): e096. https://doi .org/10.1097/EE9.0000000000000096.

West Virginia Office of Economic Development. *West Virginia Weatherization BIL State Plan 7/1/2022–6/30/2027.* State of West Virginia Development Office, 2021. https://wvcad.org /assets/files/wap/Draft-Weatherization-BIL-State-Plan.pdf.

White House. "Fact Sheet: Biden-Harris Administration Driving U.S. Battery Manu-facturing and Good-Paying Jobs." *WhiteHouse.gov,* October 19, 2022. https://www .whitehouse.gov/briefing-room/statements-releases/2022/10/19/fact-sheet-biden -harris-administration-driving-u-s-battery-manufacturing-and-good-paying-jobs/.

Willson, Miranda. "Climate Law Boost for Renewables Hits Barrier on Tribal Lands." *E&E News,* December 22, 2022. https://www.eenews.net/articles/climate-law-boost-for -renewables-hits-barrier-on-tribal-lands/.

Wishart, Ryan. "Class Capacities and Climate Politics: Coal and Conflict in the United States Energy Policy–Planning Network." *Energy Research and Social Science* 48 (February 2019): 151–65. https://doi.org/10.1016/j.erss.2018.09.005.

Wolfe, Mark. "Press Release: Gasoline and Home Heating Will Cost More Than Christmas Gifts This Winter." *National Energy Assistance Directors' Association* (blog), November 22, 2021. https://neada.org/pr-gasolineandheat/.

Wood, Matthew S., David J. Scheaf, and Sean M. Dwyer. "Fake It 'Til You Make It: Hazards of a Cultural Norm in Entrepreneurship." *Business Horizons* 65, no. 5 (September 2022): 681–96. https://doi.org/10.1016/j.bushor.2021.12.001.

World Bank Group. "Dem. Rep. Cong Data." *WorldBank.org,* 2023. https://data.worldbank .org/country/congo-dem-rep.

World Health Organization. "Ambient (Outdoor) Air Pollution." *WHO.int,* December 19, 2022. https://www.who.int/news-room/fact-sheets/detail/ambient-(outdoor)-air -quality-and-health.

World Health Organization. "Dioxins and Their Effects on Human Health." *WHO.int,* Octo-ber 4, 2016. Archived at https://web.archive.org/web/20180427100129/http://www.who .int/news-room/fact-sheets/detail/dioxins-and-their-effects-on-human-health.

Wuebbles, D. J., D. W. Fahey, K. A. Hibbard, D. J. Dokken, B. C. Stewart, and T. K. Maycock. *Climate Science Special Report: Fourth National Climate Assessment, Volume 1.* US Global Change Research Program, 2017. https://science2017.globalchange.gov/.

Yahoo! Finance. "Ford Motor Company (F)." *Finance.Yahoo.com,* accessed March 11, 2024. https://finance.yahoo.com/quote/F?.tsrc=fin-srch.

Yahoo! Finance. "General Motors Company (GM)." *Finance.Yahoo.com,* accessed March 11, 2024. https://finance.yahoo.com/quote/GM?.tsrc=fin-srch.

Yahoo! Finance. "Mercedes-Benz Group AG (MBG.DE)." *Finance.Yahoo.com,* accessed March 11, 2024. https://finance.yahoo.com/quote/MBG.DE?.tsrc=fin-srch.

Yahoo! Finance. "Tesla, Inc. (TSLA)." *Finance.Yahoo.com,* accessed March 11, 2024. https:// finance.yahoo.com/quote/TSLA?.tsrc=fin-srch.

Yahoo! Finance. "Toyota Motor Corporation (TM)." *Finance.Yahoo.com,* accessed March 11, 2024. https://finance.yahoo.com/quote/TM?.tsrc=fin-srch.

Yozwiak, Maddy, Sanya Carley, and David M. Konisky. *Clean and Just: Electric Vehicle Inno-*

vation to Accelerate More Equitable Early Adoption. Information Technology and Innovation Foundation, June 27, 2022. https://itif.org/publications/2022/06/27/electric-vehicle-innovation-to-accelerate-more-equitable-early-adoption/.

Zirogiannis, Nikolaos, Alex J. Hollingsworth, and David M. Konisky. "Understanding Excess Emissions from Industrial Facilities: Evidence from Texas." *Environmental Science and Technology* 52, no. 5 (March 6, 2018): 2482–90. https://doi.org/10.1021/acs.est.7b04887.

Zullo, Robert. "Across the Country, a Big Backlash to New Renewables Is Mounting." *Virginia Mercury* (blog), February 23, 2023. https://www.virginiamercury.com/2023/02/23/across-the-country-a-big-backlash-to-new-renewables-is-mounting/.

INDEX

Page numbers in italics refer to figures.

Abandoned Land Mines Reclamation Fund, 241n22

activism: and clean-energy transition, 9; and coal, 243n49; community, 45–46; conservative, 147; and energy justice, 11; environmental, 39; and pollution protests, 46

AES Indiana (formerly Indianapolis Power and Light), 51–52

Afrewatch, 173

African Americans, 3, 24, 52, 99, 115, 128, 179. *See also* Black Americans; communities of color and people of color

Agbogbloshie, Ghana, as e-waste disposal site, 21, 195–98

air pollution, x, 16, 197, 206; abatement, 212; conventional, 43; and energy infrastructure, 132–33, 144; and fossil fuels, 26, 34, 47–48, 50, 66–67, 240n1; and health, 112–13; from oil refineries, 23–24, 41–46; reduction, 110, 112. *See also* smog; *and specific pollutant(s)*

Alternative Motor Vehicle Credit, 119

American Battery Materials Initiative, 175

American Council for an Energy-Efficient Economy, 82, 96

American Electric Power, 213

American Lung Association, 47–48

Amnesty International, 173–74

Anadarko Petroleum Corp., 35, 63, 235n52

Anderson, Carrol, 2

Apex Clean Energy, 129–30, 149–50

appliances (household), energy-efficient, x, 10, 83, 95–96, 108–9, 114–15, 118, 251n37

Army Corps of Engineers (US), 40, 157

Arrowhead Environmental Partners, 52–53, 240n118

arsenic, 47, 50–52, 67

Asian Americans, 122–23

Attari, Shahzeen Z., 95

automotive industry: and electric vehicles, 11, 21, 163–68, 174–92, 207–9; employment statistics, 265n67; and fuel economy, 204; Mexican trade in, 265n67. *See also* electric vehicles (EVs)

autonomy, 131, 211, 215

Beckfield, Jason, 243n43

Bell, Shannon Elizabeth, 64, 68, 243n49

Biden, Joe, 11–12, 24, 39, 111, 219, 241n12, 251n37

Bipartisan Infrastructure Law, 175
Black Americans, 23–24, 42–45, 49, 52, 81, 122–23, 192, 194; and home loans and ownership, 238n100. *See also* African Americans; communities of color and people of color
Blacksmith Institute, 197
Blankenship, Don, 30
Borenstein, Severin, 125
Bosch, 189
BP, 42–44, 122
Breaking Bad (television series), 107, 250n12
bridge collapses, 255n25
Brugge, Doug, 36
Bullard, Robert D., 25
Bureau of Labor Statistics (US), on automotive-sector employment, 265n67
burning trash, for heat, 88, 246n22
Burns, Steve, 184–85
business and industry sectors, 225, 254–55n14. *See also* automotive industry; electric and gas utilities; energy sector; oil and gas industry; private sector; transportation sector
byproducts, 16–17, 50, 171. *See also* waste

cadmium, 50, 196
California Electric Vehicle Program, 119–20
Canada, 28, 35, 38, 151, 153, 176–77, 203, 270n150
Canadian Critical Minerals Strategy, The, 176–77, 264nn63–64
Cancer Alley (along Mississippi River between Baton Rouge and New Orleans, Louisiana), 4–5, 24–25, 160
Canis, Jon, 116
carbon: abatement, 165–66, 204; emissions, 4, 106, 133–39, 158, 206, 210–12; footprints, 105–6, 204–5; hydro-, 196; mitigation and reduction, 105–6, 133. *See also* net-zero emissions
carbon capture and storage (CCS), 5, 10, 20, 133–36, 139–41, 153, 158–61, 204–5, 211–12, 217

carbon dioxide: captured and stored, 20; and climate change, 6; emissions, 32, 106, 158–61, 201–2; and health threats, 6; pollution, 46–47
carbon monoxide, 43, 68
CCS. *See* carbon capture and storage (CCS)
CDC. *See* Centers for Disease Control and Prevention (US)
Centers for Disease Control and Prevention (US), 30, 93–94
CF Industries, 160
Cha, J. Mijin, 77
Chalmers Automobile Company, 178–79
Chevron, 24–25, 43–44, 63, 122
China: and carbon dioxide from coal, 46; coal reliance, 58; and electric vehicles, 163, 165, 171, 175, 177, 196; and greenhouse gas emissions, x; minerals in, 264n59, 270n150; and oil production, 203
CHIPS and Science Act, 175, 264n58
Chisholm Legacy Project, 122–23, 253n64
Chrysler Corporation, 178–82, 264n66
civil rights, 46, 53, 237n89
Civil Rights Act, 53
Clean Air Act, 26, 43–47, 58, 233n23
clean energy: activism, 9; barriers to adoption for certain communities/populations, 10–11, 105–28, 210, 218–25; benefits, x, 110–13; challenges, 26, 128, 132, 205–6, 217–18; and climate change, x, 105, 128, 153, 201–5; costs and benefits, uneven geography of, x, 37, 62, 76, 110–17, 121, 124–25, 130, 210–13; cross-sectoral solutions for, 225; and decarbonization, 128, 199, 203–5, 223; and demand-side management, 112; and disproportionate burdens, 9, 205; and economy, 20–21, 121–22, 128; electricity generated from, ix–x; embraced, 11; and energy independence, 33; and energy infrastructure, 20–21, 129–61; and energy justice, 8, 22, 222, 225; and energy systems, 22, 205–10; and equality, 41; equitable and just transition to, 22, 132, 210, 218–

25; future of, 22, 201-25; geographic constraints on, 120-21, 128; graphite for, 270n150; incentives for, access as uneven, 117-20, 124, 130; inclusivity for, 221; industrial-scale sites for, 137; inequitable and unjust outcomes of, 9, 13-14, 205, 210-11; inequities of, 13-14, 105-28; and infrastructure, 20-21, 217-18; job opportunities in, 72-73, 121-23, 244n58; and justice, 12; mandates, 33, 151; markets, 125-28; minerals for, 270n150; and natural gas, 33; partnerships, 221; and renewable energy, 116, 118, 127; shifts to, 4, 20-21; technologies, 19-20, 105-28, 199, 206; trade-offs, 169, 216-18; transition, x, 4, 8-9, 13-15, 22, 25-26, 59, 78, 105-28, 132, 149, 153-58, 201-25; as vital and urgent, 9. *See also* renewable energy

Clean Water Act, 26, 40, 47

Cleveland State University, 183-84

climate change: and CCS, 160-61; challenges of, 27, 200-205, 224; and clean energy, x, 105, 128, 153, 201-5; and climate isolationism, 11; and climate justice, 10-11, 77; and climate risks, 202-3; and climate science, ix, 201-3; and decarbonization, 7-8, 105, 110, 128, 137, 200, 203-5, 224-25; as disruptive, 224; economic costs of, 203; effects of, ix; and energy infrastructure, 161; and energy insecurity, 18, 102; and energy justice, 135; and energy system changes, 5-8, 71, 223; and energy transition, 71, 200, 218, 224-25; and equity, 225; and extreme weather, ix, 3, 5-6, 62, 69, 71, 88, 102, 202-3; and fossil fuels, 27, 46-47, 60-62, 71, 105, 150, 160-61; Gallup poll, 258-59n65; global economic damages of, 203; and health, 6; impacts of, 202-3; and inequalities, 77; international negotiations for, 61; mitigation and adaptation strategies, 6-7, 133, 224; and natural disasters, 2, 202; and new energy infrastructure, 161; and pipelines, 39; and politics, 150-53;

and renewable energy, 7-8; resilience and adaptive capacity for, 224; risks of, 202-3; severity and urgency of, 218; technologies, 10-11, 105, 128; and uninhabitable Earth, 218; unknowns of, 202-5; urgency of, 203. *See also* global warming

Clinton, Hillary, 70

coal: and activism, 243n49; addictions, 56-58; ash, 16-17, 50-53, 67-69; communities, 4, 17, 29-31, 58-59, 61, 62-65, 68, 71-72, 76-78, 206-8; companies, 58, 63, 68, 206, 241n22; consumption, 56, 203; costs and benefits of, 37; and cultural identity, 68; decline of and as dying industry, 4, 29, 56-58, 75, 136, 152-53, 203, 206-8; dependence and reliance on, 56, 203, 206-7; economic viability declining, 64, 68, 71-72, 206-7, 240n1; and energy consumption, *28*; expansion, 57; and local hegemonic masculinity, 243n49; and natural gas, 16, 28, 32, 46, 154, 234n37; and natural gas, shifts, 234n37; pollution, 16, 32, 46-47, 50-52, 67; power plants, x, 4, 50-52, 58, 62-65, 127, 207, 223, 234n36, 253n72; production, 29, 31, 56-57, 59-61, 63; products, 122; reliance on, 58; and sacrifice/abandonment, 56; shifts away from, 71-72, 203, 206-7; stranded assets of, 253n72; sulfur dioxide content, 58, 233n23; trade-offs, 67-68; waste from, 16-17, 50-53, 67-69. *See also* methane emissions

coal mining: accidents, 29-30; and acid mine drainage, 67; adverse consequences of, 16, 29-31; bankruptcies, 4; closures, 4, 29, 206-7, 210, 216-17; and cultural identity, 68; effects, 37; employment, 4, 31, 57, 61-64, 70-76, 152-53, 207; environmental issues, 31, 37; fatalities, 233n26; health issues, 30-31, 206-7; history, 29, 31; industry's overall decline, 29-30; pension funds, 241-42n22; and politics, 59; and sacrifice areas, 31; and sulfur dioxide regulation, 233n23 cobalt: blue color of, 262n27; for clean energy, 270n150; for electric vehicles, 21, 168–77; global reserves by country, 170; in history, 262n27; and lithium in batteries, 166; mining, artisanal, 171–73; mining, and auto industry, 174–77; mining, in DRC, 21, 168–77; mining, in poor locations and conditions, 198; on native lands, 175–76; sustainable mining and production of, 175

communities of color and people of color, 2–5, 16–18, 23–26, 42–44, 48–53, 81, 102, 113–15, 121–23, 128, 212–14. See also African Americans; Black Americans; disparities; Hispanic communities and populations

Community Local Energy Action Program (LEAP), 78, 245n69

Conference of the Parties (COP), 241n12

Cong, Shuchen, 87

Congo. See Democratic Republic of the Congo (DRC)

Congressional Research Service (CRS), 118

ConocoPhillips, 43–44, 63

consumer appliances, energy-efficiency standards for, 251n37

Consumer Reports, 165

cooling: Energy Star recommendations for, 83; heat pumps for, 137; and household emissions, 105; repairs for, 96. See also HVAC (heating, ventilation, and air conditioning)

COP. See Conference of the Parties (COP)

copper, 167, 175–76, 270n150

COVID-19 pandemic, 91, 101, 184, 201–2, 245n4

creosote, as carcinogenic chemical used to preserve wood, 23

cross-subsidization, 120, 126–27, 137, 253–54n73; of solar power, 253–54n73

CRS. See Congressional Research Service (CRS)

Dakota Access Pipeline (DAPL), 16, 38–42, 132

Davis, Lucas W., 127, 141

decarbonization, ix, 6–11, 17–21; challenges, 106, 199–200, 204–5, 215–16, 222–25; and clean energy, 128, 199, 203–5, 223; and climate change, 7–8, 105, 110, 128, 137, 200, 203–5, 224–25; costs of, 17–18; deep, 61, 107, 128, 137; disparities and divide, 199–200; and economy, 169, 199, 203–4; and electric vehicles, 19, 169; and energy infrastructure, 17–18, 138–39; and energy systems changes/transformations, 128, 203–4; and energy transition, 199–200; and equity, 110, 224–25; and fossil fuels, 61–62; goals, 7, 137–39, 141, 209, 224; injustice and domination patterns in, 199; policy, ix; as process and end goal, 224; and social multiplier effect trade-offs, 243n43; and solar electricity, 20; successful, 203–4; technologies, 105–8; and wind electricity, 20

Deep South Center for Environmental Justice, 160

Defense Production Act (DPA), 175

Democratic Republic of the Congo (DRC): cobalt mining in, 21, 168–77; and copper production, 270n150

Department of Agriculture (US), Energy Efficiency and Conservation Loan Program, 118–19, 252n54

Department of Energy (US), 35; Community Local Energy Action Program (LEAP), 78; and electric vehicles, 175; Office of Energy Justice and Equity, 11. See also National Renewable Energy Laboratory

Department of Interior (US), 24

Department of Transportation (US): pipeline statistics, 235n63

Devon, 63

digital economy, minerals essential to, 264n59

disadvantaged communities and populations, xii, 10–11, 13, 27, 41–42, 53, 87, 113–19, 169, 193–94, 219–22. See also disparities

disparities, x, 2–4, 7–22, 26–27, 81, 108, 110, 121–25, 128, 150, 169, 199–200, 214, 222. *See also* communities of color and people of color; disadvantaged communities and populations; low-income communities and populations; marginalized communities and populations; oppression; overburdened communities and populations; poverty; underserved communities and populations; vulnerable communities and populations

distributive and distributional justice, 9–10, 135, 220, 223

DPA. *See* Defense Production Act (DPA)

DRC. *See* Democratic Republic of the Congo (DRC)

dryer vents (household), for heat, 88, 216, 246n22

Duke Energy, 69, 122, 213, 243n52

Dumping in Dixie (Bullard), 25

DuPont, 24–25

Earthjustice, 53

economic development, 13, 71, 74–78, 178, 217–19, 223

economy: and clean energy, 20–21, 121–22, 128; and decarbonization, 169, 199, 203–4; and fossil fuels, 4, 16, 20, 25–37, 41, 46, 62–73, 218–19, 223; global, 202–3; and oil refineries, 43. *See also* infrastructure

Edwards, Sandra, 23

electric and gas utilities, 16, 111, 122, 125, 129, 139–40, 163, 213; net favorability toward, 133–34, *134*; and utility death spiral, 127. *See also* electric power; electricity; oil and gas industry

electric power, 64–65, 122, 142, 213, 234n37. *See also* electric and gas utilities; electricity

Electric Reliability Council of Texas (ERCOT), 1–2, 156–57

electric vehicles (EVs), 269n120; access to, 8, 19, 21–22, 124–25, 166, 168, 192–94; and auto industry, effects on, 11, 21, 163–68, 174–92, 207–9; autono-

mous mode in, 164; availability of, 120, 124; batteries for, 166, 168, 170, 174–76, 185, 209, 262n21; benefits of, 209, 212; charging, 18–19, 124, 182, 194; as commonplace, x; concerns over vehicle range and sticker price, 164; conditions and complications accompanying, 19, 166, 192–93, 200; consumers and owners of, 21, 163–65, 168, 192–94, 198, 204; controversies over, 221; costs and benefits of, 212–13; costs and prices of, 19, 107, 109, 114, 119, 164–66, 192; and decarbonization, 19; disenfranchisement and abandonment issues of, 192, 212–13; domestic supply chain for, 209; in electrified transportation sector, 18, 137; and environmental messages to buyers, 163; equity and justice complications of, 21–22, 166, 198–99, 212–13; and e-waste, 21–22, 168, 195–98, 212; first, in US, 163; functionality of, 164; increased sales of, 136, 204; and injustices, 21–22, 163–200, 212–13; as internal combustion engine substitute, 60–61, 114, 153, 188, 199, 207–9; and job opportunities, 185, 189, 209; marketing for, 163–64; in nascency, 207–8; next generation of, 190–91; and oil and gas companies, 260n1; performance, 164–65; plug-in hybrid (PHEV), 163; and policymaking, 165–66, 189; as renewable-energy technology, 199, 221; and social/environmental exploitation and injustices, 21–22, 163–200, 212–13; and solar energy access, 125; supply chains for, *167*, 175–92, 198; tax credits for, 19, 113–14, 118–20, 163, 165, 193, 209; as technological solutions, 10; technology life cycle of, 167–69; trade-offs, 161, 169. *See also* transportation sector

electricity: costs, 2–3, 111, 124–27, 142; demand for, 136–37; from fossil fuels, 136, 204; generation, x, 4, 56–57, 63–64, 136–37, 141–42, 153, 155, 157, 204, 206, 224; interstate transmission of, 256n36; markets, 125–28; sources of

electricity (*continued*)
in future, 154, 224; transmission of, 153–58, 256n36. *See also* electric and gas utilities; electric power; transmission lines

electronic waste. *See* e-waste

energy: affordable, 3, 9, 101–2, 215, 222; burdens and expenditures by household types, *82*; business and industry sector ratings, *134*, 254–55n14; consumption, 27–28, *28*; crises, 179–80; disparity, 4; disposal, 27; distribution, 27; domestic and global markets for, 32; efficiency, x, 10, 83, 95–96, 105–28, 251n37; extraction, 16, 27–37, 64–67; for historically disadvantaged, xii, 55–78; independence, 33; lack of, 3; life without, 17–19, 79–103, 212–16; marginalization, 214–15; markets, 11, 17, 32, 97, 125–28, 222; processing, 27; production, 66–67; profit-oriented corporations, 223–24; reliable, 15, 101–2, 214–15, 222; savings, 95, 108, 112; siting, 132–33, 141; social and economic challenges of, 4; sources and consumption, *28*; waste, 25. *See also* clean energy; energy infrastructure; energy systems; energy transition; power; renewable energy; solar energy and power; wind energy and power

Energy Efficiency and Conservation Loan Program, 118–19, 252n54

Energy Information Administration (US), 28, 82

energy infrastructure: attitudes toward, 145–46; building and deploying, 136; challenges of, 20–21, 128, 135–36, 153, 161, 217–18; and clean energy transition, 20–21, 129–61; and climate change, 161; as critical, 140; and decarbonization, 17–18, 138–39; disagreement and discourse about, 20–21, 135; and environmental injustices, 37–41; and environmental justice, 13; hosting, 133; inclusivity of, 149; industrial-scale, 37, 219; inequities of, 161; investments in, 20; location of, 224; massive scale

of, 132, 135–36, 139, 161; nuclear power in, 136; opposition to, 132, 151; and partisanship, 135, 151–53; and policymaking, 141–43; and politics, 135, 141, 150–53, 161; privately owned and operated, 139, 256n32; procedural justice for acceptance of, 149; and renewable energy, 129–32, 141–53; in sacrifice zones, 37–41; salience of, 152; siting of, 20–21, 128–61, 210, 217, 222; surveys about, 153–54; and technologies, 224; trade-offs, 161, 217; types of, 135, 145–46, 151, 153–54; in US, 136–41, 157

energy injustices: and energy transition, 12, 14; geographic, 168–69; health effects of, 14; as status quo and likely future, 12. *See also* energy justice; environmental injustices

energy insecurity, 3, 8, 17–19, 79–103; avoiding, 94–101; and climate change, 18, 102; and energy equity gap, 87; and energy poverty, 80, 86–87; and energy transition, 209–16; health consequences for, 93–94, 101; household coping strategies for, *92*. *See also* energy poverty; utility disconnections and shutoffs

energy justice: and activism, 11; aspirations for, 12; and clean energy, 8, 22, 222, 225; and climate change, 135; commitment to, 225; and disproportionate burdens, 9–10; and energy transition, 7–16, 22, 220, 225; for marginalized, 9; and policymaking, 9–12, 223; and politics, 11–12. *See also* energy injustices; environmental justice

Energy Keepers, Inc., 117

energy poverty, 3, 14–15, 80, 86–87. *See also* energy insecurity; utility disconnections and shutoffs

energy sector: net favorability toward, 133–34, *134*; political and economic dominance of, 70; trust in, lack of, 133–34, 148, 153; workforce diversity in, lack of, 121–23

energy systems: adversarial interactions in, 221; challenges of, 15–16,

215-16; and clean energy, 22, 205-10; and climate change, 5-8, 71, 223; and decarbonization, 128, 203-4; disproportionate effects of, x, xii, 7-16, 53, 62, 113; equitable and just transitions in, 218-25; hidden from plain sight, 37-38; history of, 53; inclusivity of, xi-xii; inequities and injustices of, 7-16, 25-26; infrastructure of, 16; lack of agency and opportunity to participate in, 215-16; life cycle of, 16; and lower energy bills, challenges of, 103; and marginalization, 214-15; models for, 139; opposition and skepticism toward, 215-16; and policymaking, 206; and politics, 27, 206; prioritizing people in, 225; privately owned and operated, 140; and trade-offs, 14, 210; transformations in, x

Energy Transfer Partners, 39-41

energy transition: challenges of, xi-xii, 15-16, 27, 200, 205-6, 218-19; and clean energy, 22, 201-25; and climate change, 71, 200, 218, 224-25; controversies, 221; costs and benefits, 7, 13-14, 153, 210-13, 220, 222-23; and decarbonization, 199-200; decision-making processes of, 15; disparities of, 222; disproportionate and negative affects/effects of, xi, 7-16, 53, 62, 199, 209-13, 220; and energy injustices, 12, 14; and energy insecurity, 209-10; and energy justice, 7-16, 22, 220, 225; and environmental regulations, 199; equitable and just, 22, 218-25; as fair, transparent, and inclusive, 76-77, 221-22; and fossil fuels, 17, 55-78; and future, 22, 201-25; global, 199; inclusive and fair, 76-77; inclusivity of, 76-77, 221-22; infrastructure, 19-21, 129-61, 221; and injustices, 1-22; as just, 11, 17, 77-78; and opportunities, 8, 222-23; opposition to, 68; pace and success of, 11; policymaking, 8, 77-78, 219-20, 225; politics of, 20, 56, 77, 150-53; ramifications of, 56; trade-offs, 14, 210, 216-18; unknowns of, 202-5; vul-

nerabilities of, 13, 199. See also clean energy: transition

environmental groups and organizations, 46, 53, 158, 176

environmental injustices, 16-17, 22-53, 199, 210-14. See also energy injustices; environmental justice

Environmental Integrity Project, 46, 52

environmental justice, 5, 12-13, 23-26, 42-45, 51-53, 149, 160, 168-69, 221. See also energy justice; environmental injustices

environmental problems: Gallup poll, 258-59n65. See also air pollution; climate change; global warming; water pollution

Environmental Protection Agency (EPA), 23-25

EPA. See Environmental Protection Agency (EPA)

equality, and clean energy, 41

equity: and clean energy, 22, 218-25; and climate change, 225; and decarbonization, 110, 224-25; and justice, of electric vehicle complications, 166; and justice, of energy transition, 218-25; and justice, energy-based, xi, 11-12, 22; and justice, environmental, 24. See also inequities

ERCOT. See Electric Reliability Council of Texas (ERCOT)

European Food Safety Authority, 197

Evrard, Daniel Alain, 243n43

EVs. See electric vehicles (EVs)

e-waste, 21-22, 168-69, 195-98, 212. See also wastelanding

Exelon, 122

exploitation, 21-22, 41, 169, 173-74, 206-7

Exxon Mobil, 24-25, 42-44, 46, 63, 122

Fair Housing Act (1968), 238n100

Federal Coal Mine Health and Safety Act, 30-31

Federal Energy Regulatory Commission (FERC), 156-57; jurisdiction, 256n36; winterization recommendations, 229n1

FERC. *See* Federal Energy Regulatory Commission (FERC)

Fighting King Coal (Bell), 68, 243n49

fireplaces, for heat, 2, 88, 246n22

floods and flooding, ix, 5–6, 69, 71, 202

focus groups, 17, 55–56, 71–76, 185, 245n1, 268n106

Ford, Henry, 178

Ford Motor Company, 165, 174–75, 178–82, 189, 209, 260n4, 264n66, 266nn83–84

fossil fuels: and air pollution deaths, globally each year, 240n1; benefits, 27; burdens of use, 27, 46, 48; byproducts, 50; and clean energy, 149; and climate change, 27, 46–47, 60–62, 71, 105, 150, 160–61; communities, 4–5, 17, 26, 41, 59–76, 76–78, 177–78, 210–11, 215, 219; consumption, 27–37, 53; costs of, 7; and cultural losses, 74–75; and decarbonization, 61–62; and disadvantaged, negative affects on, 27, 41–42, 48; disposal, 27; distribution, 27, 53; drawing down, commitment by countries, 241n12; and economy, 4, 16, 20, 25–37, 41, 46, 62–73, 218–19, 223; electricity from, 136, 204; and employment, 63–64, 66, 69–73, 122, 244n58; and energy consumption, 28, 203–4; and energy transition, 17, 55–78; and environmental conditions/effects, 27, 62–71, 160; environmental injustices of, 16–17, 22–53, 62–71, 212; extraction, 8, 27–37, 41–42, 53, 61, 77; fate of, 203; and health issues, 27, 53, 113, 160, 214–15; impacts of, 56; infrastructure, 6–7, 41–42; life cycle, 27; and nuclear power, 142; political dynamics of, 27; pollution from, 4–5, 29, 46–53, 66–67, 112–13, 214–15; power plants, 106, 112–13, 133, 144–45, 158–59, 210; problems with use of, 9; processing, 4–5, 27, 41–48, 53; production, 28–29, 33, 42, 53, 59–60, 66, 141; products, 38; record levels of production of, ix–x; reliance on with steep price, 26, 53, 210–11; and renewable energy, 73; and sacrifice zones, 16–17, 22–53; shifts and transition away from, 7–8, 17–18, 53, 61, 78, 112–13, 132, 153, 203, 221, 224; shut downs and loss of, 71–76; siting and operations of, 66; and social disruption/losses, 74–75, 218–19; and social identities, 78; trade-offs, 66, 69–70; and vulnerable populations, impositions on, 27, 41; waste, 46–53

Foundation for Energy Security and Innovation, 175

Fox, Josh, 34

fracking, 4, 16, 28–34, 37, 40, 50, 59, 65–66, 132, 203

France, 77, 106

Fry, Ann, 131

future: of clean energy transition, 22, 201–25; disparities, 214, 222; and energy injustices, 12; and energy transition, 22, 201–25; injustices in, 12; just and equitable, 22; and lower energy bills, challenges of, 103; and status quo, 12; as uneasy and uneven, 22, 201–25

Gallup, Inc. polls, 133–34, 148, 150, 254–55n14, 258–59n65

gas: burners, for heat, 88; prices, 65, 234n36. *See also* electric and gas utilities; natural gas; oil and gas industry

Gasland (anti-fracking film), 34

Gazmararian, Alexander, 215

General Motors (GM), 174–75, 178–88, 191, 208–9, 264n66, 266n83

geographic mobility, 8

Geological Survey (US), on minerals essential to digital economy, 264n59

geothermal operations, 10

Germany, 180, 196–97

Global North, 128, 199

Global South, 199

global warming, ix, 5–6, 202, 209; Gallup poll, 258–59n65; Pew research, 259n66. *See also* climate change

globalization, 177, 196
GM. *See* General Motors (GM)
Goble, Rob, 36
gold, stripped and repurposed, 196
graphite, for clean energy, 270n150
greenhouse gas emissions, 234n37; and
    climate change, x, 132–33, 201, 204–
    5; from human and household con-
    sumption, 105–6; increasing levels of,
    201; mitigation, 5; and policymaking,
    106; reduction, x, 106, 203–5
Greenland ice sheet, 202
Groth, Kimberly, 131, 146–47, 155
Guterres, António, 56–57, 240n1

Haaland, Deb, 24
Harvard Law Today, 118, 252n52
Hausman, Catherine, 127, 141
Hawkins, Daniel Alexander, 57
health: and air pollution, 112–13; and
    carbon dioxide, threats to, 6; and cli-
    mate change, 6; consequences, 113;
    and energy injustices, 14; and energy
    insecurity, 93–94, 101; and fossil fuels,
    27, 53, 113, 160, 214–15; and fracking,
    34; and highways, living near, 47, 49–
    50, 62; issues, 35–37, 53, 93; mental,
    66, 93, 113, 191–92; and pollution, 45–
    48; and technologies adopted, 169. *See
    also* public health; well-being
heat: Energy Star recommendations
    for, 83; fireplaces for, 2, 88, 246n22; in
    homes, 105–6; in homes, and risky
    temperature strategies, 246n22; and
    household emissions, 105; pumps, x,
    10, 12, 19, 107–8, 112, 137, 212; repairs
    for, 96. *See also* HVAC (heating, venti-
    lation, and air conditioning)
heat waves. *See* global warming
Henderson, John, Col., 40
Hispanic communities and populations,
    23, 43–44, 49, 81, 108, 115, 122–23, 194.
    See also Latino communities and
    populations
HOLC. *See* Home Owners' Loan Corpo-
    ration (HOLC)

Home Owners' Loan Corporation
    (HOLC), and redlining to demarcate
    hazardous areas, 238n100
home ownership and homeowners, 98–
    99, 115, 121, 124, 204, 238n100, 250n12.
    See also land ownership and land-
    owners
homelessness, and housing precarity, 15
Honey Creek wind farm (Crawford
    County, Ohio), 129–33, 140, 143–46,
    155, 254n2
housing precarity and homelessness, 15
Human Capital Index, 169
Hurricane Katrina, 2
hurricanes, 5–6, 71, 202
HVAC (heating, ventilation, and air con-
    ditioning), 20, 82, 95, 107, 111–12. *See
    also* cooling; heat
hydraulic fracturing. *See* fracking
hydrocarbons, 196
hydrogen sulfide, 43
hydropower, 142

identity: cultural, 68, 223–24; and self,
    75–76; social, 78
IEA. *See* International Energy Agency
    (IEA)
inclusivity: for clean energy, 221; of en-
    ergy infrastructure, 149; of energy
    systems, xi–xii; of energy transition,
    76–77, 221–22; and justice, xi–xii
Indian Reorganization Act, 116–17
Indianapolis Power and Light (now AES
    Indiana), 51–52
indigenous people, 16, 24, 35–37, 40–41,
    60, 64–65, 116–17, 123, 169, 175–77, 198–
    99, 252n47
industrialization, 4–5, 37, 49, 105–6, 120
inequalities: and climate change, 77; as
    human-made, 15; income, 8–9, 15, 77,
    225; and injustices, 225. *See also* in-
    equities; injustices
inequities: of clean energy, 13–14, 105–
    28; of energy infrastructure, 161; of
    energy systems, 7–16; and injustices,
    energy-based, 7–16, 25–26, 135, 205;

inequities (*continued*)
market dynamics and underlying conditions of, 19–20; spatial transboundary, 13–14; for technologies, access to, 10–11, 19–20, 105–28. *See also* equity; inequalities; injustices

Inflation Reduction Act (IRA), 11, 118, 152, 175, 182, 193, 219

infrastructure: and clean energy, 20–21, 217–18; and decarbonization, 17–18, 139; of energy systems, 16; and energy transition, 19–21, 129–61, 221; energy-efficient, 6–7; maintenance of, 138; and stranded assets, 253n72. *See also* bridge collapses; economy; energy infrastructure

Infrastructure Investment and Jobs Act, 140

Initiative for Responsible Mining Assurance, 176

injustices: American, 1–22; and domination, 199; and electric vehicles, 21–22, 163–200; and energy transition, 1–22; in future, 12; and inequalities, 225; and inequities, energy-based, 7–16, 25–26, 135, 205, 225; racial, and income inequality, 77, 225; and social/environmental exploitation, 21; as status quo, 12. *See also* energy injustices; environmental injustices; inequalities; inequities; justice

Innovative Technologies Loan Guarantee Program, 175

Intergovernmental Panel on Climate Change (IPCC), 5, 61–62, 71, 201

International Energy Agency (IEA), 60–61, 201–2

IPCC. *See* Intergovernmental Panel on Climate Change (IPCC)

IRA. *See* Inflation Reduction Act (IRA)

Jackson, Lisa, 24

Japan, 106, 179–80

justice: and benefits of certain government programs directed to disadvantaged, 219; and clean energy, 12, 22; climate, 10–11, 77; and energy transition, 11, 17, 77–78; and equity, of electric vehicle complications, 166; and equity, of energy transition, 218–25; and equity, energy-based, xi, 11–12, 22, 218–25; and equity, environmental, 24; and inclusivity, xi–xii; and policymaking, 223; racial, 15. *See also* distributive and distributional justice; energy justice; environmental justice; injustices; procedural justice; recognition justice

Justice40 Initiative, 11, 219

Kelley, Hilton, 45

Kennedy, Ted, 147

Kerr-McGee, 35

Keystone XL pipeline, 38–42, 132, 151

King, Martin Luther, Jr., 12

Koch, David, 147

Koch, William, 147

Koch Industries, 43–44

Krause, Dunja, 77

land dispossession, 37

land ownership and landowners, 39, 129–33, 145–49, 155, 158, 211, 215, 257n48. *See also* home ownership and homeowners

land pollution, 41, 50, 66–67, 155, 176, 197–98

Latino communities and populations, 123, 192. *See also* Hispanic communities and populations

LEAP. *See* Community Local Energy Action Program (LEAP)

LED light bulbs, 111, 120–21

Lerner, Steve, 25

LG Chem, 174, 185

light bulbs, 111, 120–21

LIHEAP. *See* Low Income Home Energy Assistance Program (LIHEAP)

lithium: for clean energy, 270n150; and cobalt in batteries, 166; in electric vehicle batteries, 166, 168, 170, 174–75; on native lands, 175–76; stripped and repurposed, 196; sustainable mining and production of, 175

Lordstown Motors Mirage, The, 184–85
Low Income Home Energy Assistance Program (LIHEAP), 84–85, 90, 98–103
low-income communities and populations, 2, 4–5, 10–19, 26, 48–50, 53, 80–85, 89–90, 96, 102, 113–15, 118–22, 128, 192–94, 212, 214. *See also* disparities; poverty
Lucid, 164, 266n85

Mallory, Brenda, 24
Marathon, 42–43, 122
marginalized communities and populations: and clean-energy transition, 8, 128; disproportionate burdens for, 132; electric vehicle access for, 21; energy justice for, 9; and energy systems, 214–15; environmental and health issues of, 27, 53, 160; environmental injustices for, 210; environmental justice for, 23–25, 160; and new technology, barriers to adoption, 212; and oil refinery locations, 43, 46; and path of least resistance, 132; in polluted locations, 24–25; quality of life impacted, 46; in redlined sections of cities, 48; as repressed, 53; social vulnerabilities of, 53. *See also* disparities
Massey Energy, 30
McKibben, Bill, 39
McSweeney, Robert, 202
mercury, 32, 47, 50
methane emissions, 29–30, 34, 67, 234n37
Mexico, 28, 177, 265n67
Mine Safety and Health Administration (US), 30, 233n26
minerals and mineral deposits: for clean-energy technologies, 270n150; and digital economy, 264n59; for electric vehicle batteries, 168, 175–76, 209, 212; major producing countries of, 270n150; rare earth, 270n150. *See also specific mineral(s)*
mobility, geographic, 8

Morena, Edouard, 77
Motor Trends, 165

NAFTA. *See* North American Free Trade Agreement (NAFTA)
NASEO. *See* National Association of State Energy Officials (NASEO)
National Academy of Engineering, 31
National Academy of Sciences, 31
National Association of State Energy Officials (NASEO), 121–22, 253n61
National Centers for Environmental Information (US), 3
National Climate Assessment (report), 6, 71
National Energy Assistance Directors' Association, 89
National Geographic, 182, 267n88
National Oceanic and Atmospheric Administration (NOAA), ix
National Renewable Energy Laboratory, 137, 144
Native Americans. *See* indigenous people
natural gas: capacity growth of, 204; and cleaner power, 33; and coal, 16, 28, 32, 46, 154, 234n37; costs and benefits of, 37, 127–28; and energy consumption, 28; and energy independence, 33; explosions, 38; and fracking, 33; infrastructure, 127–28; jobs, 59; markets, 65, 127–28; and oil, ix–x, 25–26, 31–32, 41–42, 46, 59; pollution, 16, 46–47; power plants, 1; prices, 26, 203; production, 29; and renewable resources, 29; shift from, 137
Navajo. *See* indigenous people
net-zero emissions, 10, 60–61, 105–6, 119, 126, 136, 139, 158, 201–4. *See also* zero emissions
New York Times, The, 184
NextEra Energy, 130–31
nickel: in acid coal mine drainage, 67; for clean energy, 270n150; in electric vehicle batteries, 168; on native lands, 175–76; sustainable mining and production of, 175

NIMBYism. *See* "Not-in-My-Backyard" attitudes (NIMBYism)

nitrogen oxides, 43, 47, 49

NOAA. *See* National Oceanic and Atmospheric Administration (NOAA)

Nock, Destenie, 87

Nonbusiness Energy Tax Credit program, 119

North American Free Trade Agreement (NAFTA), 177

"Not-in-My-Backyard" attitudes (NIMBYism), 145–47, 211

nuclear power: advanced, 10; and energy consumption, *28*; in energy infrastructure, 34–35, 136; and fossil fuels, 142; plants, 34–35, 37, 132, 144–45; shift from, 132

Nunes, Ashley, 118

Obama, Barack, 24, 33, 39, 59, 181

Occidental Petroleum, 35

Oceti Sakowin Power Authority, 116–17

Office of Energy Justice and Equity (US), 11

Ohio River Valley Institute, 65

oil: consumption, 203; crises, 26, 141, 179–80; and fracking, 32; jobs, 59; and natural gas, ix–x, 25–26, 31–32, 41–42, 46, 59; pollution, 41–47; production, 29, 32, 203; supply, 38. *See also* petroleum

oil and gas industry, ix–x, 17, 24–26, 33, 42–43, 59–67, 69–70, 74, 122, 160, 215, 234n37, 241n13, 260n1; net favorability toward, 133–34, *134*; and US production, 203. *See also* electric and gas utilities; energy: business and industry sector ratings; petroleum; pipelines

oil refineries, 5, 16; air pollution from, 23–25, 41–46; and clean energy, 122; and economy, 43; and energy infrastructure, 158; environmental compliance of, 45; and fossil fuels, 23–25, 38–50; and land contamination, 50. *See also* pipelines

oppression, 9, 15. *See also* disparities

ovens (household), for heat, 88, 92, 216, 246n22

overburdened communities and populations, 9, 14–17, 26–27, 41–42, 46–49, 53, 77, *82*, 103, 110, 113, 120, 132, 160, 198, 209–10, 222. *See also* disparities

oxygen concentrators, 81, 92–93, 247n32

Pacific Gas & Electric Company, 38, 122, 125, 138, 213, 255nn26–27

Paris Climate Agreement, 59–60, 77, 106, 201, 204

particulate matter, 43, 47–48, 68, 196–97

partisanship, 135, 150–53

Pastor, Manuel, 77

Peneda, Cristian Pavon, 2

people of color. *See* communities of color and people of color

Perpetua Resources (Idaho), 175

petroleum, 43, 46; and energy consumption, *28. See also* oil

Pew Research Center, 151, 259n66

PG&E. *See* Pacific Gas & Electric Company

pipelines: and climate change, 39; and energy infrastructure, 132, 151; and fossil fuels, 37–42, 66; privately owned and operated, 140; and public interest, 40; siting of, 16; statistics, 235n63. *See also* oil refineries

policymaking and policies: and decarbonization, ix; and electric vehicles, 165–66, 189; and energy infrastructure, 141–43; and energy justice, 9–12, 223; and energy systems, 206; of energy transition, 8, 77–78, 219–20, 225; and greenhouse gas emissions, 106; and justice, 223; and net-zero emissions, 10; and technologies, 19, 225. *See also* public policy

political science, 215

politics: and climate change, 150–53; and coal mining, 59; and energy infrastructure, 135, 141, 150–53, 161; and energy justice, 11–12; and energy systems, 27, 206; of energy transition, 20, 56, 77, 150–53; and fossil fuels, 27;

geo-, 169, 175; and partisanship, 135, 150–53; and pipeline protests, 40–41; and renewable energy, 151–52

pollution: coal, 16, 32, 46–47, 50–52, 67; and disparities, 26; and ecological damage, 47; in fenceline communities, 42; from fossil fuels, 4–5, 29, 46–53, 66–67, 112–13, 214–15; and health, 45–48; industrial, 4–5, 25, 37, 42–44; natural gas, 16, 46–47; from oil, 41–47; and poverty, 25; from power plants, 16, 48–49; severe impacts of, 23–25; and waste, 25, 67. *See also* air pollution; land pollution; water pollution

post-traumatic stress disorder (PTSD), 188

poverty, 64–66, 128, 181, 187, 206–7, 241n22, 245n5; in DRC, 169; and economic decline, 64; and economic opportunity, lack of, 65; and energy insecurity, 83, 87–88, 92; and environmental injustices, 38, 43–44, 52; and material hardship, 214; and pollution, 25; and public health, 31; and unemployment, 69; and vulnerability, 198. *See also* disparities; energy poverty; low-income communities and populations

power: and agency, 15, 215–16, 223; backup sources of, 112; cleaner, and energy independence, 33; hydro-, 142; lack of, 15, 93–94, 215–16; outages, 3, 138, 213, 215; shutoffs, 3, 81, 89, 94, 213–14. *See also* electric power; energy; power plants

power lines. *See* transmission lines

power plants, 18; aging, 136; coal-fired, x, 4, 50–52, 58, 62–65, 127, 207, 223, 234n36, 253n72; economic impacts of, 62–63, 68–69; and energy infrastructure, 136–38; environmental impacts of, 159–60; fossil fuel, 37, 42, 47–52, 62–69, 106, 112–13, 133, 144–45, 158–59, 210; natural gas, 1; pollution from, 16, 47–49; privately owned and operated, 140. *See also* nuclear power

Powering Past Coal Alliance Summit, 56

private sector, 17–18, 96, 116, 139–41, 157, 175, 223–24, 256n32. *See also* energy: business and industry sector ratings

procedural justice, 10, 78, 135, 149–50, 220–23

PTSD. *See* post-traumatic stress disorder (PTSD)

public health, 9, 31, 33, 36, 56, 101, 176, 198

Public Health Service (US), 36

public policy: and administrative burdens, 98; challenges, 219; and energy sources, 142; and energy systems, 206; and energy transition, xi, 11; future, 11, 206, 208, 219, 222–23, 225; and social welfare, 98; and technologies, 10–11, 206, 222–23, 225. *See also* policymaking and policies

public utility commissions, 100, 122–23, 140, 253n70

Qualified Plug-In Electric Drive Motor Vehicle Credit, 119

Qiu, Lucy, 87

racism, 25, 77, 110, 121–23, 180, 225

Reames, Tony G., 99, 120, 247n33, 249nn57–58, 253n60

recognition justice, 10, 26, 135, 220, 222–23

RECS. *See* Residential Energy Consumption Survey (RECS)

redlining, 25, 45, 48–50, 238n100

Regan, Michael, 23–24

renewable energy: and clean energy, 116, 118, 127; and climate change, 7–8; development of, 211; for electric power, 142; electricity generated from, ix–x; and energy consumption, 28; and energy efficiency, 118–19; and energy infrastructure, 129–32, 141–53; favorable support for, 143; and fossil fuels displacement, 73; impediments to, 143; jobs, 145; and low or zero fuel costs, 7; low-carbon, 199; misinformation about, 148–49; and natural gas,

renewable energy (*continued*)
29; opposition to, 141–53; and politics, 151–52; in rural landscapes, 13; shifts to, 7; siting of, 131–32, 211, 215; sources of, 27–28; technologies, 199. *See also* clean energy

Residential Energy Consumption Survey (RECS), 82

Residential Energy Efficiency Program, 119

resilience, xi, 7, 112, 125, 140, 224

Rivian, 209, 266n85

Rose, Nancy, 141

Ross, Chad, 160

rural communities, 13, 16, 19, 25, 32–33, 39, 52, 62, 66, 79, 118–20, 128–31, 144–47, 155–56, 199, 211, 215, 218

Russia, 26, 35, 58, 170, 203, 270n150

Sabin Center for Climate Change Law, at Columbia Law School, 143

Sabine National Wildlife Refuge (Louisiana), 44

sacrifice zones, 198; energy infrastructure in, 37–41, 160; environmental injustices in, 16–17, 22–53; environmental justice in, 5, 25; and fossil fuels, 16–17, 22–53; term, usage of, 5, 25, 31, 36, 230n20; and toxic communities, 25

self, and identity, 75–76

Shell, 24–25, 43–44

Sierra Club, 33, 120

smog, 32, 47. *See also* air pollution

SNAP. *See* Supplemental Nutrition Assistance Program (SNAP)

social multiplier effect, and decarbonization trade-offs, 243n43

sociodemographics, 13–14, 110–13

sociology, 64, 243n43

solar energy and power, ix–x, 27–28, 57–58; and clean energy, 204, 211; costs, 19–20, 110–12, 124–27, 231n27; costs and benefits of, 144–45; and cross-subsidization, 253–54n73; and decarbonization, 20; and energy infrastructure, 131–32, 141–49, 153; farms, 18, 116, 135–38, 144–45, 146, 175; favorable support for, 142; land required for, 144–45; panels, x, 7, 19–20, 74, 107–8, 110–11, 115–16, 124, 126, 132, 137, 219, 253n69; photovoltaic panels, peer effects in diffusion of, 253n69; project on Choctaw land in Oklahoma, 252n47; project on Moapa Band of Paiutes land, 252n47; project on Navajo land in New Mexico, 252n47; projects, 138, 145–46, 148–49, 252n47; rooftop, and world energy needs, 255n20; technologies, 10, 105, 108, 198; and utility death spiral, 253–54n73; utility-scale, 141, 146

Sostakowski, Bob, 130–31

Southern (utility company), 122, 213

Sovacool, Benjamin K., Dr., 174, 198–200

S&P Global, 122

space heaters, 87–88, 216, 246n22

Stephens, Jennie, 11

Stevis, Dimitris, 77

stranded assets, 253n72

subsidies, 105, 120, 124–27, 137, 142, 151, 158, 209, 219

sulfur dioxide: as air pollutant, 43, 47, 68; and coal mining regulations, 233n23; emissions, 58, 68; regulation of, 233n23; in smog, 47

Superfund sites, 23–24, 50

Supplemental Nutrition Assistance Program (SNAP), 90

Taylor, Dorceta, 25

technologies: availability of, 120; barriers to adoption for certain communities/populations, 10–11, 105–28, 192–94; challenges of, 27, 103, 128; for clean energy, 19–20, 105–28, 199, 206; climate change, 10–11, 105, 128; costs and benefits of, 15, 110–17, 121, 124–25; decarbonization, 105–8; and disparities, 110, 124–25; and energy infrastructure, 224; energy-efficient, 105–28; geographic constraints, 120–21, 128; government subsidies for, 105; and health, 169; incentives for, access as uneven, 117–20, 124; inequities for access to, 10–11, 19–20, 105–28, 198–99; and innovativeness-needs paradox, 110–11; life cycles of, 167–69; low-carbon, 19, 106, 110, 136, 169, 199; low-emissions, 60, 166, 205; and overburdened, 110; and policymaking, 19, 225; and public policy, 10–11, 206, 222–23, 225; renewable energy, 199; residential-energy, 106, 108, 110–12; siting, 132–33; and sociodemographics, 110–13; and wealth disparities, 124–25. *See also* clean energy: technologies

Tennessee Valley Authority (TVA), 140; Kingston power plant cave-in and coal ash slurry spill, 51–53

Tesla, Inc., x, 164–65, 169, 209, 260–61n4, 266n85

Texas Department of State Health Services, 2

thermostats, 10, 87, 95, 106–11, 114–15, 216, 251n30

Three Mile Island nuclear reactor (Harrisburg, Pennsylvania), 132

Tierney, Susan F., 57

Tingley, Dustin, 215

transmission lines, 1, 20, 132–41, 153–59, 213, 218; support or opposition to, 154–57, *154*

transportation sector, 18, 49, 136–40, 161, 204, 207, 209, 212–13. *See also* electric vehicles (EVs)

TransWest Express Transmission, 157–58

trash burning, for heat, 88, 246n22

Trump, Donald J., xi, 12, 39, 59–60, 70, 219

TVA. *See* Tennessee Valley Authority (TVA)

UAW. *See* United Auto Workers (UAW)

UMWA. *See* United Mine Workers of America (UMWA)

UN. *See* United Nations (UN)

underserved communities and populations, 9, 11, 21, 110, 120. *See also* disparities

UNICEF, 173

United Auto Workers (UAW), 264–65n66; contracts and strikes, 180–83, 189, 266n85; and electric vehicle battery manufacturing jobs, 185; grants for, 182; labor negotiations, 179, 181; membership contraction, 266n84; membership peaked, 179

United Mine Workers of America (UMWA), 241n22

United Nations (UN), 56

Upper Big Branch Mine (Raleigh County, West Virginia), 30

uranium, 16, 34–37, 172, 176

utilities. *See* electric and gas utilities

utility commissions. *See* public utility commissions

utility disconnections and shutoffs, 2–3, 14–15, 18, 79–94, 98–101, 112, 209–19, 245n1, 245n4; and households, rates of, *83*. *See also* energy insecurity; energy poverty

Valero, 42–45

Volkswagen, 174–75, 177, 182, 266n85

Voluntary Employee Beneficiary Association, 181

Voyles, Tracy, 36

vulnerable communities and populations, xi, 2, 13, 18–19, 25, 38, 41, 52–53, 64, 72, 81, 92, 96–97, 101–2, 121, 138, 198–99, 214, 225. *See also* disparities

Walker, Jason, 67–68

Wallace-Wells, David, 218

WAP. *See* Weatherization Assistance Program (WAP)

Washington Post, The, 160

waste: from coal, 16–17, 50–53, 67–69; from fossil fuels, 46–53; and health hazards, 35–36; and pollution, 25, 67; from uranium mine sites, 35–36. *See also* byproducts; energy: waste; e-waste

wastelanding, 36. *See also* e-waste

water pollution, x, 33–34, 43, 47, 50–51, 66–67, 69, 132–33, 144, 198, 206

wealth, 124–25, 224

weather. *See* climate change; global warming; winter storms

Weatherization Assistance Program (WAP), 90, 96–100

Welcome to Sodom (film), 196, 270nn135–36

well-being, 3, 8, 10–11, 14, 18, 26–27, 49, 70, 86, 89, 169, 198, 207, 215, 223; and consequences, 27. *See also* health

White House Council on Environmental Quality, 24

Whitmer, Gretchen, 182

WHO. *See* World Health Organization (WHO)

Who Killed the Electric Car? (documentary), 164, 260n1

wildfires, ix, 5, 138

Wilson, Dahvi, 149–50

wind energy and power, ix–x, 1, 27–28, 57–58; in American Midwest, 129; and clean energy, 204; costs and benefits of, 144–45, 211; and decarbonization, 20; and energy infrastructure, 129–53; farms, 18, 116, 129–40, 143–47, 151, 157, 211, 217–18, 223; favorable support for, 142; land required for, 144–45; projects, 117, 138, 140, 144–49, 157–58, 211; technologies, 10, 105, 198; turbines, 1, 37, 143–49, 155; utility-scale, 141

winter storms, 1–3, 23, 88

winterization recommendations, by FERC, 229n1

Wired UK, 197, 270n142

wood stoves, for heat, 88

World Health Organization (WHO), 47

Wright, Beverly, 160

Xing, Bo, 87

zero emissions, 7, 10, 20, 60–61, 105–6, 136, 139, 158, 165–66, 186, 202–4, 219. *See also* net-zero emissions